Seminar Nasional Fisika (SNF) Unesa 2021

Journal of Physics: Conference Series Volume 2110

Online
18 October 2021

ISBN: 978-1-7138-5081-6
ISSN: 1742-6588

Printed from e-media with permission by:

Curran Associates, Inc.
57 Morehouse Lane
Red Hook, NY 12571

Some format issues inherent in the e-media version may also appear in this print version.

Institute of Physics
Dirac House, Temple Back
Bristol BS1 6BE UK

Phone: 44 1 17 929 7481
Fax: 44 1 17 920 0979

techtracking@iop.org

Additional copies of this publication are available from:

Curran Associates, Inc.
57 Morehouse Lane
Red Hook, NY 12571 USA
Phone: 845-758-0400
Fax: 845-758-2633
Email: curran@proceedings.com
Web: www.proceedings.com

TABLE OF CONTENTS

PREFACE

Seminar Nasional Fisika (SNF) 2021, Universitas Negeri Surabaya on New Adaptations in Physics Learning and Research to Realize an Independent Learning Program

Dwikoranto, A Zainuddin, B K Prahani, N P Putri, N A Lestari, U A Deta*
Physics Department, Universitas Negeri Surabaya, Indonesia

*E-mail: utamadeta@unesa.ac.id

We want to present the National Physics Seminar or Seminar Nasional Fisika (SNF 2021), entitled *New Adaptations in Physics Learning and Research to Realize an Independent Learning Program*. SNF 2021 took place virtually by using the ZOOM meeting because of the COVID-19 pandemic. As we know, the Covid-19 pandemic has harmed almost all areas of life, including education, especially universities. SNF 2021 organized by the Physics Department, Universitas Negeri Surabaya-Surabaya on October 18, 2021. In general, there are two sessions, namely plenary and parallel. The plenary session was given to three keynote speakers: (1) Prof. Dr. Heri Sutanto, M.Si. (09.40-10.10 AM), (2) Dr. Asnawi, M.Si. (10.10-10.40 AM), (3) Prof. Dr. Budi Jatmiko, M.Pd. (10.40-11.10) which ends with a Q&A session (11.10-11.40 AM). Parallel session is for authors who present their papers and is carried out in eight groups of breakout rooms (1.00-4.00 PM).

As a scientific forum, the SNF 2021 is attended by 217 participants from students, teachers, lecturers, engineers, and practitioners who participate either as presenters or as participants. The SNF 2021 main aim is to share ideas and knowledge and build collaboration with each other. We received 56 papers that will be selected to be published in the **Journal of Physics: Conference Series** with the scope of physics and its related area. In this volume, we accept 32 papers for publication after meeting the scope, checking originality, and performing peer review.

The technology needed to deliver the meeting successfully is zoom meeting, laptop, sound system, Wi-Fi (internet signal source), and the backup of electrical power. In this particular case, we have two times electricity outages (± 3 minutes). It is not a big problem because we have the backup of electricity and internet signal. We are hoping that the event will become a foothold of thinking in welcoming global developments. In addition, publications packaged in this seminar will further strengthen the existence and reputation of the physics department as an educational institution that excels in scientific publication at national and international levels. The results of these studies need to be disseminated so that they are scientifically helpful and provide practical value for the community.

The organizing and editorial committee of the SNF 2021 hopes you enjoy reading this volume of the JPCS. Besides, we would like to fully respect keynote speakers and all the presenters for providing their valuable contributions for this proceeding and the reviewers for their constructive recommendations and criticism aiding to improve the presented articles. We, on behalf of the committee of the SNF 2021, would like to thank all parties for their participation in supporting this publication. We want to invite the participants back to participate in SNF 2022, which will be held by the Physics Department, Universitas Negeri Surabaya.

Seminar Nasional Fisika (SNF) Unesa 2021 Committee

Executive Chairman
Prof. Dr. Munasir, M.Si. (Head of the Department of Physics, Unesa)

Chairman of the Seminar Committee
Dr. Dwikoranto, M.Pd. (Lecturer, Department of Physics, Unesa)

The Organizer
Universitas Negeri Surabaya (Unesa), Indonesia

Descriptions	**Deadlines**
Abstract Submission Duedate:	August 9th, 2021
Notification of Acceptance:	September 27th, 2021
Final Revision Due:	October 3th, 2021
Conference Day:	October 18th, 2021

Keynote Speakers
Prof. Dr. Heri Sutanto, M.Si. (Universitas Diponegoro, Semarang, Indonesia)
Dr. Asnawi, M.Si. (Universitas Negeri Surabaya, Surabaya, Indonesia)
Prof. Dr. Budi Jatmiko, M.Pd. (Universitas Negeri Surabaya, Surabaya, Indonesia)

Committee Boards

Executive Chairman	: Prof. Dr. Munasir, M.Si.
Chairman	: Dr. Dwikoranto, M.Pd.
Vice Chairman	: Abu Zainuddin, M.Pd.
Secretary	: Dr. Binar Kurnia Prahani, M.Pd.
Treasurer	: Nugrahani Primary Putri, M.Si.

Secretariat and Manuscripts:
Nurita Apridiana Lestari, M.Pd.
Utama Alan Deta, M.Pd., M.Si
Meta Yantidewi, M.Si.
Dr. Rohim Aminullah Firdaus, M.Si.
Riana Ambarsari
Fatihatun Nikmah
Ekaningtyas Wahyu Epriliyani
Ermia Fadilata Khoir
Satya Cantika Agustinur

Reviewer:
Prof. Dr. Heri Sutanto, M.Si.
Prof. Dr. Budi Jatmiko, M.Pd.
Prof. Dr. Madlazim, M.Si.
Prof. Dr. Munasir, M.Si.
Prof. Dr. Wasis, M.Pd.
Prof. Tjipto Prastowo, Ph.D.

Dr. Dwikoranto, M.Pd.
Dr. Zainul Arifin Imam Supardi, M.Si.
Dr. Asnawi, M.Si.
Dr. Titin Sunarti, M.Si.
Nadi Suprapto, Ph.D.
Mita Anggaryani, Ph.D.
Dr. Binar Kurnia Prahani, M.Pd.
Utama Alan Deta, M.Pd., M.Si.
Dr. Rohim Aminullah Firdaus, S.Pd., M.Si.
Dr. Muhimmatul Khoiro, S.Si.

Journal of Physics: Conference Series **2110** (2021) 011001 doi:10.1088/1742-6596/2110/1/011001

Documentations of SNF Unesa 2021

Plenary Session

Master of Ceremony

Welcoming Speech
Chairman SNF Unesa 2021

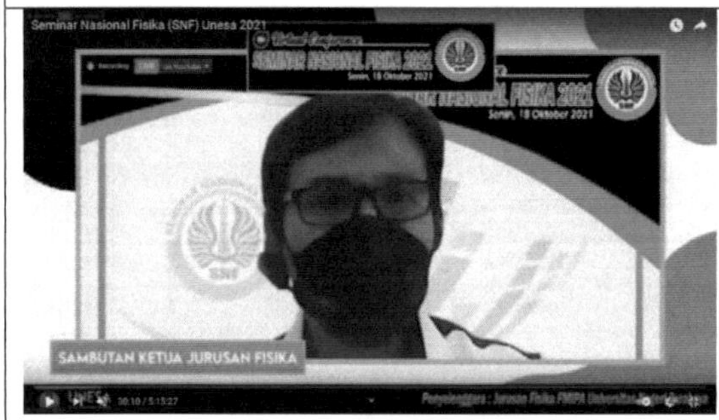

Welcoming Speech
Head of Department of Physics

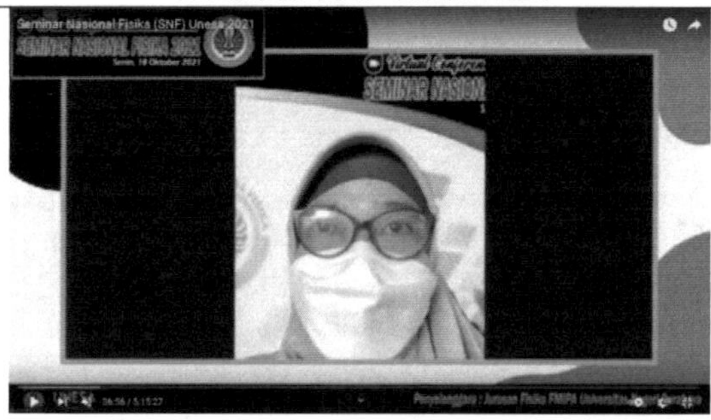

Welcoming Speech
Head of Physical Society Indonesia
(Surabaya Chapter)

Welcoming Speech
Dean of Faculty of Mathematics and Natural Sciences
Universitas Negeri Surabaya

Moderator of Plenary Session

Keynote Speaker Presentation
Prof. Dr. Heri Sutanto, M.Si.

Keynote Speaker Presentation
Dr. Asnawi, M.Si.

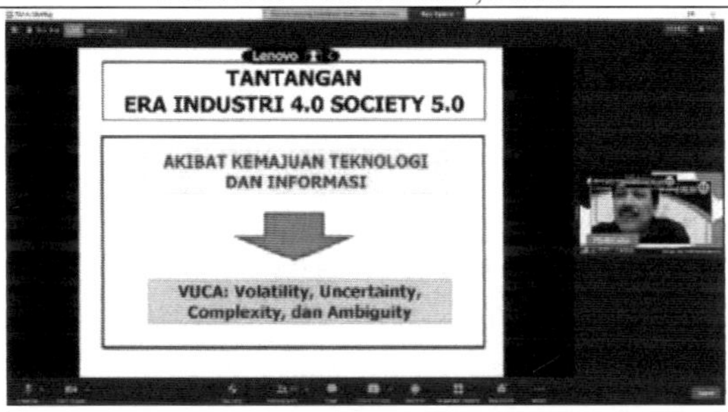

Keynote Speaker Presentation
Prof. Dr. Budi Jatmiko, M.Pd.

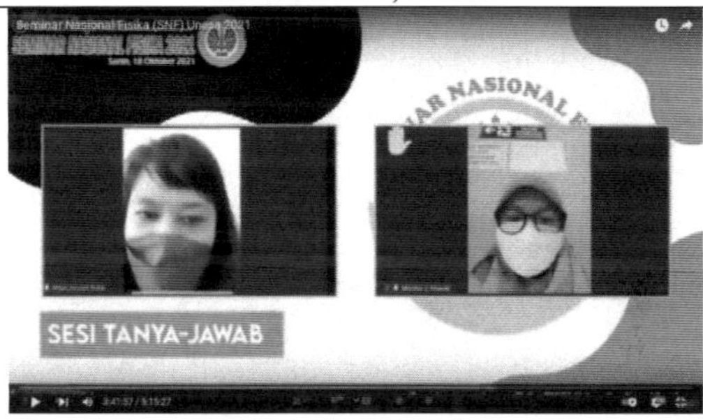

QnA Session (1)

QnA Session (2)

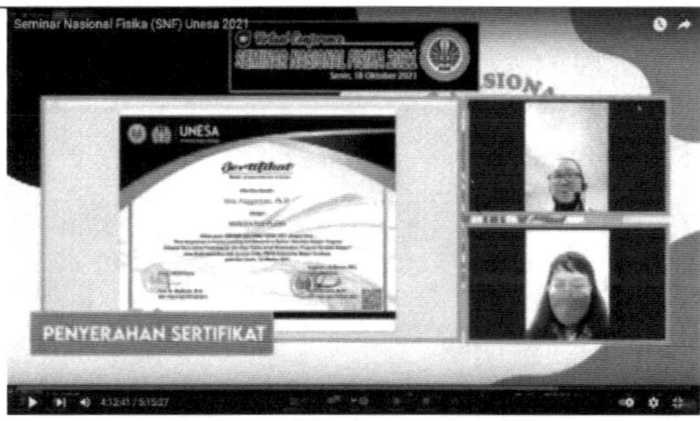

Token of Appreciation (1)

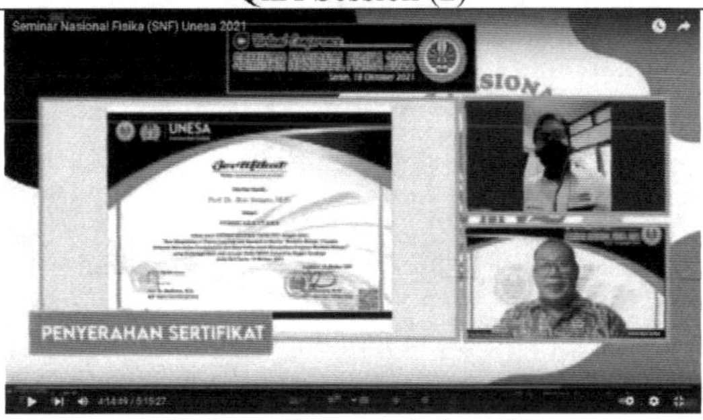

Token of Appreciation (2)

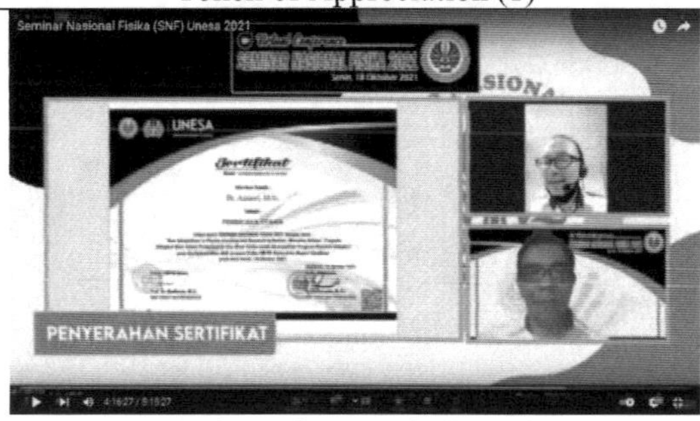

Token of Appreciation (3)

Journal of Physics: Conference Series **2110** (2021) 011001 doi:10.1088/1742-6596/2110/1/011001

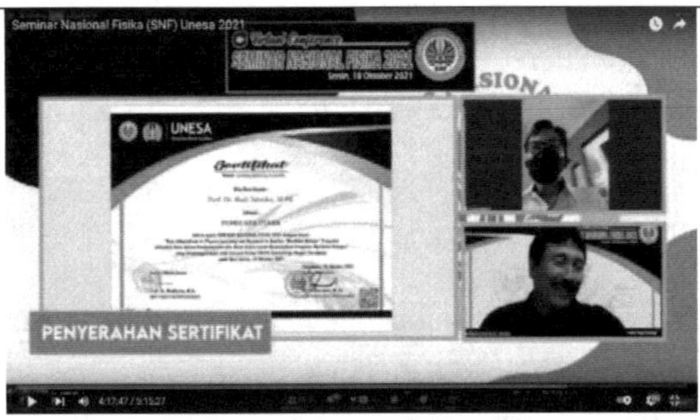

Token of Appreciation (4)

Students Performance

Full video of SNF Unesa 2021 Plenary Session: https://youtu.be/pJDSt5Jt-lQ

Parallel Session

Room 1

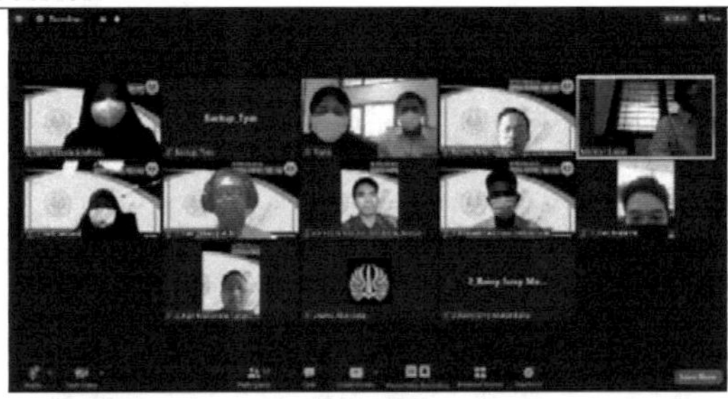
Room 2

Room 3

Room 4

Room 5

Room 6

Journal of Physics: Conference Series **2110** (2021) 011001 doi:10.1088/1742-6596/2110/1/011001

Room 7	Room 8

Full video of SNF Unesa 2021 Parallel Session:
https://drive.google.com/drive/folders/1BbHGVH65WqSpg1-VAF7IjLl4lZNa7Z_T?usp=sharing

Journal of Physics: Conference Series **2110** (2021) 011002 doi:10.1088/1742-6596/2110/1/011002

Peer review declaration

All papers published in this volume of Journal of Physics: Conference Series have been peer reviewed through processes administered by the Editors. Reviews were conducted by expert referees to the professional and scientific standards expected of a proceedings journal published by IOP Publishing.

- **Type of peer review: Double-blind**
- **Conference submission management system: Morressier**
- **Number of submissions received: 56**
- **Number of submissions sent for review: 46**
- **Number of submissions accepted: 32**
- **Acceptance Rate (Number of Submissions Accepted / Number of Submissions Received X 100): 32/56 x 100 = 57.14 %**
- **Average number of reviews per paper: 1**
- **Total number of reviewers involved: 15**
- **Any additional info on review process:**
  There are three steps of the editorial and review process in SNF Unesa 2021. First, Editor selects the paper based on the topics, language, manuscript structure, and similarity check. Next, the Editor assign a Reviewer for each paper to review the manuscript. Last, the Editor rechecked the revised manuscript based on the Reviewer's Comments, proofread the manuscript, edited the manuscript based on the JPCS template, and checked the similarity before submitting the manuscript in IOP.
- **Contact person for queries:**
  Name : Utama Alan Deta
  Affiliation: Department of Physics, Universitas Negeri Surabaya
  Email : utamadeta@unesa.ac.id

Identification of Crustal Thickness in Central Part of Sumatra Using Teleseismic Receiver Function Method

P Ariyanto[1], M D Atthonthowi[1]*, B Pranata[2], and B S Prayitno[2]

[1] Program Studi Geofisika, Sekolah Tinggi Meteorologi Klimatologi dan Geofisika, Indonesia
[2] Badan Meteorologi Klimatologi dan Geofisika, Indonesia

*Email: ditto.atthonthowi@gmail.com

Abstract. The Central part of Sumatra is a region that has a high potential for earthquakes. This research intended to determine the crustal thickness of the earth, P and S wave velocity profiles, and *vp/vs* value in the Central part of Sumatra using stacking H-k and inversion techniques based on the analysis of receiver function. This study utilized teleseismic earthquake data with a distance of 30° to 90° from the station and magnitude more than 6 (M>6). The stations used in this study were 3 BMKG broadband stations located in 3 zones, the fore arc ridge zone (SISI), the volcanic zone (PLSI) and the back arc zone (TPRI). The crustal thickness varies in the fore arc ridge zone (SISI) estimated 17.8 km, volcanic zone (PLSI) reaches 29.7 km and the back arc zone (TPRI) reaches 34 km. The crustal thickness is quite thick under the PLSI and thicker beneath TPRI station. These possibly due to the effect of topography and isostatic compensation in the station. However, whether there is a correlation between crustal thickness and topography needs further research using more stations. The highest *vp/vs* value was found in the volcanic zone of 1.9, that might be associated with the presence partial melting beneath the station. Meanwhile, the *vp/vs* value in the back arc zone is 1.72, indicating a relatively more homogeneous structure.

1. Introduction

The Island of Sumatra is part of the convergence zone between two plates, namely the Eurasian plate and the Indo-Australian plate. The gathering zone between the plates forms a trough known as the collision zone. The trench integrates the northward movement of the Indo-Australian plate against the Eurasian plate. The movement of the Indo-Australian plate and the Eurasian plate in the subduction zone has caused earthquakes [1].

The central part of Sumatra is a continental plate boundary which consists of two fault systems, namely a strike-slip fault system with a dextral orientation and a subduction dip-slip interface which has a greater influence [2]. Central Sumatra is an area that has a high potential for earthquakes and tsunamis. This region has three main sources of active earthquake generators, namely the subduction line (Megathrust), the Mentawai fault and the Sumatran fault. Based on historical records, the central Sumatra region has experienced several destructive earthquakes in 1822, 1835, 1981, 1991, 2005, and 2009 in Padang, 2003 in Agam, 2007 in Bukitinggi, and an earthquake accompanied by a tsunami in 1861 in Mentawai, 1904 in Sori-sori, and in 2010 it happened again in Mentawai [3]. Historically segments in the Sumatra region can generate earthquakes with a magnitude of 7 SR. [4]. Seismicity map in Central Sumatra can be shown in Figure 1.

Journal of Physics: Conference Series **2110** (2021) 012001 doi:10.1088/1742-6596/2110/1/012001

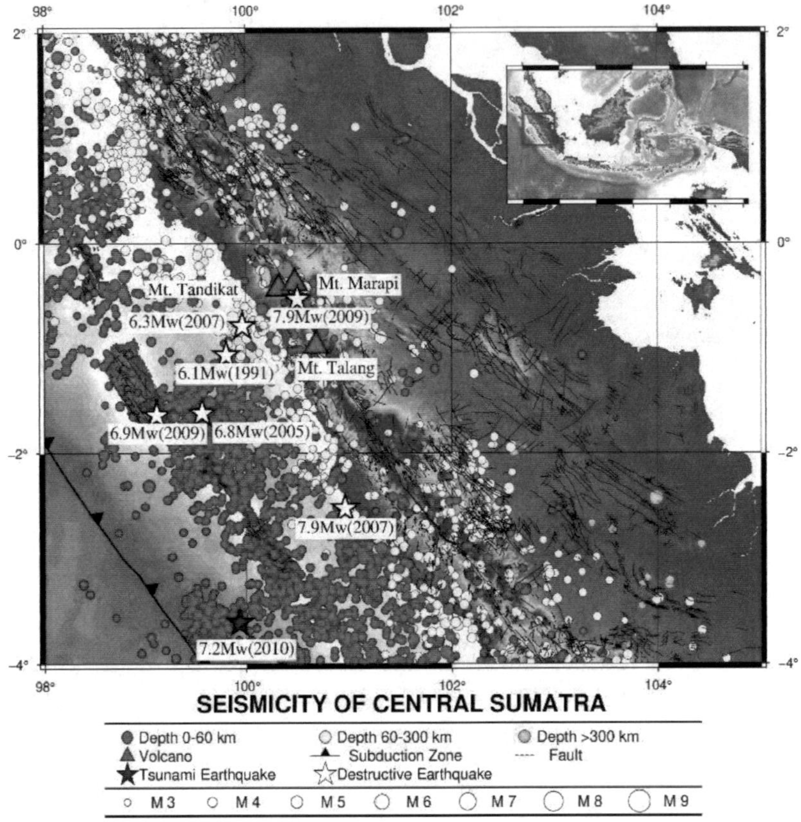

Figure 1. Central part of Sumatra seismicity map

This research uses the receiver function method to analyze the structure of the earth's crust in central part of Sumatra. In Indonesia, several studies about crustal structure using receiver function analysis have been carried out by several researcher [5-8]. It is necessary to analyze the structure of the earth's crust in this area given the high level of vulnerability to earthquakes and tsunamis. The receiver function is a method that can interpret the structure of the earth and its discontinuity plane by utilizing a teleseismic waveform on a three-component seismic sensor. Teleseismic P waves will be converted into S waves in the discontinuity plane below the receiving sensor and the difference in arrival time between the two can be used to determine the crustal thickness [9]. In addition, the local velocity model from the inversion of the receiver function of this study is expected to improve accuracy in determining hypocenters and tsunami early warnings in central part of Sumatra.

2. Method

The data utilized in this research is teleseismic waveform data with a magnitude of more than 6 (M>6) and an epicenter distance of $30^\circ - 90^\circ$ from the receiver station. This research is located in central Sumatra and the data from this study comes from 3 broadband 3 component seismometer sensors belonging to the Meteorology, Climatology and Geophysics Agency (BMKG). The study area, epicenter of earthquake and the seismic stations used in this study can be shown in Figure 2.

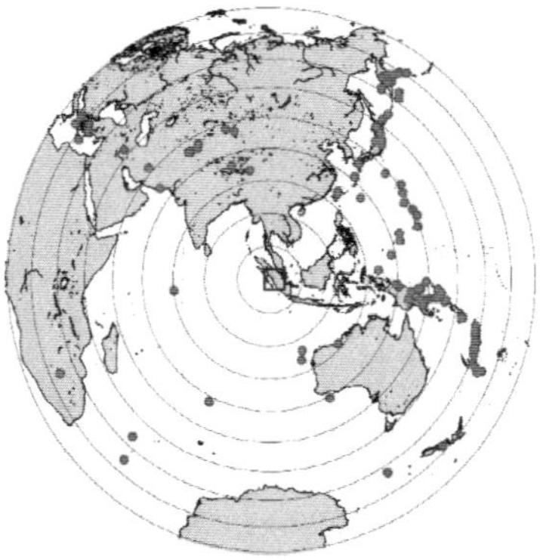

Figure 2. Distribution map of earthquake epicenters used in this research

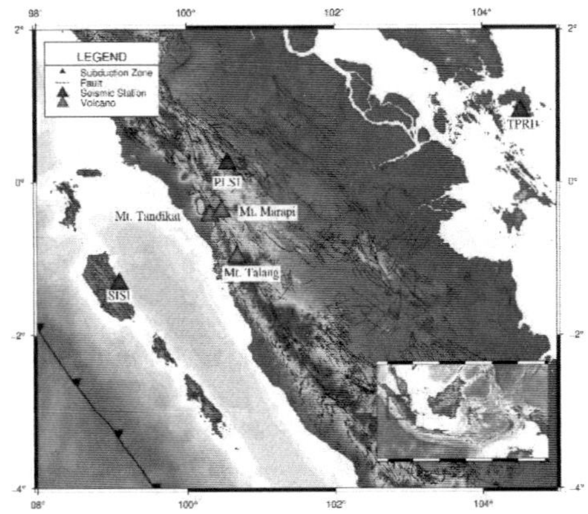

Figure 3. Distribution map of the seismic stations used in this research

2.1. Receiver function method

The receiver function is a technique used to map the structure of the earth layer underneath the receiver by utilizing teleseismic waves and is a time series that represents the conversion of P to S waves at discontinuities in response to the structure of the earth beneath the receiver. The response comes from the propagation of teleseismic waves that pass through a medium with different characteristics. If the propagation of a body wave such as a P wave passes through a discontinuity plane, then the P wave energy will be divided into several phases such as being transmitted directly to the surface, some of the energy is reflected back, and can be converted into an S wave. The first wave amplitude and clearly visible in the receiving function is a wave P directly and then followed by phase conversion of Ps waves. The distinction in arrival time between the P wave phase and the Ps conversion phase can be utilized to calculate the estimated crustal thickness if the velocity model of the area is identified [10].

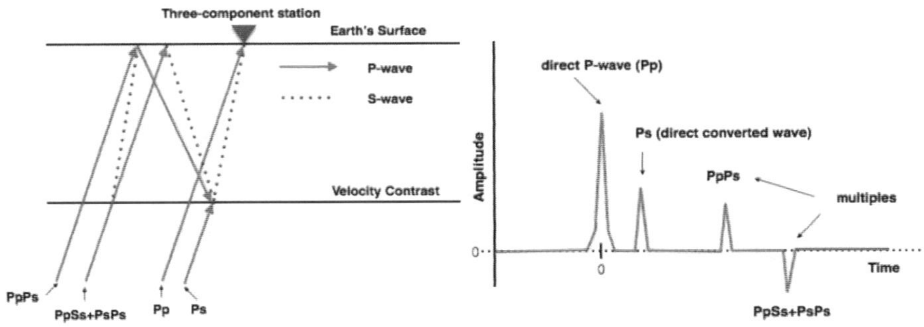

Figure 4. Illustration of the phase conversion of seismic waves and their reflected phases as they pass through the contrasting velocity and layers of the earth's surface [11]

In this research we only calculated the radial component of receiver function using the iterative time domain deconvolution method developed by Ligorria and Ammon [11] which is a time domain deconvolution approach based on the difference between synthetic seismograms and observational seismograms which have advantages in estimating the receiver function on noisy data, based on the following formula.

$$R(t) = E_{iR}(t) * Z(t) \tag{1}$$

Denoted *R(t)* is the signal from the radial component in the time domain, E_{iR} is the iteratively obtained radial component receiving function model. *Z(t)* is the vertical component seismogram and * is the convolution operator [11].

2.2. Stacking H-k

The aim of stacking *H-k* is to calculate the estimated crustal thickness (*H*) and the v_p/v_s (*k*) value from the time series of the receiver function. This method calculates the distinction in the propagation time of Ps, PpPs, and PpSs+PsPs waves to direct P waves.

Estimation of the crustal thickness is done by stacking the receiver function data in the *H-k* domain, using the equation.

$$s\,(H,k) = w_1 r(t_1) + w_2 r(t_2) - w_3 r(t_3) \qquad (2)$$

Denoted w is the weighting factor provided that $\Sigma W = 1$ and r is the radial receiver function at t_1, t_2, and t_3 with t being the estimated time difference between the arrival times of the Ps, PpPs, and PpSs+PsPs [10].

3. Results and Discussion

In this study, we used 3 stations which represent 3 zones, fore arc ridge zone (SISI), the volcanic zone (PLSI) and the back arc zone (TPRI). The receiver functions generated by each station has a fit >90% of deconvolution.

3.1. Fore Arc Ridge Zone

The teleseismic earthquake data used at this station was recorded from 2018 to 2020 and produced as many as 20 radial component receiving functions as shown in Figure 5. Based on the results that are plotted in quadrant 4 there is no receiver function signal, this is because the data in quadrant 4 does not have good quality so that the receiver function cannot be calculated. A direct P phase or Pp is observed at ~0 s. The Ps phase used to estimate the depth of the moho layer was observed at ~3 s. Furthermore, the PpPs phase can be observed at ~6 s.

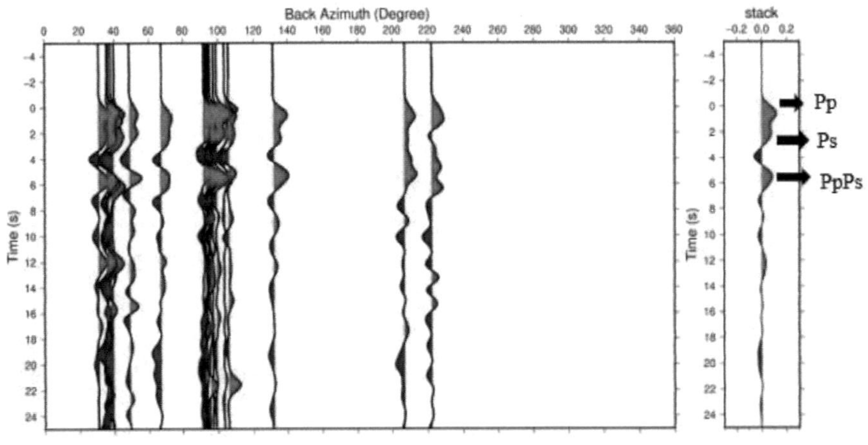

Figure 5. Radial component receiver functions at SISI station based on back azimuth

The P wave velocity profile can be seen in Figure 6. The P wave velocity has a minimum value of 3.06 km/s and a maximum value of 9.81 km/s. The estimation of the crustal thickness is approximately 2,022 km, which is indicated by an increment in the P wave velocity from 6.4 km/s to 6.8 km/s. The S wave velocity at SISI station has a minimum value of 1.7 km/s and a maximum of 5.36 km/s. The crustal thickness is estimated around 20-22 km marked by an increment in the velocity of the S wave from 3.2 km/s to 4 km/s.

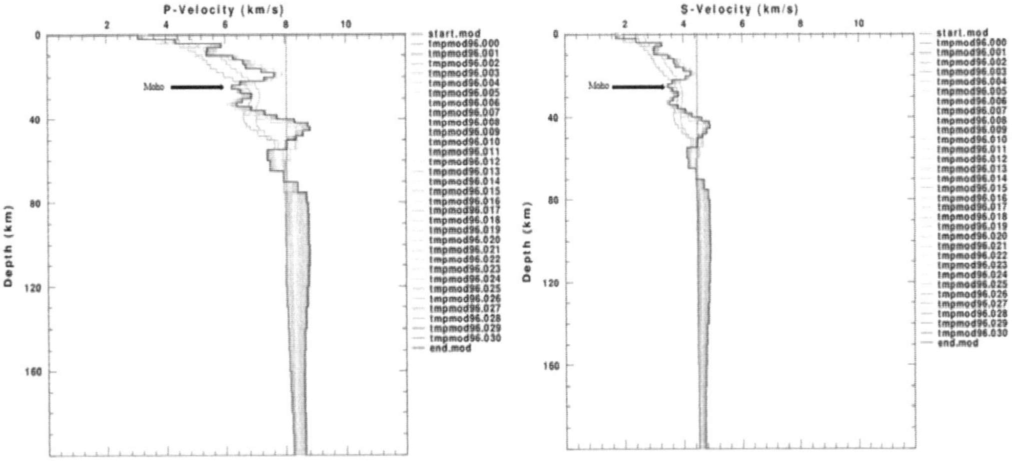

Figure 6. P and S wave velocity profiles at SISI station

Figure 7 shows the results of the stacking *H-k* calculation at the SISI station. Based on the results of these calculations, the crustal thickness is 17.8 km and the *vp/vs* value is 1.2. The fore arc ridge zone in Sumatra is a seismic anisotropy zone [12]. The *vp/vs* value generated at SISI station is very small, this is because the use of the stacking *H-k* method to estimate the *vp/vs* value in the anisotropy zone is less accurate [13].

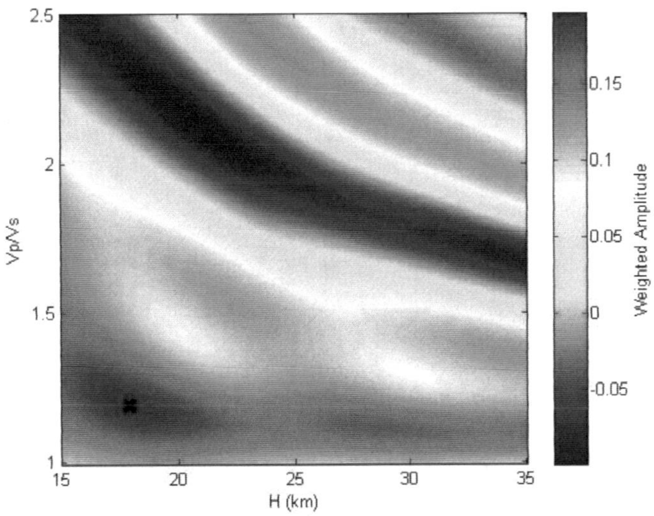

Figure 7. Stacking *H-k* at SISI station

3.2. Volcanic Zone

Based on 22 radial component receiver functions as shown in Figure 8, the direct P phase or Pp is seen at ~0 second and the P wave phase converted to S wave is seen at ~5 second and the PpPs phase is seen at ~13 second.

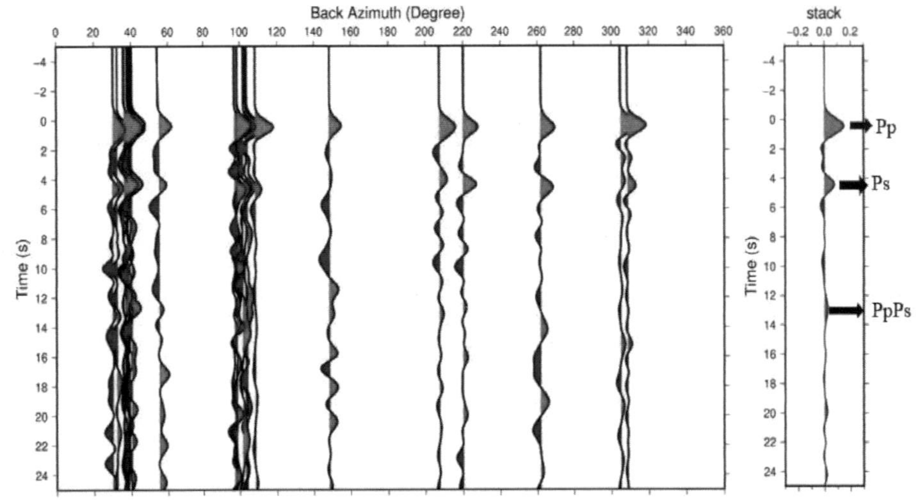

Figure 8. Radial component receiver functions at PLSI station based on back azimuth

The P wave velocity profile under the PLSI station is shown by Figure 9. The P wave velocity beneath this station has a minimum value of 4.46 km/s and a maximum of 10.4 km/s. The estimation of crustal thickness is approximately 28-32 km, which is indicated by an increment in P wave velocity from 7.2 km/s to 8.4 km/s. The velocity of the S wave at the PLSI station has a minimum value of 2.48 km/s and a maximum of 5.8 km/s. The crustal thickness is estimated around 28-32 km which is indicated by an increment in the velocity of the S wave from 4 km/s to 4.8 km/s.

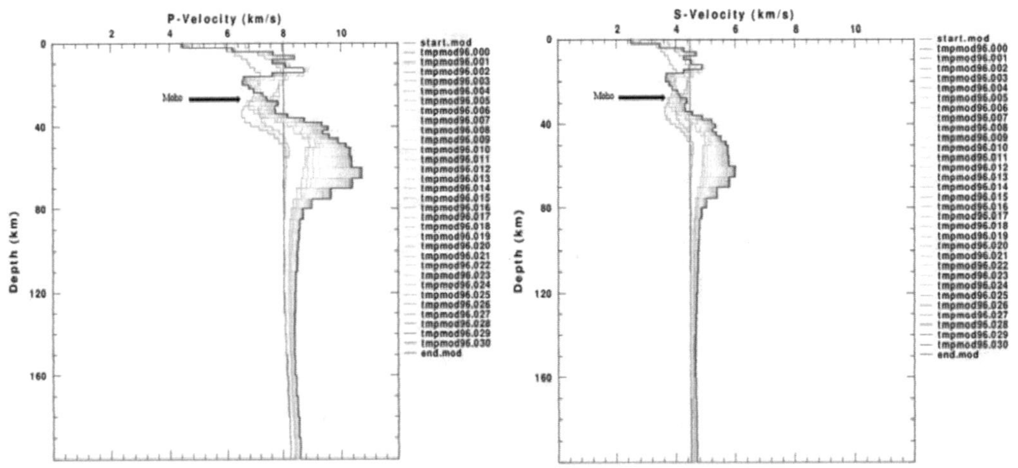

Figure 9. P and S wave velocity profiles at PLSI station

The estimated crustal thickness obtained based on the calculation of stacking *H-k* is 29.7 km and the *vp/vs* value is 1.9. The high *vp/vs* value at the PLSI station that might be associated with the presence of partial melting so that the S wave velocity when passing through the fluid medium is more attenuated than the P wave. The results of *H-k* stacking can be seen in Figure 10. This result is in agreement with the receiver function study conducted in the Toba Caldera (volcanic zone) using the PSI station which produced a crustal thickness of 30 km and the *vp/vs* value was 1.87 [14].

 2110 (2021) 012001 doi:10.1088/1742-6596/2110/1/012001

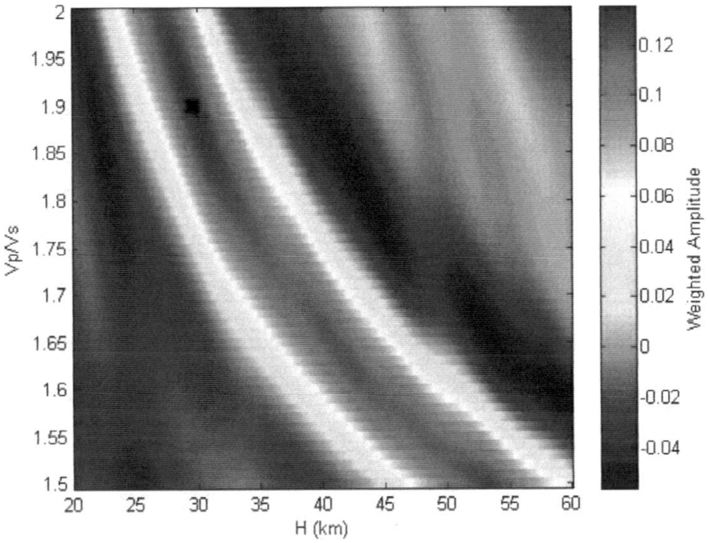

Figure 10. Stacking *H-k* at PLSI station

3.3. Back Arc Zone

The receiver function of the radial component generated at the TPRI station is 48 as shown in Figure 11. Based on the stacking results of the receiver function, the resulting phases are quite simple, indicated by the fact that there are only two phases at this station. This condition is appropriate because the TPRI station represents a back arc zone where seismic activity is low in this zone so that the resulting phase is very clear and not complex. The direct P phase or Pp is seen at ~0 s and a P wave phase converted to an S wave is seen at ~4 s.

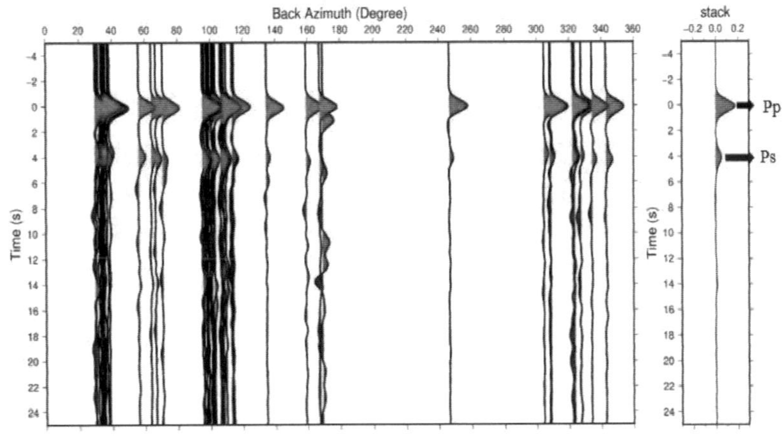

Figure 11. Radial component receiver functions at TPRI station based on back azimuth

Figure 12 shows the P wave velocity profile below the TPRI station. The P wave velocity below this station has a minimum value of 6.1 km/s and a maximum of 9.8 km/s. The estimation of crustal thickness is approximately 32-36 km, which is indicated by an increase in the velocity of the P wave from 8 km/s to 8.4 km/s. The velocity profile of the S wave at TPRI station has a minimum value of 3.41 km/s and a maximum of 5.46 km/s. The crustal thickness is estimated around 32-36 km which is indicated by an increase in the velocity of the S wave from 4.4 km/s to 4.8 km/s.

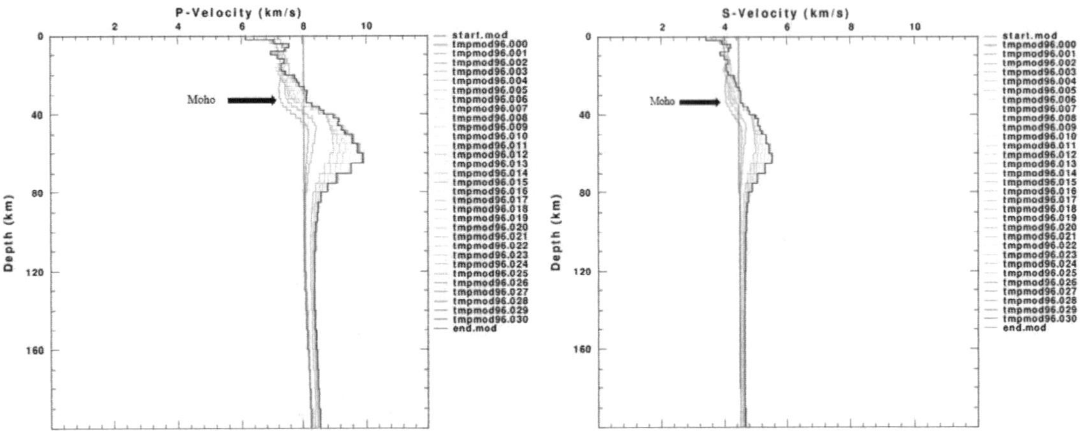

Figure 12. P and S wave velocity profiles at TPRI station

Figure 13 shows the results of the stacking *H-k* calculation at the TPRI station. Based on the results of this calculations, the estimated crustal thickness is 34.1 km and the *vp/vs* value is 1.72. This result is supported with the study of receiver function by Bora et al. [15] using the inversion Neighborhood Algorithm (NA) method in the back arc zone with a crustal thickness of 27-33 km. Bora et al. used the BKNI and PMBI stations to calculate crustal thickness in the back arc of Sumatra [15].

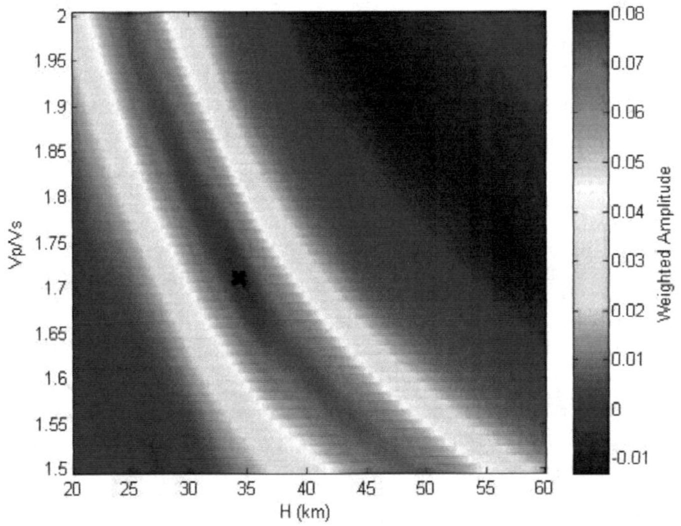

Figure 13. Stacking *H-k* at TPRI station

4. Conclusion

Analysis of the P and S wave velocity profile and the crustal thickness in central Sumatra has been carried out using the inversion method and stacking *H-k*. Based on the results of the receiver function analysis, we estimated the crustal thickness in the fore arc ridge zone (SISI) is around 17.8 km, in the volcanic zone (PLSI) reaches 29.7 km and the back arc zone (TPRI) reaches 34 km. The crustal thickness is quite thick under the PLSI and TPRI station, possibly due to the effect of isostasy in the zone. Stations with high topography may have deeper Moho as an effect of isostatic compensation. The highest *vp/vs* value was found in the volcanic zone of 1.9, that might be associated with the presence of magmatic material or partial melting beneath the station. Furthermore, the normal *vp/vs* value in the back arc zone (*vp/vs* ~ 1.72) probably related to the simple and relatively homogeneous structure beneath this station.

Acknowledgements
We would like to thank Indonesian Agency for Meteorology, Climatology, and Geophysics (BMKG) for providing the teleseismic data in this research. We also acknowledge Earthquake and Tsunami Study Committee in BMKG for supporting this research.

References

[1] McCaffrey R 2009 *Annu. Rev. Earth Planet. Sci.* **37** 345.
[2] Lasitha S, Radhakrishna M, and Sanu T D 2006 *Curr. Sci.* **90** 690.
[3] BMKG 2019 *Katalog Gempabumi Signifikan dan Merusak 1821 -2018* (Jakarta: BMKG)
[4] Sieh K and Natawidjaja D 2000 *J. Geophys. Res.* **105** 28295
[5] Ariyanto P, Syhada, Rosid S, Anggono T, and Januarti Y 2018 *J. Phys. Conf. Ser.* **985** 012018
[6] Ariyanto P, Ariwibowo S, Wijaya A, Pratama R 2020 *J. Phys. Conf. Ser.* **1491** 012055
[7] Ariyanto P, Rosid, M S, Syuhada, Januarti Y 2021 *J. Phys. Conf. Ser.* **1805** 012018
[8] Pratama R, Ariyanto P, Wijaya A, and Ariwibowo S 2020 *J. Phys. Conf. Ser.* **1491** 012052
[9] Vinnik L P 1977 *Phys. Earth Planet. Inter.* **15** 39.
[10] Zhu L and Kanamori H 2000 *J. Geophys. Res.* **105** 2969.
[11] Ligorría J P and Ammon C J 1999 *Bull. Seismol. Soc. Am.* **89** 5.
[12] Collings R, Rietbrock A, Lange D, Tilmann F, Nippress S, and Natawidjaja D 2013 *J. Geophys Res. Solid Earth* **118** 5372.
[13] Syuhada, Anggono T, and Febriani F 2020 *AIP Conf. Proc.* **2256** 090004.
[14] Sakaguchi K, Gilbert H, and Zandt G 2006 *Geoph. Res. Lett.* **33** L20305.
[15] Bora D K, Borah K, and Goyal A 2016 *J. Asian Earth Sci.* **121** 127.

Crustal Structure Study in Lampung Region Using Teleseismic Receiver Function Analysis

A R Puhi[1]*, P Ariyanto[1], B Pranata[2], and B S Prayitno[2]

[1] School of Meteorology, Climatology, and Geophysics (STMKG) Banten, Indonesia
[2] Agency for Meteorology, Climatology and Geophysics (BMKG), Indonesia

*Email : aan.rahardji.puhi@stmkg.ac.id

Abstract. Lampung region is seismically and volcanic active because located in subduction zone of Indo-Australian and Eurasian plate. We applied receiver function and stacking *H-k* analysis to estimate the crustal structure in Lampung region. We used teleseismic earthquake data (epicenter distance 30°-90°) and M>6 recorded at 3 seismic broadband stations owned by Agency for Meteorology Climatology and Geophysics (BMKG). Those stations are PSLI (located on Sebesi Island approximately 20 km from Anak Krakatau) represented volcanic arc zone, KASI (located on Kota Agung, Lampung) represented Sumatran Fault Zone and KLI (located on Kotabumi, Lampung) represented back-arc basin. Crustal thickness estimated at PSLI station 32-36 km, KASI station 36-40 km, and KLI station 30-36 km. Furthermore, in 3 stations P wave velocity estimated 4.1-11 km/s, S wave velocity 2.2-6.2 km/s, while *vp/vs* value estimated 1.7-2.05. We estimated Anak Krakatau volcano's magma chamber beneath PSLI station in depth 16-30 km, Great Sumatran Fault structure in depth about 8-14 km beneath KASI station, and thick sediment layer about 4 km near surface beneath KLI station. This study result is expected to explain more detail crustal of Lampung region and can be useful for developing of BMKG's seismic monitoring systems and other geophysical fields in future.

1. Introduction

Lampung region located in the southern part of Sumatra Island has complex tectonic. This makes Lampung region seismically active that we can see on Lampung seismicity map at Figure 1. This condition is caused by Lampung located in subduction zone of Indo-Australian plate to Eurasian plate that move with velocity 7 cm/year [1]. Earthquake source in Lampung region divided in two, those come from subduction zone and from Great Sumatran Fault (GSF). There are 3 segments of Great Sumatran Fault crossed Lampung region, those are Sunda segment, Semangko segment, and Kumering segment [2]. This Kumering segment was recorded to caused earthquake in Liwa, West Lampung on 1994 with M6.6 and hipocenter depth 23 km. This earthquake caused 217 people died, more than 2,000 injured, more than 6,000 house, shops and buildings were permanently damaged [3].

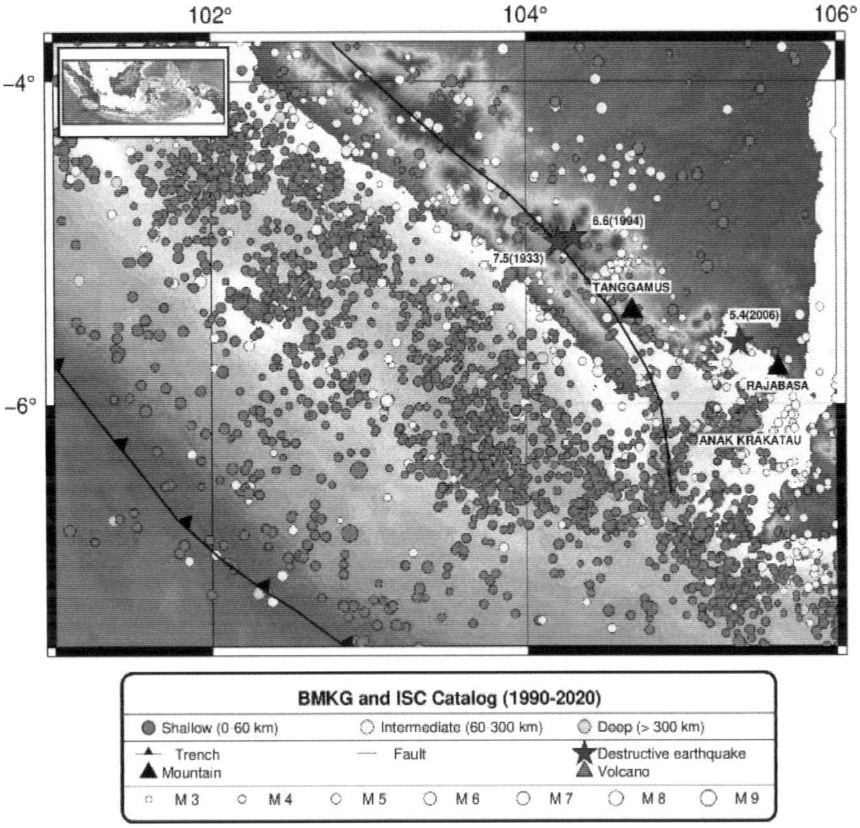

Figure 1. Lampung Region Seismicity Map (BMKG [4] and ISC [5])

Lampung region not only has an active seismically, but also has a very active volcanic condition. This condition is caused by the presence of Krakatau volcano which now being Anak Krakatau volcano in Sunda Strait. Anak Krakatau Volcano was recorded to have erupted in December 22nd, 2018 and caused volcano tsunami with maximum run up 13.5 m and inundation of 330 m [6].

With the high level of seismicity and volcanic activity in Lampung region as well as the history of destructive earthquake and volcanic eruption have occurred, so it is very important to study about crustal structure in Lampung region. This study aim is to explain more detail crustal structure of Lampung region using receiver function analysis. Previous study using receiver function method such as [7-11]. We expect this study result can be useful for developing the capability of BMKG's seismic monitoring systems and research other geophysical fields in the future.

2. Method

In this study, we used time series earthquake seismogram data recorded at 3 broadband stations owned by Meteorology Climatology and Geophysics Agency (BMKG) which can be accessed in webdc BMKG earthquake catalog [4]. Those stations are PSLI represented volcanic arc zone (located on Sebesi Island, Sunda Strait approximately 20 km to Anak Krakatau volcano), KASI represented Great Sumatran Fault zone, and KLI stations represented back-arc basin. Earthquake data criteria used are teleseismic earthquake events with epicenter distance of 30°-90° from broadband stations and earthquake magnitude more than 6 (M>6) with high signal to noise ratio (SNR) and clear P arrival phase.

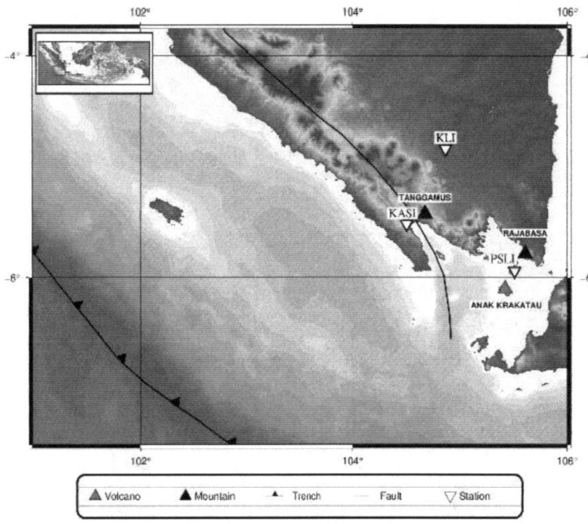

Figure 2. Seismic Broadband 3 Components Distribution

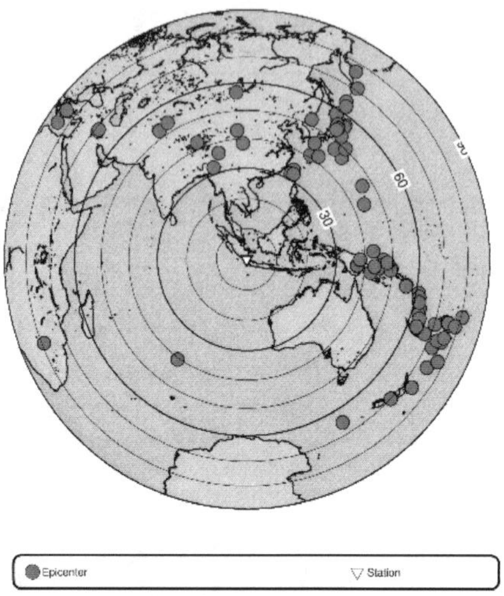

Figure 3. Teleseismic Data Used in this Study. Data Obtained from Agency for Meteorology Climatology and Geophysics (BMKG) Catalog.

2.1. Receiver function method

Receiver function is a method to estimate earth structure using teleseismic earthquake recorded at 3 components broadband receiver station. This method introduced by Langston [12]. This method utilizes the phase conversion of seismic waves that occurs when the seismic waves pass through different impedances. By eliminating the influence of source effect and instrument response, the seismogram remains information about medium of seismic wave in its propagation from source to receiver. The diagram of receiver function can be seen on Figure 4.

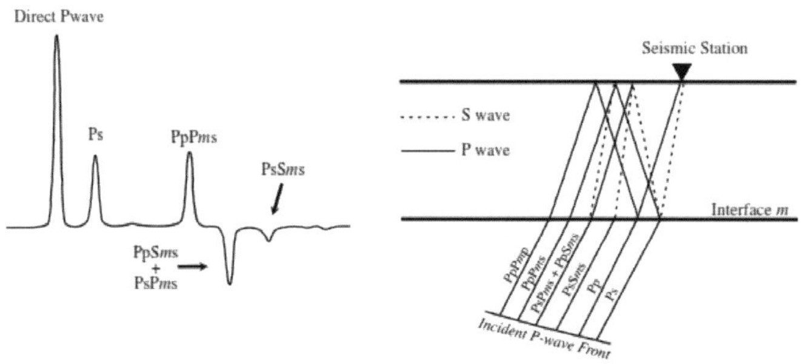

Figure 4. Receiver Function Diagram (left) and Schema of Receiver Function Come from Source to Receiver (right) [13]

After data preparation, the processing start with removed instrument response to eliminate the influence of seismograph amplifier to get the original signal from the earthquake event. Then we rotated the seismogram to change seismogram orientation from ZNE coordinate system (vertical, north-south, and east-west) to ZRT coordinate system (vertical, radial, and transverse) to clarify receiver function signal with following equation:

$$\begin{bmatrix} R \\ T \\ Z \end{bmatrix} = \begin{bmatrix} -\cos\alpha & -\sin\alpha & 0 \\ \sin\alpha & -\cos\alpha & 0 \\ 0 & 0 & 1 \end{bmatrix} \begin{bmatrix} N \\ E \\ Z \end{bmatrix} \quad (1)$$

Where α is back-azimuth, Z is vertical component, T is transverse component, and R is radial component.

Then iterative time domain deconvolution [14] is applied to produce receiver function with time window 5 second before and 25 second after P arrival. To produce receiver function signal with low frequency, we applied low pass gaussian filter with width of 1,5 which able to separate frequencies higher than 0.75 Hz. Furthermore, to modelling 1D P and S wave velocity, an inversion processing is applied to the stacked receiver function with the initial model using AK135 [15]. The process of removed instrument response, seismogram rotation, iterative time domain deconvolution, and inversion used program CPS330 [16].

2.2. Stacking H-k

To estimate crustal thickness and *vp/vs* value, we also used stacking *H-k* method in Matlab program. This method is developed by [17]. This method estimates crustal thickness and *vp/vs* value by calculating difference time arrival of receiver function phase Ps, PpPs, PpSS+PsPs to arrival of direct P phase. Stacking *H-k* method is matematically formulated as following equation.

$$s(H,\kappa) = \sum_{j=1}^{N} w_1 r_j\,(t_1[H,\kappa]) + w_2 r_j(t_2[H,\kappa]) - w_3 r_j(t_3[H,\kappa]) \quad (2)$$

Where N is number of receiver function stacked, $\sum W_i$ weighting factor with total 1, in this study we applied weight 0,7, 0,2, 0,1, $r_j(t)$ is receiver function amplitude, and t_1, t_2, and t_3 is the difference time arrival of Ps, PpPs, and PpSs phase against direct P phase.

3. Result and Discussion

3.1. PSLI Station

PSLI station also known as Rajabasa station is located in Sebesi Island, Sunda strait approximately 20 km from Anak Krakatau volcano. We got total 13 traces of receiver function fit>85% from earthquake event data recorded from 2020 to April 2021. Receiver function radial component of PSLI station is shown in the Figure below.

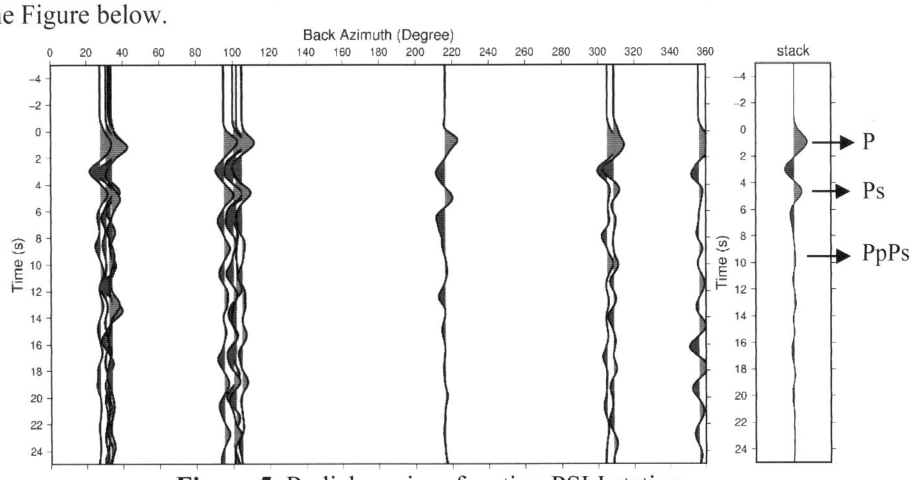

Figure 5. Radial receiver function PSLI station

From Figure 5, we observed direct P phase delayed and arrived in 1s. This delay time according to [18] this delay time is caused by thick sediment layer near surface and it is confirmed by [19] that identified 5 km sediment near surface of Anak Krakatau volcano. We also observed the strong negative phase consistent from all back-azimuth and according to [20] this negative phase indicated the low velocity zone associated with Anak Krakatau volcano's magma chamber. Then, PpPs observed in 10s.

To see the layering structure under the PSLI station, then we reprocessing the receiver function with high resolution (gaussian filter width 5) and migrated into the depth domain as we can see in Figure 6. The results obtained is fit with the model from [19] with slightly different depths. See at

 2110 (2021) 012002 doi:10.1088/1742-6596/2110/1/012002

green ellipse on Figure 6. The first positive phase after direct P wave arrival was observed at depth of 5 km that showed discontinuity between the sedimentary layer and granitic basement. Then at depth of 10 km, the next positive phase is observed that showed a discontinuity between granitic basement layer and diorite/gabbro layer. Furthermore, at depth 29-40 km, a positive Ps phase was observed which indicated a discontinuity between the diorite/gabbro layer and Moho discontinuity in the upper mantle (shown by yellow dotted line). A strong negative phase is also observed which estimated to be a low velocity zone from Anak Krakatau's magma chamber at depth about 20 km (shown by black dotted line).

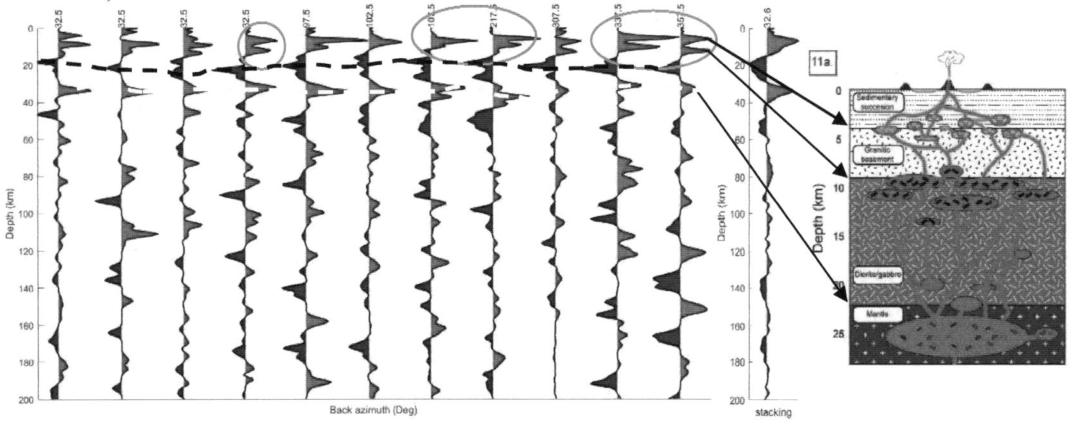

Figure 6. Receiver Function of PSLI Station Migration (left), and Schematic Illustration of Magma Plumbing System of Anak Krakatau Volcano [19]

The result of P and S wave velocity modelling from inversion processing are shown in the following Figure. We observed that the P and S wave velocity beneath PSLI is fluctuated. P wave velocity estimated in interval 4.1 km/s-11 km/s, while S wave velocity in interval 2.3 km/s-6.2 km/s. On the surface to a depth of 4 km, a low S wave velocity 2.3 km/s-3.2 km/s is observed and it is associated with a sediment layer. The P wave velocity is observed decrease at depth of 14-30 km from 11 km/s to 6,6 km/s, while S velocity also observed decrease at depth 16-30 km from 6.2 km/s to 3.6 km/s. This depth interval estimated as the low velocity of Anak Krakatau magma chamber. This result also fit with [21] using the seismic tomography method to identified the magma chamber at depth about 10-20 km, while [22] with local tomography identified magma chamber about down to 4 km. Furthermore, the most significant P and S wave velocity increation was observed at depth about 32-36 km/s which estimated to be Moho discontinuity.

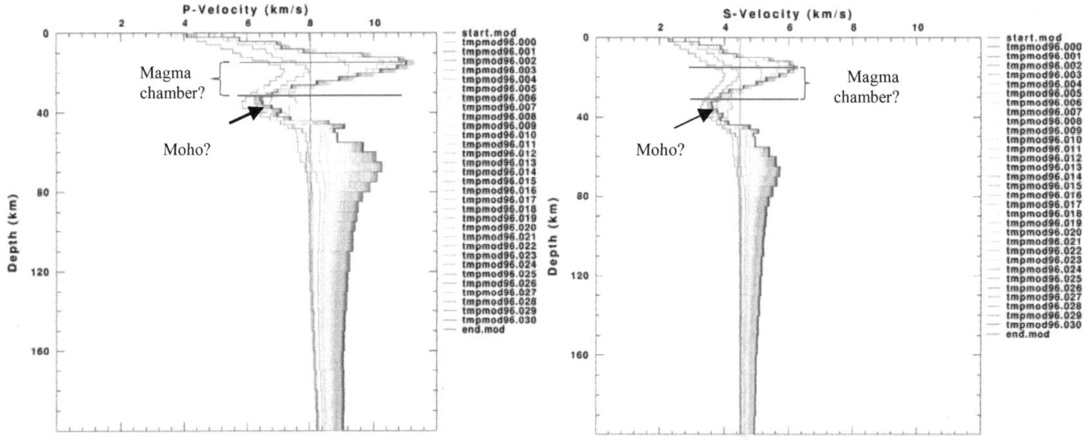

Figure 7. P and S wave velocity model of PSLI Station

Crustal thickness estimation in PSLI station is about 26.59 and *vp/vs* value 2.05 using stacking *H-k* method can be seen at Figure below. The crustal thickness estimation about 26.59 km using stacking *H-k* method is different with crustal thickness identification by velocity modelling and receiver function migration. According to [23], the stacking *H-k* method application in anisotropic zone is less accurate, so crustal thickness estimation in PSLI is about 32-36 km by contrast increation of velocity modelling. This depth interval is not much different with several previous study such as, [24] identified Moho depth about 22 km beneath Anak Krakatau volcano using seismic method, and [25] using seismic reflection in Sunda strait identified Moho depth at 28 km.

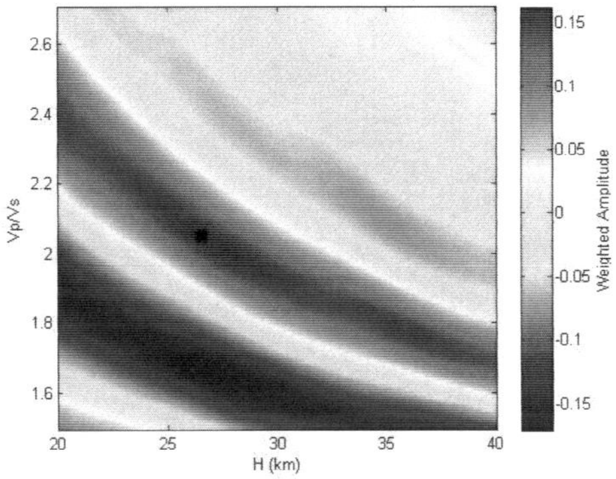

Figure 8. Stacking *H-k* of PSLI station.

3.2. KASI Station

KASI station also known as Kota Agung station is located in Kota Agung, Tanggamus district. We got total 44 trace receiver function fit>90% from earthquake event 2016-2018. Receiver function radial component of KASI station is show in the Figure 9.

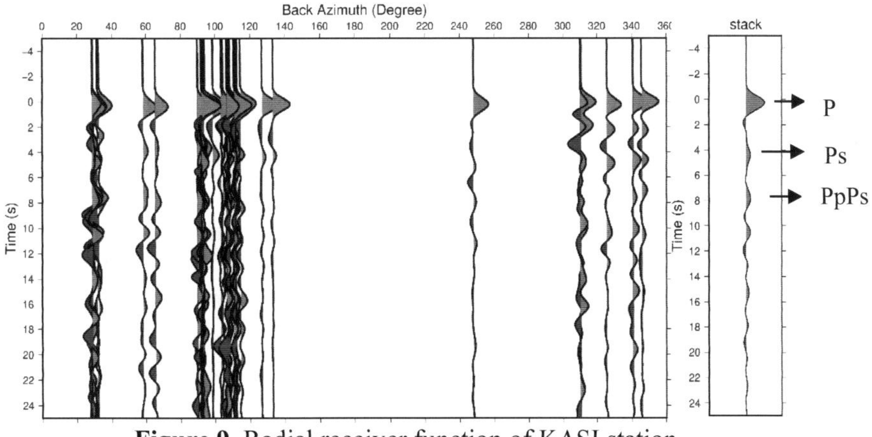

Figure 9. Radial receiver function of KASI station

From Figure 9, we observed direct P phase at 0 s, followed by a small negative phase at 2s, Ps phase which is the conversion phase of P wave into the S wave at the Moho discontinuity is observed at 4s, and PpPs phase at 8s.

Figure 10 shows P and S velocity model of KASI station. We observed that P and S wave velocity fluctuated, P wave velocity is in interval 4.1 km/s-8 km/s, while S wave velocity is in interval 2.2-4.4 km/s. P wave velocity is observed decrease from 6 km/s while S velocity decrease 3.2 km/s to 2.8 km/s at depth of 8-14 km. This depth interval estimated as the structure of Great Sumatran Fault. According to [26] who identified fracture zone of fault has relatively small density, and [27] identified the linear corelation between velocity and density. We also observed significant increase of P and S wave velocity at depth about 36-40 km estimated as Moho discontinuity.

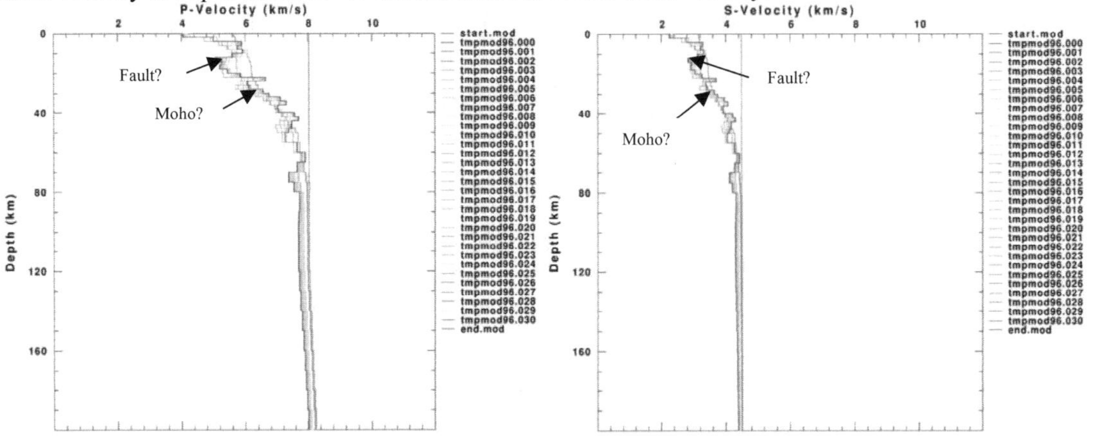

Figure 10. P and S wave velocity Model of KASI station

Crustal thickness estimation in KASI station is about 37.18 km and *vp/vs* value 1.7 using stacking *H-k* method shown at Figure below. This depth also fit with the identification of depth Moho by P and S wave velocity about 36-40 km.

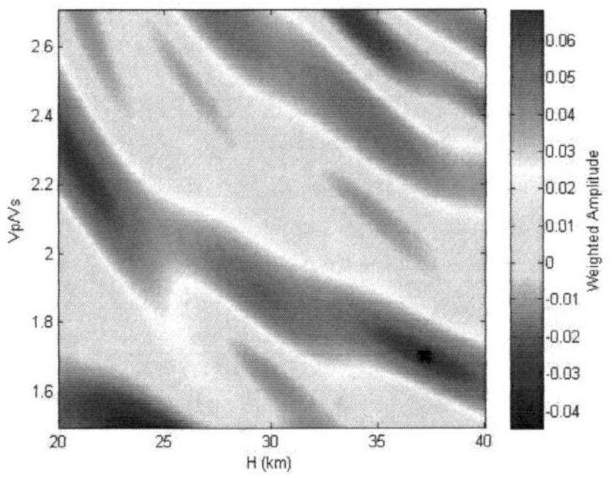

Figure 11. Stacking *H-k* of KASI station

3.3. *KLI Station*

KLI station also known as Kotabumi station is located in Kotabumi, North Lampung. We got total 47 trace receiver function fit>90% from earthquake event 2016-2017. Receiver function radial component of KASI station is shown in the Figure 12. From Figure 12, we observed direct P phase delay and arrived in 1s. according to [18], this delay time is caused by thick sediment layer. Then Ps phase is observed at 3s, while PpPs in 5s.

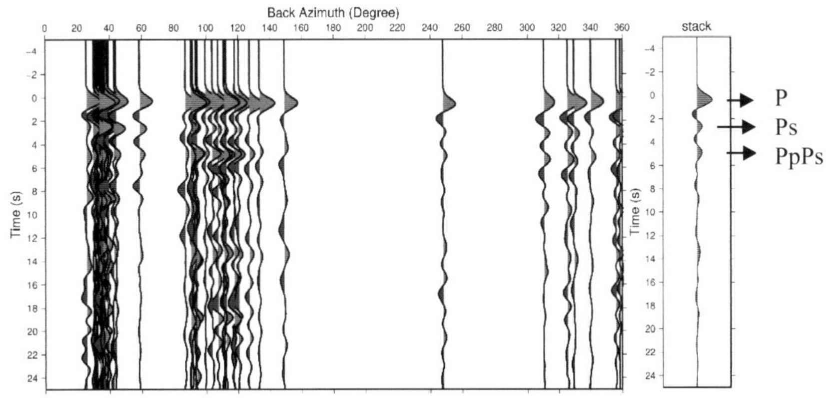

Figure 12. Radial receiver function of KLI station

Figure 13 shows P and S wave velocity model by inversion processing of KLI station. P and S wave velocity is observed fluctuated, P wave velocity is in interval about 4.6-9 km/s while S wave velocity is in interval about 2.6-5 km/s. A low S wave velocity observed from surface to a depth of 4 km associated with sediment layer. A significant P and S wave velocity increase was observed at depth about 30-36 km estimated as Moho discontinuity.

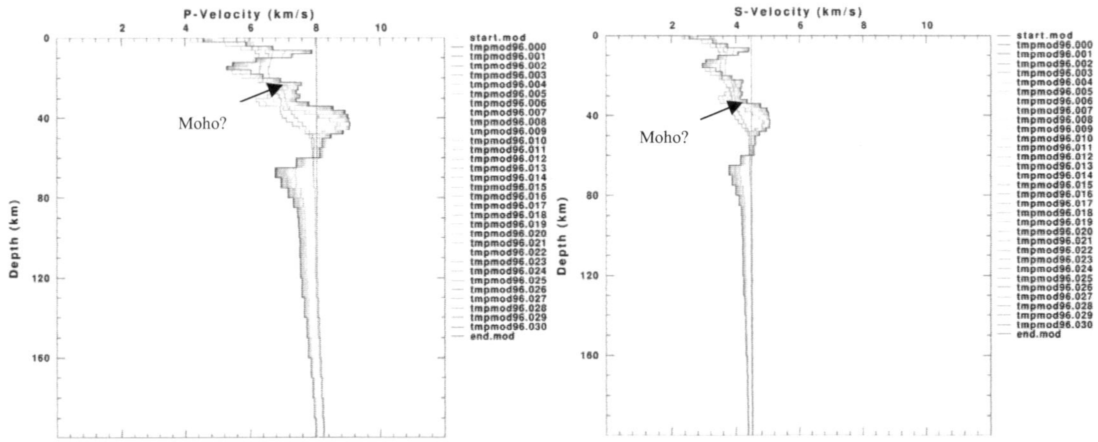

Figure 13. P and S wave velocity model of KLI station

Crustal thickness estimation in KLI station is about 28.14 km and *vp/vs* value 2.05 using stacking *H-k* method can be seen at Figure 14. This depth result by stacking is different with crustal thickness estimation by velocity modelling. According to [23], the stacking *H-k* method application in anisotropic zone is less accurate, so crustal thickness estimation in KLI station is about 30-36 km by contrast increation of velocity modelling. This depth interval is not much different with several previous study such as [28] and [29] in Southern Sumatran Basin zone using receiver function. The crustal thickness in this zone is about 27-33 km.

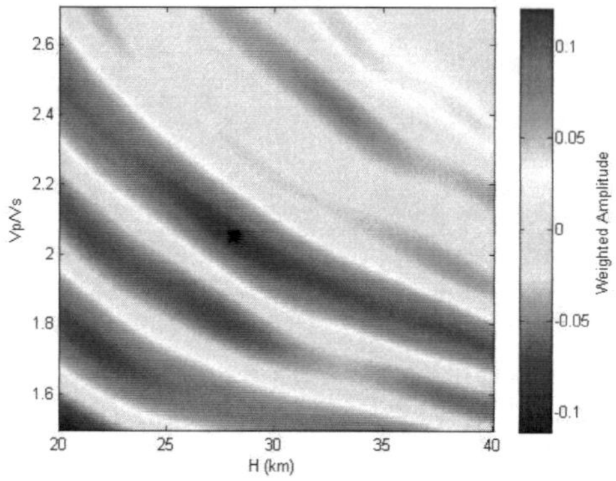

Figure 14. Stacking *H-k* of KLI station

4. Conclusion

We estimated crustal thickness, P and S velocity modelling, and *vp/vs* value in 3 seismic broadband station in Lampung. Crustal thickness estimated at PSLI station 32-36 km, at KASI station 36-40 km, and at KLI station is 30-36 km. Then P wave velocity in 3 seismic stations estimated in interval 4.1-11 km/s, S wave velocity estimated 2.2-6.2 km/s, and *vp/vs* value estimated 1.7-2.05. We observed high *vp/vs* value 2.05 and decrease of P and S wave velocity in depth 16-30 km beneath PSLI station associated with presence of Anak Krakatau volcano's magma chamber. Then we observed the decreasion of seismic velocity at KASI station in depth about 8-14 km associated with Great Sumatran Fault structure. Furthermore, in KLI station we also observed high *vp/vs* value 2.05 and low velocity of S wave near surface to depth 4 km association as sediment layer. We expect this study result can be useful for developing the capability of BMKG's seismic monitoring systems and research other geophysical fields in the future.

Acknowledgements

We acknowledge Agency for Meteorology, Climatology, and Geophysics for providing us earthquake data used in this study. We also thank Earthquake and Tsunami Study Committee for supporting this study.

References

[1] Barber A, Crow M J, and Milsom J S 2005 *J. Geol. Mag.* **143** 933
[2] Sieh K and Natawidjaja D 2000 *J. Geophys. Res.* **105** 28295
[3] Setyono U, et al 2019 *Katalog Gempa Bumi Signifikan dan Merusak 1821-2018* (Jakarta: Pusat Gempabumi dan Tsunami Kedeputian Bidang Geofisika Badan Meteorologi Klimatologi dan Geofisika)
[4] BMKG 2021 *Data Archives* Retrieved from: https://geof.bmkg.go.id/webdc3/
[5] ISC-EHB Bulletin 2020 *Bulletin Search 2020* Retrieved from: http://www.isc.ac.uk/iscbulletin/search/catalogue/
[6] Muhari A, Heidarzadeh M, Susmoro H, Nugroho H D, Kriswati E, Supartoyo, Wijanarto A B, Imamura F, and Arikawa T 2019 *Pure Appl. Geophys.* **176** 5219
[7] Ariyanto P, Ariwibowo S, Wijaya A, and Pratama R 2020 *J. Phys. Conf. Ser.* **1491** 012055
[8] Ariyanto P, Rosid, M S, Syuhada, and Januarti Y 2021 *J. Phys. Conf. Ser.* **1805** 012018
[9] Pratama R, Ariyanto P, Wijaya A, and Ariwibowo S 2020 *J. Phys. Conf. Ser.* **1491** 012052
[10] Syuhada, Anggono T, and Febrian F 2019 *J. Phys. Conf. Ser.* **1191** 012017
[11] Anggono T, Syuhada, Febrianti F, and Amran A 2020 *J. Phys. Conf. Ser.* **1568** 012030
[12] Langston C A 1979 *J. Geophys. Res.* **84** 4749
[13] Ligorría J P 2000 *An Investigation of the Mantle-Crust Transition Beneath North America and Poisson's Ratio of the North America Crust* Thesis (America: Saint Louis University)
[14] Ligorría J P and Ammon C J 1999 *Bull. Seismol. Soc. Am.* **89** 1395

[15] Kennett B L N, Engdahl E R, and Buland R *Geophys. J. Int.* **122** 108
[16] Herrmann R B 2013 *Seismol. Res. Lett.* **84** 1081
[17] Zhu L and Kanamori H 2000 *J. Geophys. Res.* **105** 2969
[18] Yu Y, Song J, Liu K H, and Gao S S 2015 *J. Geophys. Res. Solid Earth* **120** 3208
[19] Dahren B, Troll V R, Andersson U B, Chadwick J P, Gardner M F, Jaxybulatov K, and Koulakov I 2012 *Contrib. Mineral. Petrol.* **163** 631
[20] Cassidy J F 1995 *Can. J. Earth Sci.* **32** 938
[21] Rosalia S, Widiyantoro S, Nugraha A D, and Supendi P 2019 *Earthq. Sci.* **32** 12
[22] Jaxybulatov K, Koulakov I, Seht M I von, Klinge K, Reichert C, Dahren B, and Troll V R 2011 *J. Volcanol. Geotherm. Res.* **206** 96
[23] Syuhada, Anggono T, and Febriani F 2020 *AIP Conf. Proc.* **2256** 090004
[24] Harjono H, Diament M, Nouaili L, and Dubois J 1989 *J. Volcanol. Geotherm. Res.* **39** 335
[25] Lelgemann H, Gutscher M A, Bialas J, Flueh E R, Weinrebe W, and Reichert C 2000 *Geophys. Res. Lett.* **27** 3545
[26] Setianingsih, Efendi R, Kadir W G A, Santoso D, Abdullah C I, and Alawiyah S 2013 *Procedia Environ. Sci.* **17** 248
[27] Barton P J 1986 *Geophys. J. Royal Astronom. Soc.* **87** 195
[28] Macpherson K A, Hidayat D, and Goh S H 2012 *J. Asian Earth Sci.* **46** 161
[29] Bora D. K, Borah K, and Goyal A 2016 *J. Asian Earth Sci.* **121** 127

Comparison Study of Crustal Structure in Aceh Region based on Volcanic Arc System using Receiver Function Method

R I Mahardiana[1]*, P Ariyanto[1], B Pranata[2], and B S Prayitno[2]

[1] School of Meteorology, Climatology, and Geophysics (STMKG), Indonesia
[2] Agency for Meteorology, Climatology and Geophysics (BMKG), Indonesia

*Email: rainy.ismy.mahardiana@stmkg.ac.id

Abstract. Aceh region has a very complex crustal structure from the forearc ridge to the backarc basin. This study aims to determine the velocity model of P and S waves and the depth of Moho discontinuity. This research was conducted using teleseismic earthquake data (30°-90° from the station) with M>6 from four seismic stations belonging to the BMKG in Aceh region. The stations are qualified based on the volcanic arc system zone. Furthermore, the velocity model determined by result of forward modelling, while the depth of the Moho layer estimated by migrated receiver function from time domain to the depth domain. At station SNSI that represented the forearc ridge zone, the depth of Moho is ±28 km, at station TPTI represent the forearc basin is ±16 km, while at zone with higher topography, namely volcanic arc zone represented by station KCSI, the Moho depth was identified at ±38 km, and the backarc basin represented by station LASI with ±40 km depth of Moho. This variation occurs because the composition of the earth's layers below the station is diverse also different topography for each station.

1. Introduction

Aceh region is located at the northern tip of Sumatra Island and it has characteristics of a tectonic structure that is generally the same as other regions in Sumatra. One of tectonic feature which is a continuation fault from the southward is the Great Sumatran Fault (GSF) lineament which plays an important role in the number of seismicity in the Aceh region. The lineage of the Sumatran Fault in Aceh Province is divided into 3 segments, namely the Tripa Segment, the Aceh Segment, and the Seulimeum segment. These segments stretch from 120 to 200 km with significant seismicity in the past [1] and active until today. In addition, the Sumatra region in western Indonesia is considered part of the Sunda Arc, which results from the confluence of the Indo-Australian and Eurasian Plates [2]. Plate subduction in the southwest of Aceh also plays a very important role in the formation of tectonic system and seismic activity in the Aceh area, including the potential to cause tsunamis. According to two complex geological conditions, Aceh Province is one of the areas with high seismic activity in Indonesia.

Because of the influence of high seismic activity, it is deemed necessary to take a deeper look at the structure of the subsurface crust, in the form of slabs, the depth of the Moho discontinuity, and the determination of the velocity profile in order to understand the geodynamic mechanism of plate movement that occurs under the earth's surface in the Aceh region. This can be reviewed through the receiver function method. This method uses a teleseismic signal as input data which will then be processed with a radial or transverse component deconvolution technique from a vertical component seismogram consisting of P-to-S conversion waves from various layers below the seismic station [3].

Research using this method was first applied by Langston [4] to describe the subsurface structure of Mount Reiner with teleseismic data recorded at the LON Longmire station, Washington. This research was conducted by rotating the wave vector horizontally after applying the deconvolution technique to eliminate the function of time and instrument response. The deconvolution technique is applied because the seismogram used is earthquake data from various depths. Furthermore, Macpherson conducted research in the Sumatra region [5]. This study estimates the speed structure under GE network seismic stations and several broadband seismic stations, namely LHMI, GSI, BKNI, and MNAI stations. Time domain iterative deconvolution is used in this study to minimize the noise present in the waves. The average Moho depth around the island of Sumatra is obtained using a linear inversion technique. From this study, the results obtained with a significant difference from the CRUST2.0 speed model. This difference is considered a representation of the diversity of the geological environment of the station used, so the results of this study have a more detailed value.

Beside to these two studies, a similar study conducted in Indonesia identified layers of the Earth's crust in the Simeuleu area using the combined inversion of the receiver and surface wave dispersion functions with this method and found a change in velocity associated with the subduction of the Indo-Australian Plate [6]. In addition, in the Simeuleu area, research using the tomography method was also carried out which resulted in a velocity model showing the intermediate velocity anomaly zone between the oceanic crust related to the uplifted topography above the oceanic crust [7].

2. Method

This study was conducted using teleseismic earthquake data with the provisions of the earthquake distance of 30°-90° from the station location and magnitude of more than 6 (M>6) from 4 seismic stations belonging to the BMKG (Agency for Meteorology, Climatology and Geophysics) which are qualified in the Aceh region. The stations used are based on the volcanic arc zone of the Aceh region, namely the forearc ridge zone represented by the SNSI station, the forearc basin zone represented by the TPTI station, the volcanic arc zone represented by the KCSI station, and the backarc zone represented by the LASI station as can be seen in Figure 1.

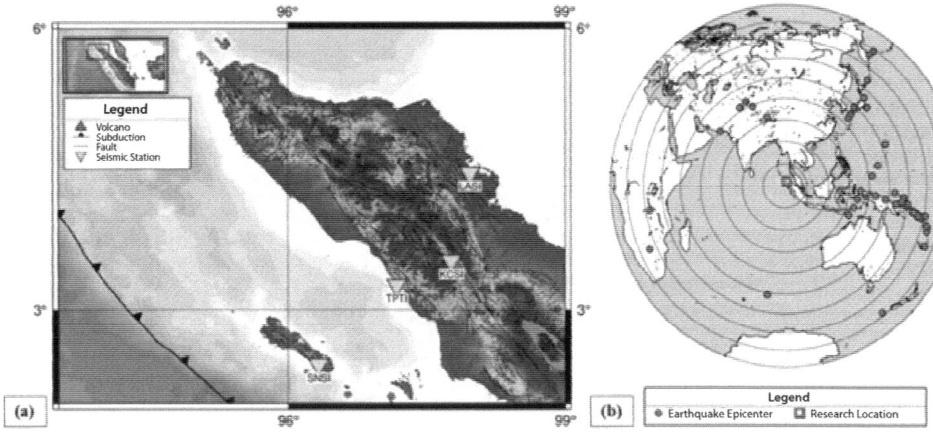

Figure 1. (a) Distribution of stations used and (b) Distribution of used teleseismic earthquakes (BMKG)

The receiver function method utilizes the propagation of body waves in distant earthquake events received by the receiving station by eliminating the source time and instrument response functions from the seismogram at the receiving station. The P wave that strikes an inclined object at the discontinuity boundary is refracted as a P wave and partly converted into an S wave and several other conversion wave phases which will later provide information about the depth of the discontinuity boundary and impedance contrast [8] as can be seen in Figure 2.

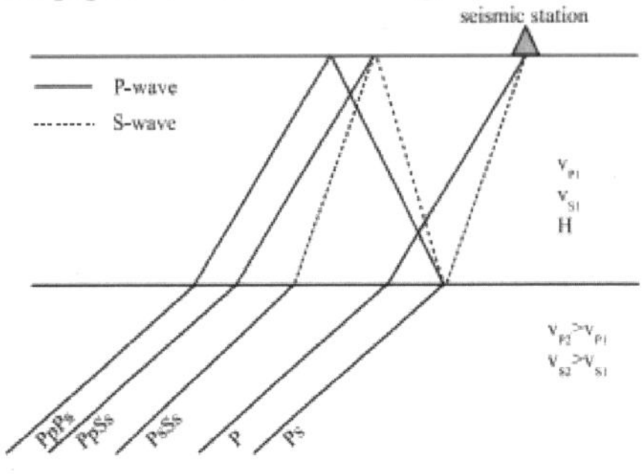

Figure 2. Receiver function diagram [8]

Qualification of teleseismic earthquake data is based on a fairly high and intact SNR (signal to noise ratio). Furthermore, the appropriate waveform needs to be recorrected using the CPS (Computer Program for Seismology) program. Seismic waves recorded in broadband seismometers are composite waves generated by many events; multi-path propagation, conversion, and splitting of waves on the earth which are heterogeneous and laterally anisotropic [9] and recorded in ZNE coordinates (vertical, north-south, east-west). Therefore, it is necessary to make corrections in the form of rotation of the seismogram components from ZNE coordinates (vertical, north-south, east-west) to ZRT coordinates (vertical, radial, transverse). The horizontal component of the original signal from the seismometer is rotated to the radial and transverse coordinates to match the direction of the earthquake wave propagation [10]. Furthermore, the correction process is continued with the deconvolution method which aims to eliminate the effect of the earthquake source so that only the receiving function is obtained. This stage is carried out on the radial component which is carried out repeatedly (iteratively) [11] until the best value is obtained (fit 90 %). In addition, at this stage filtering is also carried out to remove noise using a Gaussian filter with a width of $a = 1.5$ which can filter noise with a frequency > 0.75 Hz. In this study, the determination of the depth of the Moho layer was carried out based on the output of the inversion process through forward modelling [12] and was strengthened through the migration output of the receiver function [13]. In addition, through the output of the inversion results can also determined local velocity variations under the observation station. The formulation used in the inversion process is Equation (1).

$$y = F[x] \tag{1}$$

where y is the data from the observation that the receiver function has not been updated and x is the wave velocity vector. Next, Equation (1) is iteratively linearized to Equations (2) and (3).

$$\delta y = \nabla F|_{x_n} \cdot \delta X_n \tag{2}$$

$$x_{n+1} = x_n + \delta X_n \tag{3}$$

where $\delta X_n = x - x_n$ is the model correction vector and $\delta y = y - F[x_n]$ is the residual data vector [14], and in the iterative calculation x_0 using the AK135 global velocity model.

3. Results and Discussion

In this study, the output of the processing results is presented in the form of velocity profiles of P and S waves and the migration of the velocity model into the depth domain. The results of this processing have been arranged based on the division of the Volcanic Arc System group for each station.

3.1. SNSI Station

At the SNSI station which represents the forearc ridge zone, a Gaussian filter width of 5.0 is used because it is considered to be able for interpret the subsurface layer. At this station, 9 signals of the receiver function were obtained which were then performed inversion of forward modeling and migration to determine the depth of the Moho layer. Through the forward inversion modeling process, the fitting value between the inversion signal and the observation signal is 96.83 % as can be seen in Figure 3(c). In the P wave velocity model and the S wave velocity model at SNSI stations, it is found that there is a significant variation in velocity wave. The variation of the P wave velocity profile is seen from ~3.4–8.8 km/s from the surface to a depth of 100 km, while the variation of the S wave velocity shown is from ~1.8-4.9 km/s from the surface to depth of ~100 km, as can be seen in Figures 3(a) and 3(b). Furthermore, from this velocity profile, it is observed that there is a significant increase in velocity and it is suspected that the Moho discontinuity layer is at depth of about 24 km. This is reinforced by the migration output (Figure 3(d)) which shows a significant Ps phase at a depth of about 28 km below the top slab of the Indo-Australian continental plate subducting beneath the Eurasian continental plate.

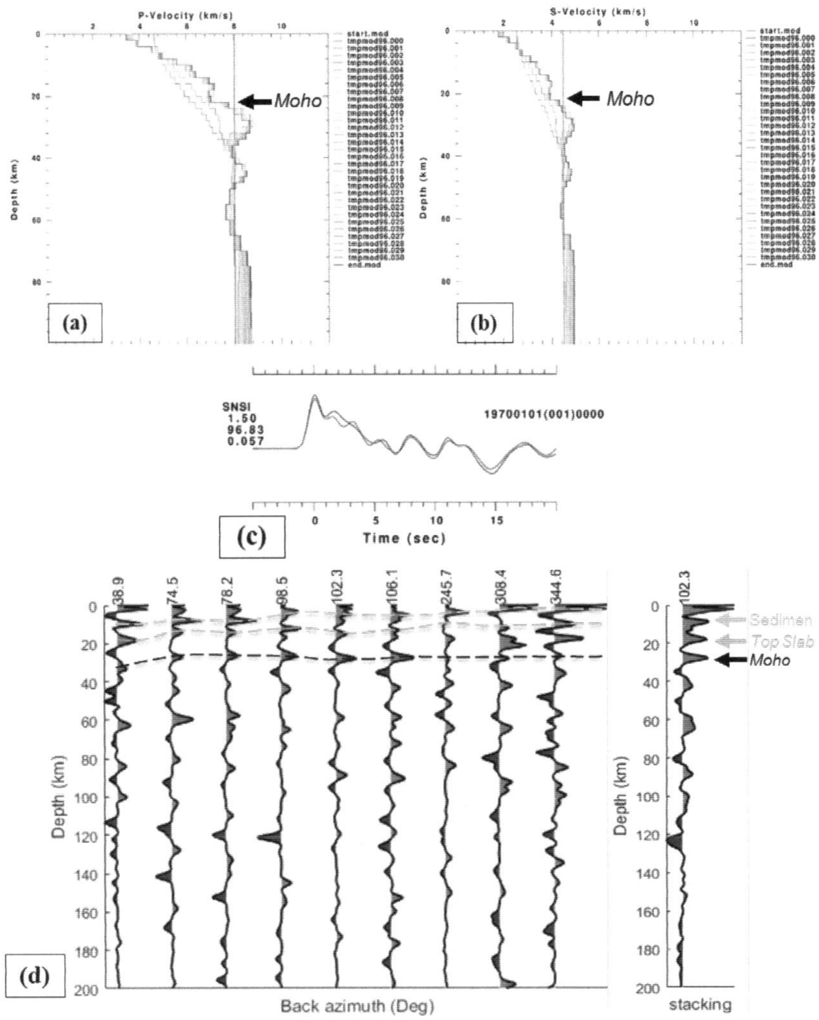

Figure 3. (a) P wave velocity profile, (b) S wave velocity profile, (c) Fitting value, and (d) Receiver function migration on SNSI Station

3.2. TPTI Station

The TPTI station in this case represents the forearc basin zone with 21 receiver function signals originating from various back-azimuths. Through the inversion process, the fitting value between the inversion signal and the observation signal is 96.03 % as can be seen in Figure 4(c). In the P-wave and S-wave velocity models of the TPTI station, it is found that there is a significant variation in wave speed. The variation of the P wave velocity profile at the TPTI station is seen at ~5.6-9.6 km/s from the surface to a depth of 100 km, while the variation of the S wave velocity shown is from ~3.1-5.2 km/s from the surface to a depth of ~100 km, as can be seen in Figures 4(a) and 4(b). From these two velocity profiles, it can also be seen that there is a significant increase in velocity which can be considered as a change in the Moho discontinuity layer, which is at a depth of about 16-24 km. This is consistent with the migration output which shows a significant Ps phase suspected to be the Moho layer at a depth of ~16 km as shown in Figure 4(d).

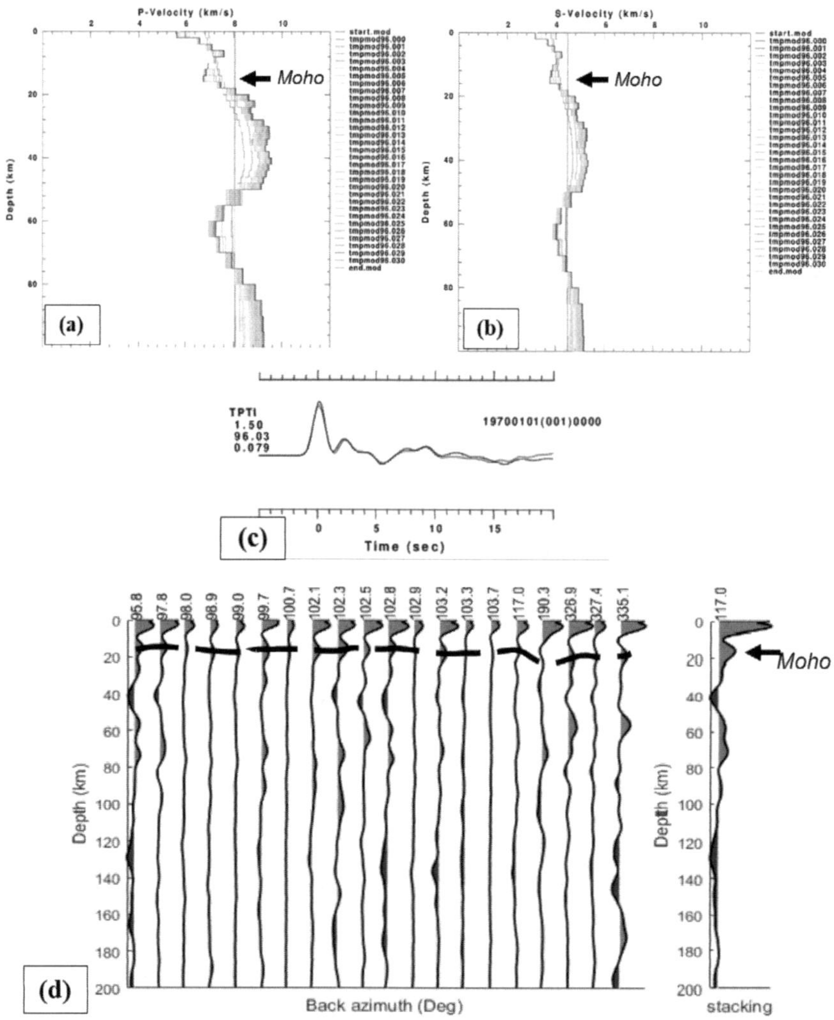

Figure 4. (a) P wave velocity profile, (b) S wave velocity profile, (c) Fitting value, and (d) Receiver function migration on TPTI Station

3.3. KCSI Station

At the KCSI station which represents the volcanic arc zone, there are 30 receiver functions with fitting value between the inversion signal and the observation signal of about 97.90 % as can be seen in Figure 5(c). The variation of the P wave velocity profile at the KCSI station starts from ~4.4-10.3 km/s from the surface to a depth of 100 km, while the variation of the S wave velocity shown starts from ~2.5-5.6 km/s from the surface to a depth of 100 km, as can be seen in Figures 5(a) and 5(b). The Moho depth of these two velocity profiles can be observed at a depth of 34-38 km. Reinforced by the migration output which shows the Moho layer is at a depth of ~40 km as shown in Figure 5(d).

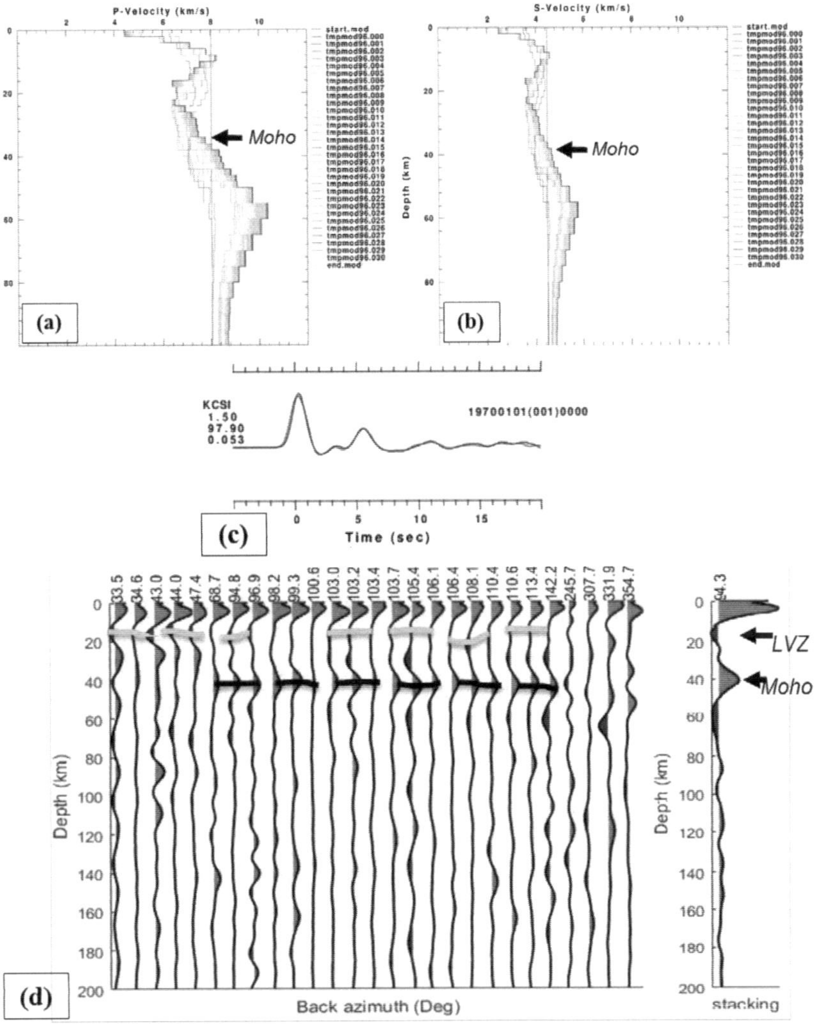

Figure 5. (a) P wave velocity profile, (b) S wave velocity profile, (c) Fitting value, and (d) Receiver function migration on KCSI Station

3.4. LASI Station

In the back arc basin zone represented by the LASI station, 14 receiver function signals were obtained from various back-azimuths. Through the inversion process, the fitting value between the inversion signal and the observation signal that does not qualify the data criteria is obtained, which is 76.37 % as can be seen in Figure 6(c). The lack of fitting values in this zone is thought to be the result of the layer below the station which is composed of thick sediment. In the P-wave and S-wave velocity models at the LASI station, it was found that there was a significant variation in velocity wave. The variation of the P wave velocity profile at the LASI station is seen at ~1.6-9.2 km/s from the surface to a depth of 100 km, while the variation of the S-wave velocity is shown from ~0.9-5.4 km/s from the surface to a depth of ~100 km, as can be seen in Figures 6(a) and 6(b). Through this velocity profile, it is observed that there is a significant increase in velocity which is considered a change in the Moho discontinuity layer, which is at a depth of 42-46 km. Then, based on the migration output, it is shown that the Moho layer is at a depth of about 40 km as shown in Figure 6(d).

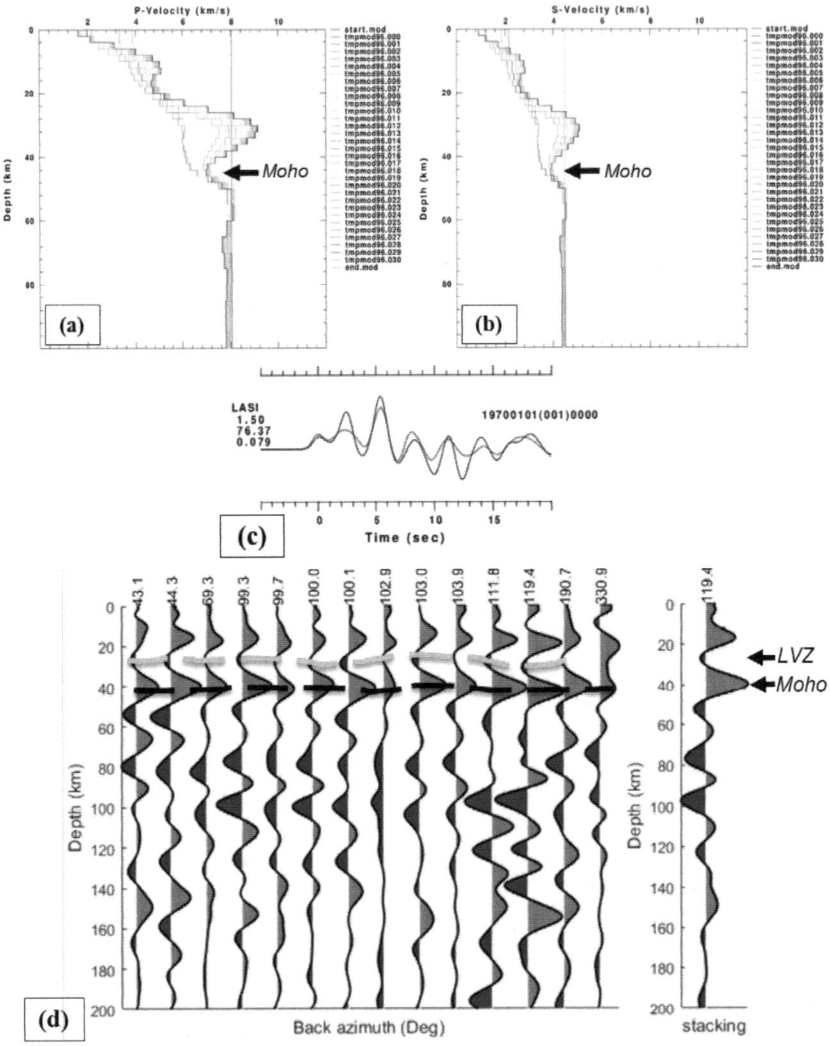

Figure 6. (a) P wave velocity profile, (b) S wave velocity profile, (c) Fitting value, and (d) Receiver function migration on LASI Station

From four stations that have been processed, the receiver function provides the Moho depth value and the velocity model with varying results. The recapitulation of the Moho depth and the values of *Vp* and *Vs* are summarized in the Table 1.

Table 1. Recapitulation of the Moho depth, *Vp*, and *Vs*

Station			RF Signal	Moho		
Zone	**Sensor**	**Elevation (m)**		**Moho Depth (km)**	**V_p (km/s)**	**V_s (km/s)**
Forearc ridge	SNSI	14.5	9	28 ± 2	8.37	4.67
Forearc basin	TPTI	9.0	21	16 ± 2	7.03	3.92
Volcanic arc	KCSI	204.7	27	40 ± 2	8.12	4.53
Back arc basin	LASI	13.0	14	40 ± 2	8.36	4.66

Based on the recapitulation of the values listed in Table 1, in the forearc ridge zone, the Moho depth value is around 28 ± 2 km. This is supported by the subduction scheme of Singh et al [15], where the Moho oceanic below the Simeuleu plateau is at a depth of about 25-28 km. Furthermore, at the TPTI station with the station elevation at 9 m above sea level, the depth of the Moho layer is around 16 ± 2 km. This depth value is supported by research by Anggono et al who have researched the receiver function using the H-k stacking method at the TPTI station, where in this study the depth of Moho was about 20 km [16]. This research also supports the results of an observation from volcanic arc zone represented by the KCSI station which states that the depth of Moho is at a depth of about 38 km. Then, in the last zone, namely the backarc basin represented by the LASI station, the Moho depth is found at a depth of 40 ± 2 km. This value was confirmed by the research of Bora et al who measured the receiver function using the Neighbourhood Algorithm (NA) inversion method at three stations located in the backarc basin zone, namely LHMI, PMBI, and BKNI, in which the thickness of the crust is up to 33 ± 2 km as the beginning of Moho layer discontinuity boundary [3]. These four stations are located in four volcanic zones and sequentially have a deeper Moho discontinuity layer associated with an increasing topography.

4. Conclusion

The depth of the Moho layer varies from the forearc ridge to the backarc associated with the topography of the measuring station. Moho's depth ranges from 16-42 km below the surface. From the forearc ridge, the depth of the Moho ranges from 26-30 km. Then, in the forearc basin zone, the Moho depth is around 16-25 km. Then, in the zone with increased station topography, namely the volcanic arc zone, the Moho depth was found to be around 37-40 km. Finally, at stations in the back arc zone, the depth of the Moho layer is deeper, at a depth of 38-40 km.

Local velocity models of P and S waves in the Aceh region also vary based on the volcanic zone. In the forearc ridge zone, the values of Vp and Vs are 8.37 km/s and 4.67 km/s, in the forearc basin zone the values of Vp and Vs are 7.03 km/s and 3.92 km/s. Furthermore, in the volcanic arc zone, the values of Vp and Vs are 8.12 km/s and 4.53 km/s, and in the volcanic backarc zone, Vp is 8.36 km/s and Vs is 4.66 km/s.

Acknowledgements

The author would like to thank all those who have contributed to this research for all constructive comments on the manuscript, Agency for Meteorology, Climatology and Geophysics (BMKG) especially for Earthquake and Tsunami Study Committee for providing data for this research, and the School of Meteorology, Climatology and Geophysics (STMKG) as a facilitator.

References

[1] Sieh K and Natawidjaja D 2000 *J. Geophys. Res.* **28** 295
[2] Puspito N T and Gunawan I 2005 *ISET J. Earthq. Technol.* **42** 111
[3] Bora D K, Borah K, and Goyal A 2016 *J. Asian Earth Sci.* **121** 127
[4] Langston C A 1979 *J. Geophys. Res.* **84** 9
[5] Macpherson K A, Hidayat D, and Goh S H 2012 *J. Asian Earth Sci.* **46** 161
[6] Anggono T, Syuhada, Febriani F, Amran, Soedjatmiko B, Prasetio A D, and Dewi C N 2020 *J. Phys. Conf. Ser.* **1568** 012030
[7] Tang G, Barton P J, McNeill L C, Henstock T J, Tilmann F, Dean S M, Jusuf M D, Djajadiharja Y S, Permana H, Klingelhoefer F, and Kopp H 2013 *Geophys. Res. Lett.* **40** 48
[8] Kieling K, Roessler D, and Krueger F 2010 *J. Seismol.* **15** 235
[9] Plesinger A, Hellweg M, and Seidl D 1986 *J. Geophys.* **59** 129
[10] Pratama R 2019 *Identifikasi Variasi Kedalaman Diskontinuitas Moho dan Slab Subduksi di Sumatra Bagian Utara dengan Metode Fungsi Penerima, Sekolah Tinggi Meteorologi Klimatologi dan Geofisika* Dissertation (Jakarta: Pertamina University)
[11] Ligorría J P and Ammon C J 1999 *Bull. Seismol. Soc. Am.* **89** 5

[12] Julià J, Ammon C J, Herrmann R B, and Correig A M 2000 *Geophys. J. Int.* **143** 99
[13] Krisnayadi M N 2011 Migrasi Pre-Stack Domain Kedalaman Menggunakan Model Kecepatan Inversi Tomografi Gelombang Normal Incidence Point (NIP) Thesis (Jakarta: Universitas Indonesia)
[14] Menke W 1989 *Geophysical Data Analysis: Discrete Inverse Theory* (San Diego: Academic Press)
[15] Singh S C, Hananto N, Mukti M, Robinson D P, Das S, Chauhan A, Carton H, Gratacos B, Midnet S, Djajadiharja Y, and Harjono H 2011 *Nat. Geosci.* **4** 308
[16] Anggono T, Syuhada, Soedjatmiko B, and Amran A 2017 *J. Phys. Conf. Ser.* **817** 012075

Geothermal Reservoir Identification in Way Ratai Area Based on Gravity Data Analysis

M Sarkowi, R C Wibowo*, and Karyanto

Geophysical Engineering, Engineering Faculty, Universitas Lampung, Indonesia

*Email: rahmat.caturwibowo@eng.unila.ac.id

Abstract. Gravity research in the Way Ratai geothermal prospect area was conducted to determine geothermal reservoirs, heat sources, and the structure of the geothermal reservoir. The research carried out includes 3D inversion modeling of gravity data. The Bouguer anomaly in the study area has 50 mGal to 120 mGal with low anomalies located in the southeast (Ketang and Kelagian), Northeast (Gedong Air, Sungai Langka, Gunung Betung) areas, and in the Pesawaran mountain area. The high anomaly is in Merawan – Hanuberak – Padang Cermin, Sumbersari and Kaliawi. The horizontal gravity gradient map analysis shows a pattern of fault structure trending northwest-southeast and southwest-northeast, according to the main fault structure in the area. 3D inversion modeling obtains a density distribution between 1.8 g/cc to 3 g/cc with a low-density distribution in the south, Mount Pesawaran/Ratai, Gunung Betung, and Sidoharum. The location of the manifestation is 9 km southeast of the Mount Ratai/Pesawaran summit. The existence of geothermal reservoirs is estimated to be in the Lubuk Badak and Miwung Hills areas which are located between the peaks of Mount Ratai/Pesawaran and geothermal manifestations. This is supported by the low-density distribution in the area and the resistivity map from audio-magnetotelluric data.

1. Introduction

The Way Ratai Geothermal area is located in a volcanic complex which has two volcanic cones side by side, namely Mount Pesawaran in the southwest and Mount Betung in the northeast. Mount Pesawaran has peaks, namely: Pematang Petai Peak, Kambing Neck Peak and, the highest, Tugu/Ratai peak (Figure 1). Mount Pesawaran and Mount Betung are part of the many volcanoes on Sumatra that form a model of the volcanic hydrothermal system or liquid-dominated high standing terrain [1,2].

The Way Ratai geothermal prospect area is geographically located in Pesawaran Regency and Bandar Lampung City. Several studies have been conducted to determine the geothermal system in the Way Ratai area. This area has an estimated reserve of 105 MW [3]. Magnetic research around the Way Ratai geothermal manifestation found a low magnetic anomaly associated with manifestations in the area. The area is estimated to be an outflow zone [4].

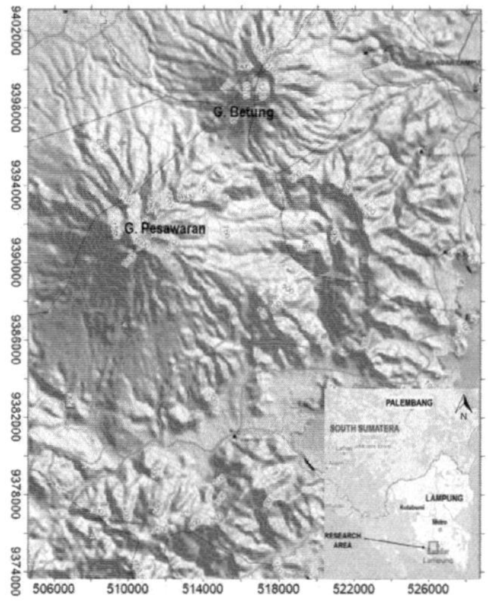

Figure 1. Research area location, Way Ratai geothermal

Radon's research shows that geothermal manifestations in the Way Ratai geothermal area (Kali Asin – Bambu Kuning – Padok – Kali Tigas) are correlated with the presence of SW-NE trending faults [5]. Taufiq conducted a gravity study only in the Way Ratai geothermal manifestation area, which found a graben to the north of the geothermal manifestation [6]. The Way Ratai geothermal manifestation area has a reasonably high temperature, which is between 85°C – 98°C. Measurement of the thermal conductivity of the area around the manifestation to get the value of thermal conductivity: 0.056 to 0.644 W/mK. This thermal conductivity value is influenced by geological structures such as the presence of faults, the presence of alteration zones, and the presence of manifestations in the area [7].

Research that previous researchers have carried out has concentrated on the area around the geothermal manifestation. In contrast, research in the north, between Mount Pesawaran and the manifestation area, has not been carried out. For this reason, in this study, processing and analysis of gravity data will be carried out in the Way Ratai geothermal prospect area, which covers 12 x 15 km and includes the area of Mount Pesawaran, Mount Betung, and geothermal manifestation areas. This study aims to obtain the structure and location of the geothermal reservoir prospect area and the heat source of the Way Ratai geothermal system.

Rock dominates the geology of the study area from young volcanoes (Qhv) both from Mount Pesawaran/ Ratai and Mount Betung, which consist of lava rock (andesite-basalt), breccia, and tuff (Figure 2). Stratigraphy of the Way Ratai geothermal field is grouped into four groups [8], namely:

a. Tertiary rocks, sedimentary rocks from the Ratai Formation, are composed of conglomerates, sandstones, breccias lava, and claystones sometimes associated with andesite tuff, which are scattered on the south-southwest side.
b. Pre-Eruption Volcanic Rocks of Mount Betung and Ratai. Groups of volcanic rocks from old to young are volcanic Gebang rock (GV), ignimbrite Gebang rock (GI), lava flow Gebang rock (GL), debris deposit (ED), and Banjarmeger volcanic rock (BV).
c. Volcanic Rocks Eruption of Mount Betung and Ratai. Quaternary volcanic rocks erupted into two eruption sources, namely the eruption of Mount Betung and Ratai at the base of the Gebang Caldera.
d. Surface Deposits, which consist of Lava (LH) and Alluvium Deposits.

Normal fault structures dominate the fault structure in the Way Ratai geothermal field and its surroundings with SE-NW and NE-SW directions. The study area is also characterized by the main trending lineaments NE-SW and SE-NW. Lineaments are present quite a lot, especially in the west, southwest, south, and slightly in the center of the study area. Geothermal manifestations are scattered in the southeastern part of the study area, namely: Kali Tiga (85°C), Padok (96°C), Bambu Kuning (90°C), Margodadi (98°C), Way Asin (85°C).

2. Method

The data used is satellite gravity data of 238 gravity points spread over Mount Pesawaran - Mount Betung and its surroundings with an area of 12 km x 15 km [9]. Gravity data is carried out by Boguer correction so that the Bouguer anomaly is obtained. Spectrum analysis was carried out on several paths so that the average depth of the regional Bouguer anomaly was carried out. The power spectrum of the Bouguer anomaly is carried out by transforming the Fourier gravity data into the frequency domain [10]:

$$F(g) = 2\pi\gamma m \frac{e^{|k|(z_0 - z_1)}}{|k|} \tag{1}$$

The energy spectrum of the equation is:

$$E(k) = \frac{4\pi^2\gamma^2\rho^2}{|k|^2} e^{-2|k|z} \tag{2}$$

$$\log E(k) = \log(4\pi^2\gamma^2\rho^2) - 2|k|z - 2\log|k| \tag{3}$$

$$\log E(k) = \log A - 2|k|z \tag{4}$$

where: Z_0 (depth of point measurement), Z_1 (center of mass depth) $Z_1 > Z_0$, $k = 2\pi/\lambda$ ((wave number), λ (wavelength), g (gravity anomaly), and ρ (density).

From these equations, it can be determined the wavelength and depth of anomaly objects. First-order horizontal gravity gradient analysis ($FHDx = \frac{\Delta g}{\Delta x}, FHDy = \frac{\Delta g}{\Delta y}, SFHD = FHDx + FHDy$) [11,12] and the second order ($SHDx = \frac{\Delta^2 g}{\Delta x^2}, SHDy = \frac{\Delta^2 g}{\Delta y^2}, SSHD = SHDx + SHDy$) [13-15] carried out to obtain fault structures/intrusions/lithological boundaries as well as sources of anomalies originating close to the surface [10-16]. Gradient analysis correlates gravity gradient with geological and geochemical maps to obtain fault structure patterns or lithological boundaries in the area [16].

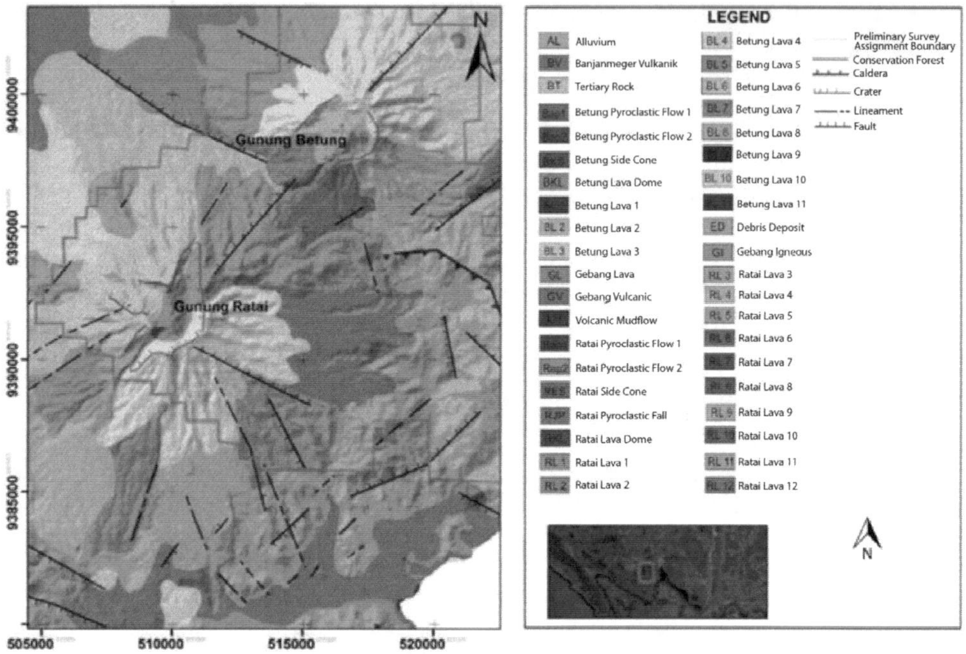

Figure 2. Research area location, Way Ratai geothermal

Bouguer anomaly 3D inversion modeling was conducted to obtain a subsurface density distribution model [17,18] using the GRAV3D 2.0 program [19]. Based on the density distribution model and correlation with geological, geochemical, and other geophysical data are obtained, the fault structure, geothermal reservoir, and heat source of the Way Ratai geothermal system.

3. Results and Discussion

The Bouguer anomaly in the study area has a value of 45 mGal to 120 mGal, with high anomalies in the western part (Merawan, Sidomulyo, Sentongan, Bunut, Hanuberak, and Padang Cermin areas which correlate with Tertiary rocks), Malawi, Sumbersari, and Sanggi areas. Low anomalies are in Mount Pesawaran, Mount Betung, and the southeastern part of the study area (Figure 3).

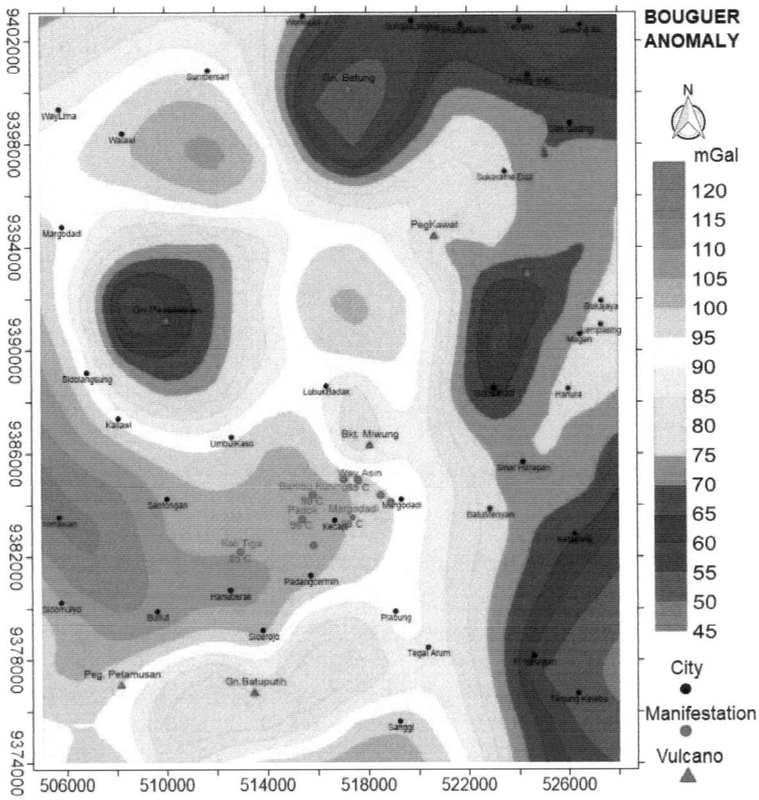

Figure 3. Bouguer anomaly map of the research area and geothermal manifestations, Mount Pesawaran, and Mount Betung location

Spectrum analysis was carried out by making a cross-sectional trajectory of the Bouguer anomaly to get the depth of the regional anomaly. The spectrum analysis results for 6 cross-sectional paths of the Bouguer anomaly with an average depth of 3,293 meters (Figure 4). The depth of this regional anomaly correlates with the average depth of the basement rocks in the area.

3.1. Bouguer anomaly horizontal gradient

To support the Bouguer anomaly analysis in identifying fault structures, lithological boundaries and to generate shallow effect anomalies, a vertical gradient analysis of the Bouguer anomaly was carried out. Theoretically, the horizontal gradient method of the Bouguer anomaly is the derivative of the Bouguer anomaly for the horizontal direction, which can be written as first-order horizontal gradient x-direction = $FHDx = \frac{\Delta g}{\Delta x}$ and first-order horizontal gradient y-direction = $FHDy = \frac{\Delta g}{\Delta y}$, while second-order horizontal gradient x-direction = $SHDx = \frac{\Delta^2 g}{\Delta x^2}$ and second-order horizontal gradient y-direction = $SHDy = \frac{\Delta^2 g}{\Delta y^2}$.

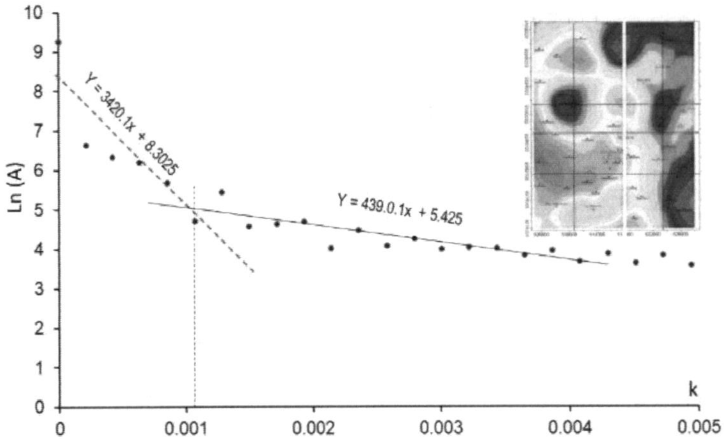

Figure 4. Spectrum analysis of the West-East trajectory, which obtained the regional anomaly boundary and residual 3,420.1 m

In general, fault structures, intrusions, lithological boundaries, and boundary models of anomaly objects are not always in the x or y-direction. The sum of the horizontal gradients in the x and y directions is used in the horizontal gradient analysis. That analysis of the sum of SFHD ($SFHD = FHDx + FHDy$) and the sum of SSHD ($SSHD = SHDx + SHDy$) are applying in this study.

Horizontal gradient analysis of the Bouguer anomaly is carried out to reveal anomalies that originate close to the surface and also to find out: boundaries of density changes, lithological boundaries, intrusions, presence of faults, and others [12,15,20]. The first-order and second-order horizontal gradient analysis in the Way Ratai geothermal area was carried out to obtain the presence of fault structures or lithological rock boundaries. The analysis is carried out by correlation or comparison with geological data and geothermal manifestations. The first-order horizontal gradient map of the study area is shown in Figure 5. The presence of a fault, intrusion, lithological boundary, and density change limit is indicated by the maximum or minimum value of the first-order horizontal gradient anomaly map [13,21]. The contour of the horizontal gradient anomaly of the order of the Bouguer anomaly shows that there is a dominant maximum and minimum anomaly pattern with a southeast-northwest trending which indicates a structure that has a southeast-northwest trending. This structural pattern is in accordance with the pattern of the main fault structure trending southeast–northwest in the area [8].

The second-order horizontal gradient map of the Bouguer anomaly in the Way Ratai geothermal area is shown in Figure 6. The presence of fault structures, lithological boundaries, intrusion structures, caldera, and density changes limits are shown by the horizontal gradient value of order-2 Bouguer anomaly = 0 (zero) [19]. Analysis of the presence of fault structures, lithological boundaries, intrusion structures, calderas, and density changes boundaries is carried out by analyzing the contours of the horizontal gradient anomaly of order 2 Bouguer anomaly with a value of 0 (zero). The second-order horizontal gradient anomaly contour of the Bouguer anomaly in the Way Ratai geothermal prospect area has a dominant 0 anomalous contour pattern with the direction of southeast-northwest, which indicates a fault with a direction southeast-northwest in the area. This follows the structural pattern of the main fault structure trending southeast-northwest in the area [8]. Besides that, there is also a horizontal gradient anomaly pattern of second-order Bouguer anomaly with a zero value trending southwest-northeast, which indicates a fault with a southeast-northwest trending in the area following the geological structure in the area [8].

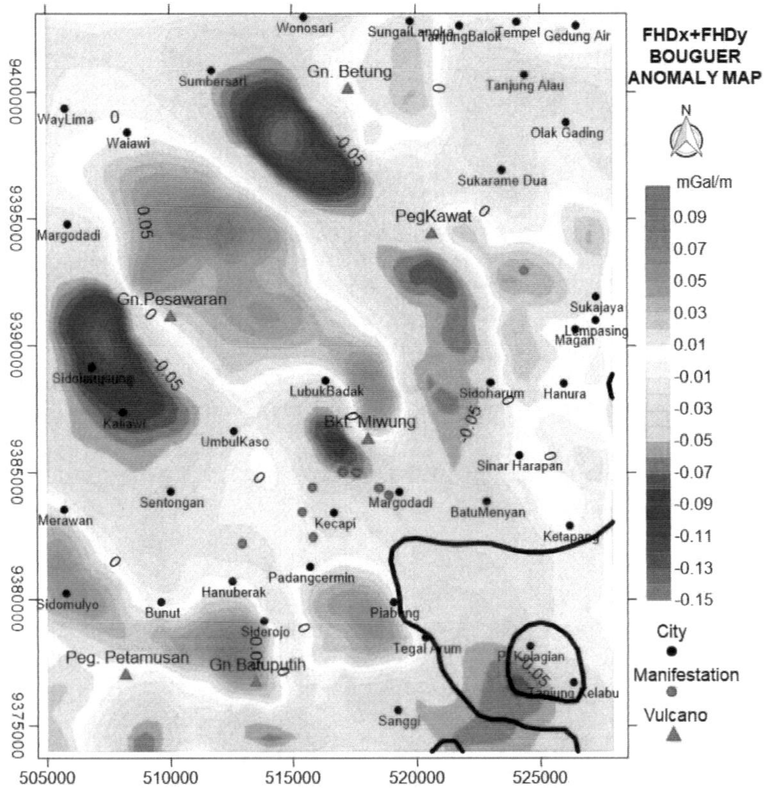

Figure 5. The first order gradient map of the Bouguer anomaly of the research area and the location of geothermal manifestations and Mount Pesawaran and Mount Betung

The second-order horizontal gradient contour of the Bouguer anomaly with a value of 0 (zero) which forms a closing closure around Mount Pesawaran and Mount Betung, is probably related to the lithology of the volcanic products of the mountain and the presence of magmatic rock under Mount Pesawaran and Mount Betung.

3.2. 3D modeling

Modeling the 3D inversion of the Bouguer anomaly is done to get a subsurface structure model of the Way Ratai geothermal field. The equation used and the calculation of the 3D inversion modeling is the subsurface model approach composed of prisms with the number of adjusting to measurement area, data grid, thickness, and depth of objects. Calculation of the gravity response of each prism block using the Plouff equation [22]:

$$g = G\Delta\rho \sum_{i=1}^{2}\sum_{j=1}^{2}\sum_{k=1}^{2}\mu_{ijk}\left[z_k \arctan\frac{x_i y_i}{z_k R_{ijk}} - x_i \log\left(R_{ijk} + y_i\right) - y_i \log\left(R_{ijk} + x_i\right)\right] \quad (5)$$

where: $R_{ijk} = \sqrt{x_i^2 + y_j^2 + z_k^2}$ and $\mu_{ijk} = (-1)^i (-1)^j (-1)^k$

Journal of Physics: Conference Series **2110** (2021) 012004 doi:10.1088/1742-6596/2110/1/012004

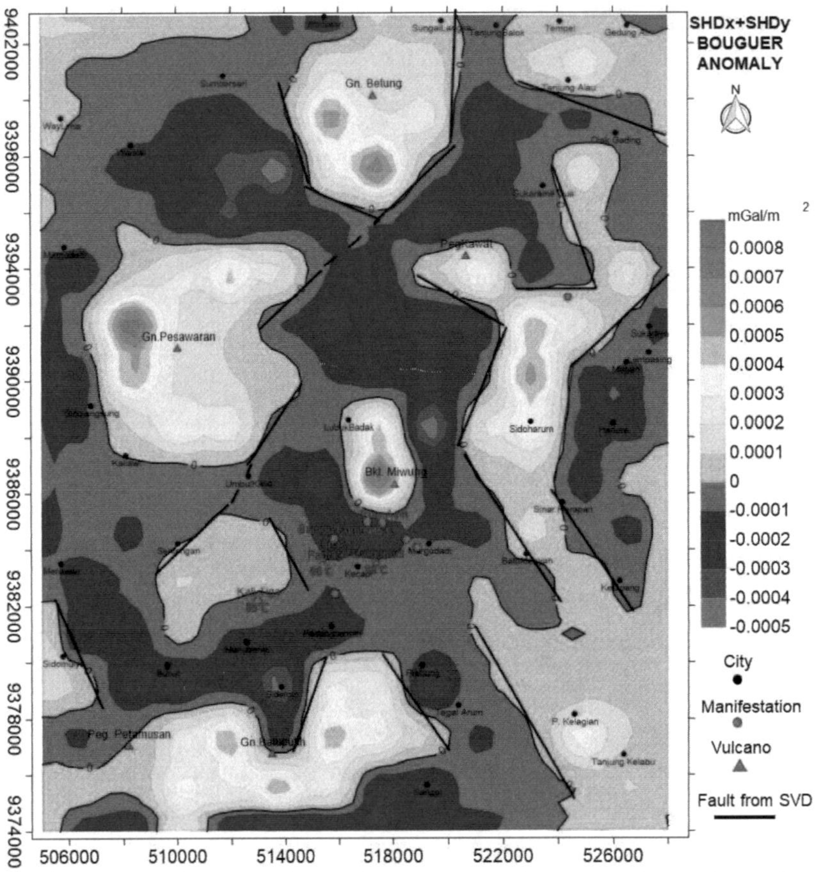

Figure 6. Second-order gradient map of the Bouguer anomaly and geothermal manifestations, Mount Pesawaran and Mount Betung location

The 3D inversion modeling of the Bouguer anomaly using the 3D Grav program is carried out by creating a subsurface object model composed of a box model with specific dimensions and depths. The box model is arranged in a mesh with the same dimensions as the area of the Bouguer anomaly data. The depth is adjusted according to geological information and depth results from spectrum analysis. Meanwhile, the input is the Bouguer anomaly data grid and the topographic data grid. The results of the 3D inversion modeling are shown in Figure 7.

The density distribution model from the 3D inversion model obtained a density value between 1.8 g/cc to 2.9 g/cc. To obtain a detailed model of the subsurface density distribution, slicing is done either horizontally or vertically so that the distribution of the subsurface density model can be known in detail [23]. To interpret the model, comparisons are made with geological data and other geophysical data [24,25]. The cross-sectional model of the subsurface density distribution in the S–N direction correlates manifestations. The fault structure model identified from the horizontal gradient analysis of first-order and second-order is shown in Figure 8. In this path, there are low-density models in the south-central and northern parts. The presence of geothermal manifestations in the middle that comes out through the fault line indicates a geothermal reservoir near the area. The possibility of a low-density model in the middle is related to the presence of a geothermal reservoir.

The cross-sectional model of the subsurface density distribution in the NW–SE direction, which correlates with the fault structure model identified from the first-order and second-order horizontal gradient analysis, is shown in Figure 9. Faults in this path limit low-density models on the left, middle, and right sides. The low density on the left under Mount Pesawaran is probably the result of former volcanic activity. In contrast, the high density on the left and under Mount Pesawaran may be related to the remaining magma that can function as a heat source for the geothermal system in the area. The low density in the middle, precisely in the Lubuk Badak and Miwung hills, can be interpreted as a geothermal reservoir supported by geothermal manifestations in the south of the area.

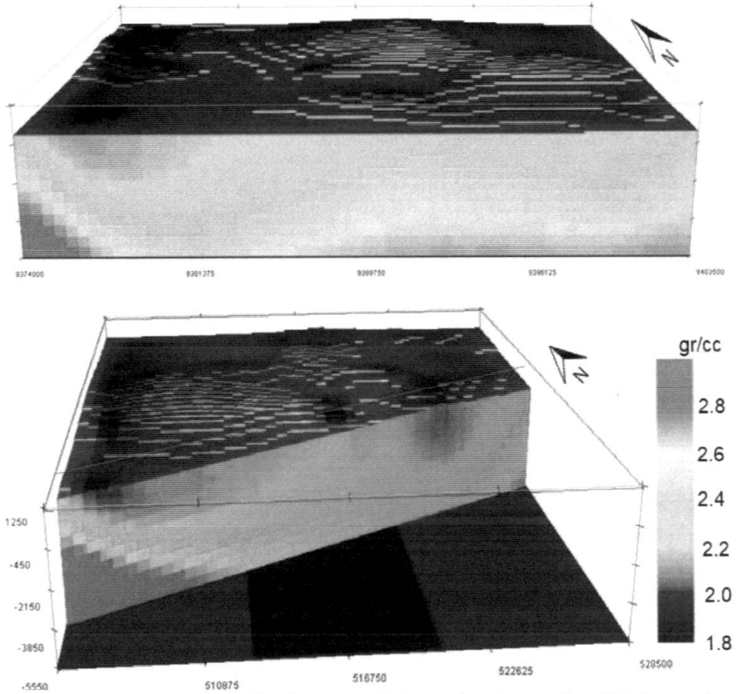

Figure 7. The subsurface density distribution model results from the 3D inversion modeling of the Bouguer anomaly in the Way Ratai geothermal area

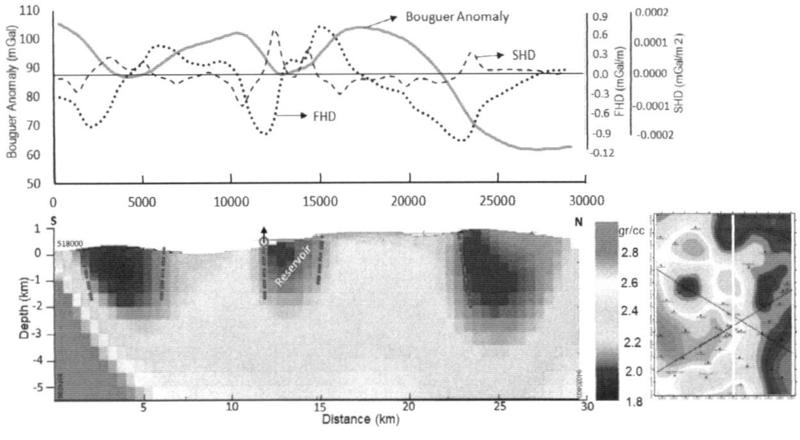

Figure 8. The cross-sectional model of the subsurface density distribution in the S–N direction correlates with manifestations. The fault structure model was identified from the analysis of horizontal gradients of first-order and second-order

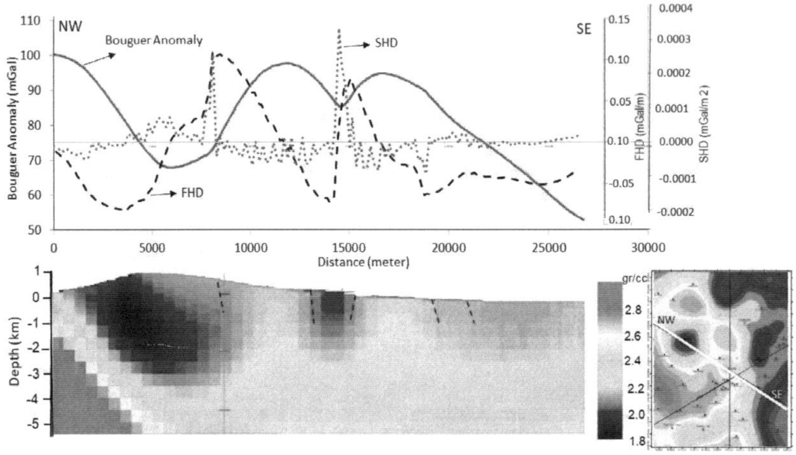

Figure 9. The cross-sectional model of the subsurface density distribution in the NW–SE direction is correlated with the fault structure model identified from the analysis of horizontal gradients of first-order and second-order

The cross-sectional model of the subsurface density distribution of the SW–NE direction correlated with the manifestation, the fault structure model identified from the first and second order horizontal gradient analysis, and the resistivity model of the AMT [23] is shown in Figure 10. In this path, there is a low-density model in Figure 10: the left (Bunut) and right (Sidoharum), which are limited by faults. The correlation of the density model in the middle (manifestation area) with the resistivity model from AMT shows a good correlation, namely the presence of manifestations in Padok and Kaliasin in the fault area, which is clearly illustrated from the AMT model and the gravity gradient analysis. The presence of caprock in the manifestation area indicates that underneath there is a geothermal reservoir around the manifestation area [27].

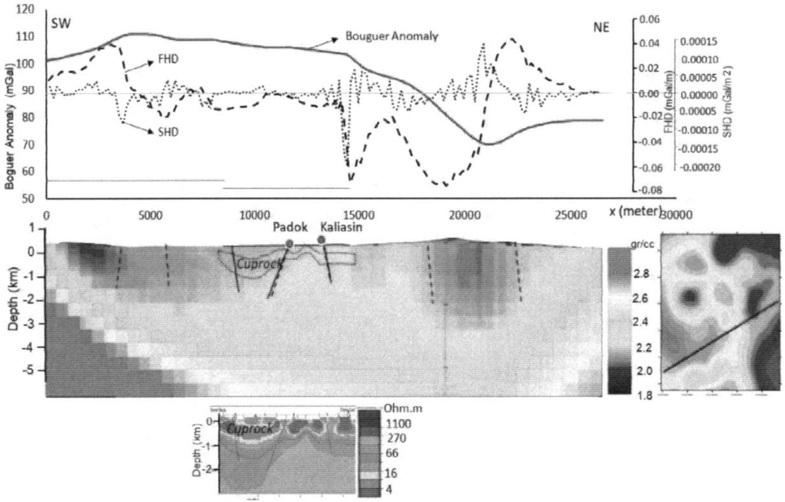

Figure 10. The cross-sectional model of the subsurface density distribution in the SW–NE direction is correlated with the resistivity model from AMT [23], the presence of manifestations, and the fault structure model identified from the horizontal gradient analysis of first-order and second-order

The cross-sectional density distribution model at a depth of 1,000 meters below MSL is correlated with contour 0 second-order horizontal gradient map, presence of manifestations, Mount Pesawaran and Mount Betung (Figure 11).

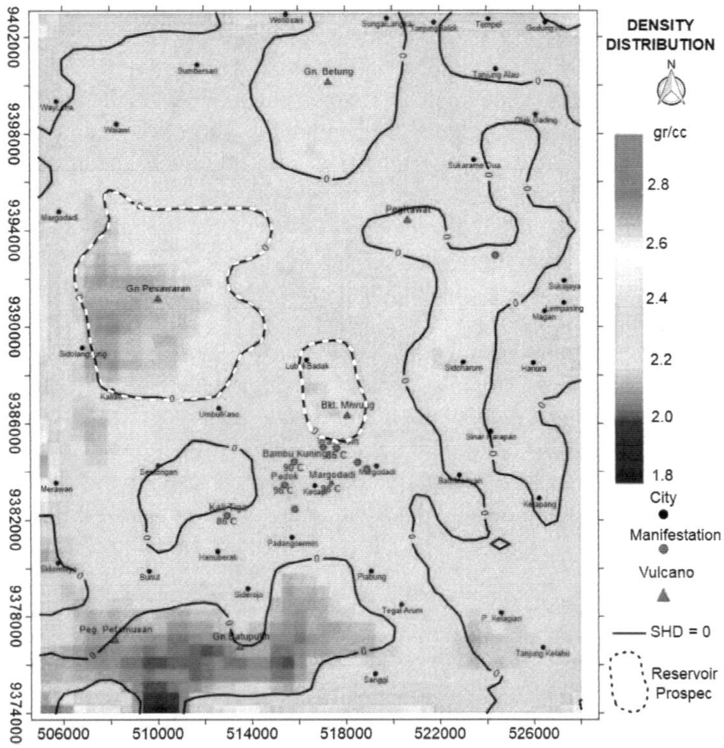

Figure 11. The cross-sectional density distribution model at a depth of 1000 meters below MSL is correlated with: contour 0 (zero) horizontal gradient map of second-order, presence of manifestations, and Mount Pesawaran and Mount Betung

The low-density distribution model (ρ < 2.2 g/cc) is located in the following areas: Gunung Pesawaran, Gunung Betung, Petamusan mountain, Mount Batuputih, Lubuk Balak, Miwung Hill, Tanjung Alu, Sidoharun and around the island of Kelagian. This low-density model is at a low-order second horizontal anomalous gradient ($SHDx + SHDy < 0$ mGal/m^2). Reservoirs in good geothermal systems generally have high porosity and permeability, and low density [24].

Based on this, areas with low density will function as good reservoirs, but not all areas with low density can function as reservoirs. The existence of a reservoir in an area is also strongly influenced by caprock, which will trap the hot fluid in the geothermal reservoir to not come out to the top. From the resistivity map, the results of AMT data processing in the area around the manifestation show a low resistivity zone which is interpreted as caprock at depths below 500 meters [23]. The caprock in the manifestation area is likely to continue to the North and Northwest.

The prospect of the Way Ratai geothermal system is a volcanic geothermal system where the heat source comes from the presence of a volcano. The heat source from the Way Ratai geothermal system may come from the Mount Pesawaran area. This is also supported by the results of the inversion modeling, which found a high density around the Mount Pesawaran area. The location of the heat source in Mount Pesawaran and the emergence of geothermal manifestations that appear in the southeast can be interpreted as that the existence of the geothermal reservoir is located between Mount Pesawaran and the manifestation area, namely in the Miwung Hills, Lubuk Badak, Umbul Kaso, and around the peaks of Mount Pesawaran. The Lubuk Badak and Bukit Miwung reservoir areas are the most prospective reservoirs of the several prospective geothermal reservoir areas. These areas have low density, and in the south, there are geothermal manifestations with high temperatures (98°C).

4. Conclusion

The Bouguer anomaly in the study area has a value of 50 mGal to 120 mGal with low anomalies located in the Southeast area (Ketang and Kelagian), the Northeast area (Gedong Air, Sungai Langka, Gunung Betung), the middle area (Bukit Miwung and Lubuk Badak) and in the middle area (Bukit Miwung and Lubuk Badak). Pesawaran Mountain. The high anomaly is in Merawan – Hanuberak – Padang Cermin, Sumbersari and Kaliawi. The horizontal gravity gradient map analysis shows a pattern of fault structure trending northwest-southeast and southwest-northeast, which is following the main fault structure in the area.

The Way Ratai geothermal prospect area is a volcanic geothermal system with a heat source located on Mount Pesawaran, supported by high anomalies around the area. The existence of geothermal potential in the Way Ratai area is indicated by the presence of geothermal manifestations in the study area spread over 6 areas, namely: Kalitiga (85°C), Padok (96°C), Margodadi (98°C), Bambu Kuning (90°C), Way Asin (85°C), and the other side of the island. West Sukajaya. The location of the manifestation is 9 km to the southeast of the summit of Mount Ratai/Pesawaran (heat source).

3D inversion modeling obtained a density distribution between 1.8 g/cc to 3 g/cc with a low-density distribution (ρ < 2.2 g/cc) located in Gunung Pesawaran, Gunung Betung, Petamusan Mountain, Mount Batuputih, Lubuk Balak, Miwung Hill, Tanjung Alu, Sidoharun and around the island of Kelagian. Based on the cross-sectional model of density distribution and horizontal gradient anomaly profile, both first-order and second-order, it is found that there is a low-density area bounded by a fault structure to the north of the manifestation. The low-density model in the Bukit Miwung and Lubuk Balak areas which are located between Mount Pesawaran and geothermal manifestations, can be interpreted as the most prospective reservoir locations in the area.

Acknowledgments

We thank the Faculty of Engineering, the University of Lampung, which has funded this research to complete this article.

References

[1] Sudarman S, Suroto, Pudyastuti K, and Aspiyo S 2000 *Proc. World Geotherm. Congr. 2000* 455
[2] Hochstein M P and Sudarman S 2008 *Geothermics* **37** 220
[3] Saefulhak Y 2017 *Potensi Panas Bumi Indonesia Jilid 1* (Jakarta: Direktorat Panas Bumi Direktorat Jenderal Energi Baru, Terbarukan dan Konservasi Energi Kementerian Energi dan Sumber Daya Mineral)
[4] Zaenudin A, Karyanto, Damayanti L, and Sarkowi M 2011 *11th Annu. Indones. Geotherm. Assoc. Meet. Conf.* 324
[5] Haerudin N, Karyanto, and Kuntoro Y 2016 *Asian Res. Publ. Netw. J. Eng. Appl. Sci.* **11** 4804
[6] Taufiq 2020 *IOP Conf. Ser. Mater. Sci. Eng.* **982** 012049
[7] Karyanto, Haerudin N, Mulyasari R, Suharno, and Manurung P 2020 *J. Earth Space Phys.* **45** 98
[8] Mangga SA, Amirudin, Suwarti T, Gafoer S, Sidarta 1993 *Peta Geologi Lembar Tanjungkarang, Sumatera, skala 1:250.000* (Jakarta: Perpusatakaan Nasional Indonesia)
[9] Anonym 2021 *Extract XYZ Grid - Topography or Gravity* Retrived from: https://topex.ucsd.edu/cgi-bin/get_data.cgi
[10] Blakely R J 1995 *Potential Theory in Gravity and Magnetic Applications* (Cambridge: Cambridge University Press)
[11] Al-Khafaji W M S 2017 *Baghdad Sci. J.* **14** 625
[12] Ming Y, Niu X, Xie X, Han Z, Li Q, and Sun S 2021 *IOP Conf. Ser. Earth Environ. Sci.* **660** 012057
[13] Sumintadireja P, Dahrin D, and Grandis H 2018 *J. Eng. Technol. Sci.* **50** 127
[14] Muhammad Y, Faisal A, Yenny A, Muzakir Z, Abubakar M, and Nazli I 2021 *J. Teknol.* **83** 145
[15] Elkins T A 1951 *Geophys.* **16** 29
[16] Sarkowi M and Wibowo R C 2021 *J. Appl. Sci. Eng.* **25** 392
[17] Witter J B, Stelling P, Knapp P, and Hinz N H 2016 *Geotherm. Resour. Counc. Trans.* **40** 647
[18] Rosid M S and Sibarani C 2021 *J. Phys. Conf. Ser.* **1816** 012083

[19] Department of Earth and Ocean Science 2005 *GRAV3D Version 2.0 A Program Library for Forward Modelling and Inversion of Gravity Data over 3D Structures* (Columbia: University of British Columbia)
[20] Cevallos C 2018 *Geophys.* **82** G115
[21] Sarkowi M 2010 *J. Sains MIPA* **16** 111
[22] Plouff D 1976 *Geophys.* **41** 727
[23] Hafidah AD, Daud Y, and Usman A 2019 *E3S Web Conf.* **125** 14008
[24] Daud Y, Rosid M S, Pati G P, Maulana M R, and Khoiroh M 2018 *AIP Conf. Proc.* **2023** 020263
[25] Puspita O D, Juniarto R, Larasati S S, Hamdalah H, and Takodama I 2020 *J. Phys. Sci. Eng.* **5** 22
[26] Suryadi, Haerudin N, Karyanto, and Sudrajat Y 2019 *J. Geofis. Eksplor.* **3** 85
[27] Maithya J, Fujimitsu Y, and Nishijima J 2020 *Geothermics* **85** 101795
[28] Siratovich P A, Heap M J, Villenueve M C, and Cole J W Reuschle T 2014 *Geotherm. Energy* **2** 10

Optical Power Splitter Based on Single Mode Directional Coupler Waveguide Using SnO_2 Nanomaterial

M Khoiro[1*], Asnawi[1], R A Firdaus[1], and M A Bustomi[2]

[1] Departement of Physics, Universitas Negeri Surabaya, Indonesia
[2] Departement of Physics, Institut Teknologi Sepuluh Nopember, Indonesia

[*]Email: muhimmatulkhoiro@unesa.ac.id

Abstract. Optical power splitter based on waveguide had been simulated numerically using Finite Difference Beam Propagation Method (FDBPM). Proposed waveguide was designed in the form of simple directional coupler waveguide. The waveguide was contained SnO_2 nanomaterial as film or the guide part and the other supporting material as cladding with lower refractive index such as flint glasses. The waveguide used 2 μm of width to establish single-mode waveguide. The structure of waveguide is divided into three parts such as input, coupling and output part. While the waveguide was modified with angle in input and output parts to avoid coupling between waveguides. Furthermore, the proposed waveguide was analysed by varying the angle and coupling length. The analysed result shows that the waveguide has best performance in angle of 0.5 degrees and coupling length of 300 μm when the propagation loss was around 0.53%. Using the parameter, the output distribution percentage of waveguide approached 55%:44.5%. This performance indicated that the proposed waveguide can be used as optical power splitter. The application is very useful for optical telecommunication networking development.

1. Introduction

During the current COVID-19 pandemic, long-distance telecommunication has become very important. The telecommunication device is necessary to continue some activities without having to meet each other. Optical device is one of the best alternatives to support development of the telecommunications. Optical telecommunication can produce ultra-fast and more effective communications, especially in the data transfer.

The process data transfers in the current telecommunication has already used optical fiber or optical waveguides. However, the transfer process is still obtained losses due to various factors. One of them occurs in branching due to the wider network. Many researchers are still looking for solutions to reduce these losses. The most popular methods are modifying the branching design of the Y waveguide [1], changing the coupling parameters of the directional coupler waveguide [2,3], and making use of the waveguide containing optical nonlinear materials [4]. Previous research has shown that nonlinear waveguides can switch signals at speeds up to terabits per second [5]. This high speed is one of the reasons why nonlinear waveguides are still the concern of many researchers until now.

In this study, the data transfer process was represented by power splitter will be investigated on a directional coupler-based waveguide. The waveguide contains SnO_2 nanomaterial as an optical nonlinear material. SnO_2 material was chosen because of many advantages such as high refractive index (2.006-2.486), high transparency (43.8-75% in visible light region), high conductivity $(\sim 10^3\ cm)^{-1}$, and wide bandgap (≥ 3.6 eV) and high nonlinearity coefficient [6,7]. SnO_2 has been developed and applied in research on waveguides such as optical logic gates [8,9] and optical divider [10]. The waveguide in this study was also developed by modifying its parameters. The waveguide was generated to be single-mode waveguide to obtain lower losses. In the future, it is hoped that the design and simulation of waveguides in this research can help the development of optical telecommunications.

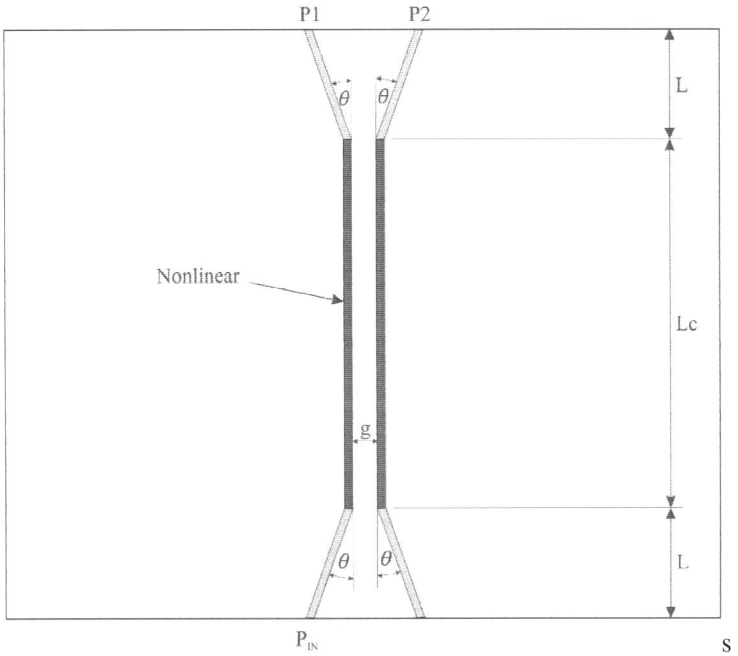

Figure 1. The proposed waveguide design

Table 1. Parameter of proposed waveguide design

Waveguide Parameter	Unit symbol	Value
Input/output length	L	200 μm
Coupling length	Lc	500 μm
Film/core width	d	2 μm
Gap	g	1.4 μm
Angle	θ	1°
Refractive index of film/core	n_{co}	2.006
Refractive index of cladding	n_{cl}	2.000
Wavelength of input signal	λ	632.8 nm
Propagation constant	β/k_0	2.0038445086913765
Nonlinearity coefficient of Kerr	α	1.7×10^{-11} m^2/V^2

2. Waveguide Design and Method

Figure 1 illustrates the proposed waveguide that designed in parallel straight waveguide as based of directional coupler. The structure of waveguide consists of three parts that are input part, central part, and output part. The input and output part of structure were arranged with a certain angle (θ) to avoid coupling effect. While the coupling phenomena was expected to occur in central part to generate power splitter function. Therefore, the part was built with two straight waveguide which separated a slight gap (g). In this study, the gap was chosen as 1.4 μm. The waveguide consists of SnO_2 nanomaterial as film or core and the other material with lower material such as flint glass where the refractive index of core (n_{co}) and cladding (n_{cl}) are 2.006 and 2.000 respectively. According to the previous research, SnO_2 has high nonlinearity coefficient which is α = 1.7 x 10^{-11} m^2/V^2. While the wavelength of input signal is chosen by λ = 632.8 μm. nonlinear material was placed in the central part where coupling was happened to increase obtained output power [11].

As single mode directional coupler, the proposed waveguide width (*d*) was arranged to ensure that the propagated signal was in basic mode TE_0. The mode of propagated optical signal was represented by effective refractive index or better known as propagation constants ($N = \beta/k_0$) which can calculate using dispersion relation equation that showed in equation 1.

$$f(x) = k_0 h\sqrt{n_f^{\,2} - N^2} - \tan^{-1}\sqrt{\frac{N^2 - n_s^{\,2}}{n_f^{\,2} - N^2}} - \tan^{-1}\sqrt{\frac{N^2 - n_k^{\,2}}{n_f^{\,2} - N^2}} - m\pi \quad (1)$$

Numerical calculation of equation results dispersion relation graph was described in Figure 2. It represented how many modes that propagated according to the film/core width. As the results, basic mode was occurred while the core width in range of 0.9 – 2 μm. The widest size was chosen to make the fabrication process easier. While *d* = 2 μm, the propagation constant was obtained on, β/k_0 = 2.0038445086913765. The overall parameter of waveguide summarized in Table 1.

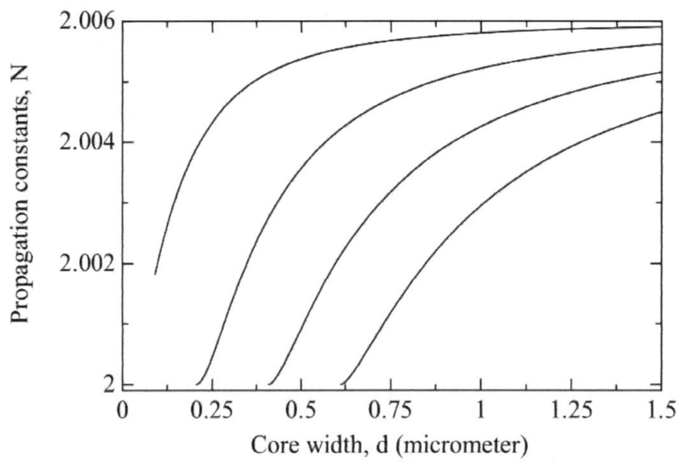

Figure 2. Dispersion relation of propagated optical signal mode in waveguide

Furthermore, signal propagation in the proposed waveguide was simulated using Finite Difference Beam Propagation Method (FDBPM) which as one of numerical calculation tools to solving differential equation for wave function during propagate in nonlinear waveguide that written by equation 2.

$$\frac{\partial^2 u}{\partial x^2} + \frac{\partial^2 u}{\partial z^2} - 2jk_0 N\frac{\partial u}{\partial z} + k_0^{\,2}\left(n^2 - N^2\right)u + k_0^{\,2}\alpha(x,z)|u|^2 u = 0 \quad (2)$$

In the study, the problem of waveguide simulation was reduced to have one-dimensional index profile in the cross section by defining effective indices to various parts of practical two-dimensional structures. The propagation of signal was built in cartesian area with $|x| < x_{max}$ (= 25μm) by calculating of the homogeneous grid with interval of the transverse coordinate as Δx = 0.05 μm and Δz = 1 μm in the propagation direction. For analysis process, input signal was given just on one the input port. The output port in the same line with input signal was labelled with P1 since the other output port is P2. Both of output ports was comparing to achieve better power splitter application.

3. Result and Discussion

The study analysis was carried out by parameter optimization to obtain best performance of waveguide, especially as power splitter. Firstly, parameter of angle in input and output part should been optimized. Angle was varied in three value which are 0.5^o, 1°, and 1.5^o. Table 2 shows output power distribution from the simulation results by varying the angle. According to the results, the smallest value of angle produces the best performance which is obtained smallest propagation loss. It is because bent in waveguide with angle can cause signal scattering as one of reason of propagation loss. Smaller angle parameter can avoid the scattering effect and obtain smaller loss [1]. Therefore, 5° of angle was chosen as fixed parameter of waveguide for further analysis.

Table 2. Output power distribution by varying in angle parameter

Angle (deg.)	Output Power in P1 (%)	Output Power in P2 (%)	Propagation Loss (%)
0.5	30.1	68.7	1.16
1.0	28.2	66.8	5.05
1.5	21.7	67.7	10.6

Coupling length parameter is one of important parameter that should be considered while perform directional coupler. In the next analysis, wave propagation in the waveguide was analysed by varying coupling length parameter in the central part. The coupling length was varied in 300 μm, 400 μm, and 500 μm. Table 3 shows output power distribution from simulation results by varying coupling length. The results indicated that power distribution had been divided almost stable while coupling length of 300 μm and 400 μm was used. It clarifies that the wave function of signal has propagated on a half beat cycle in spatial length between 300 μm and 400 μm. While in coupling length of 500 μm, power distribution of output signal tends to P2. It due to the signal begin to transmit into another waveguide in that length. The best coupling length value of power splitter waveguide is 300 μm. The value results almost balance distribution of power splitter and the lowest propagation loss. The simulation results of signal propagation in the proposed waveguide can be observed in Figure 2. The results confirm that longer the coupling length parameter between 300 μm to 500 μm, more periodic the deflection of the signal in coupling part. It can be reason why the propagation loss became higher.

Table 3. Output power distribution by varying in coupling length parameter

Coupling length (μm)	Output Power in P1 (%)	Output Power in P2 (%)	Propagation Loss (%)
300	55.0	44.5	0.53
400	43.4	55.8	0.84
500	30.1	68.7	1.63

Table 4. Comparison of previous studies with the proposed waveguide

No	Reference	Waveguide Type	Material	Coupling ratio	Losses
1	[12]	Multi-mode DC	Optical Fiber	50.2:49.2	14.6%
2	[10]	Multi-mode DC	SnO_2	44.4:44.14	0.78%
3	Antenna proposed	Single-mode DC	SnO_2	55:44.5	0.53%

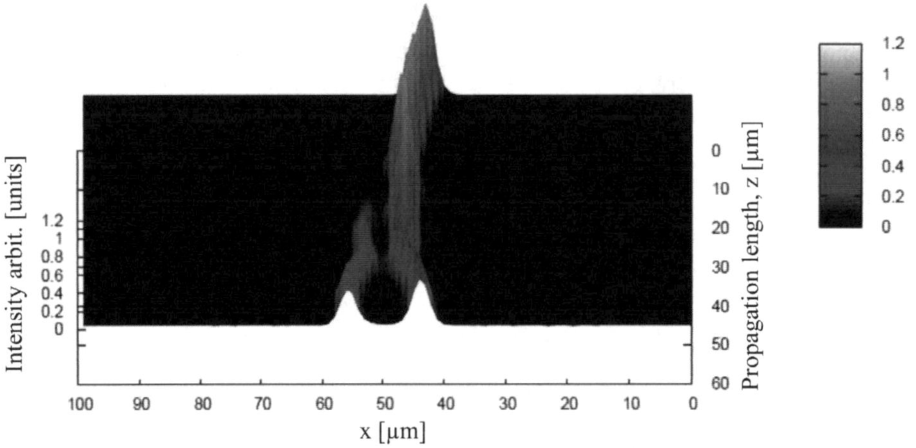

(a)

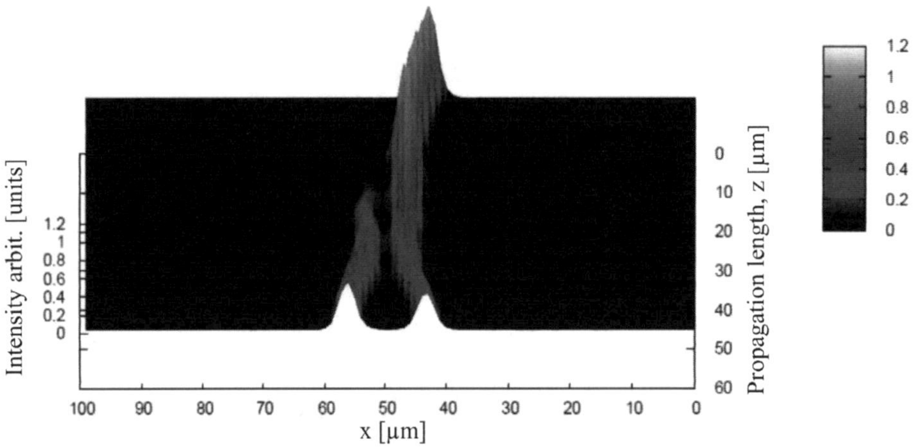

(b)

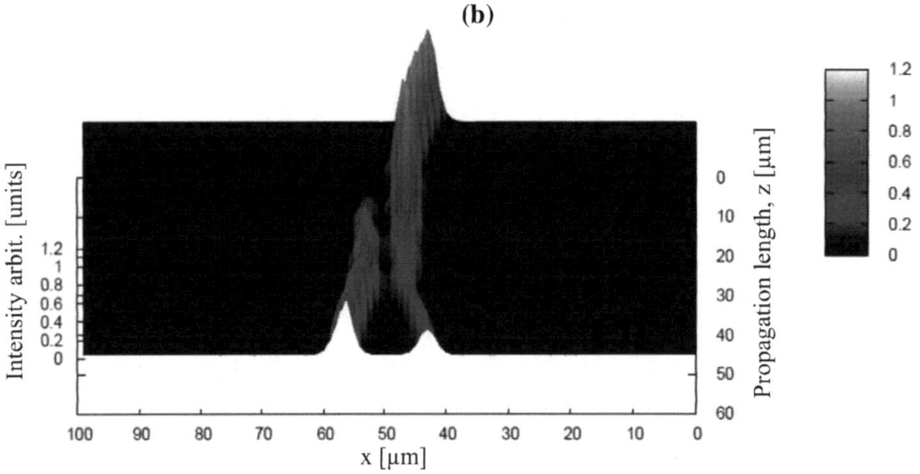

(c)

Figure 3. Propagation of signal on the proposed waveguide with coupling length of (a) 300 μm; (b) 400 μm; and (c) 500 μm

According to Table 4, the proposed waveguide has advantage by comparison with previous studies. The waveguide in this study obtains the smallest losses, that is 0.53%. It is due to the waveguide is a single-mode waveguide and contain nonlinearity from SnO_2 material which can improve the output power percentage. The power distribution also quite balance and can be considered as power splitter for telecommunication networking applications.

4. Conclusion

A power splitter based on single-mode directional coupler waveguide using SnO_2 nanomaterial had been simulated using FDBPM. On the coupling part, the core of waveguide contains nonlinearity of SnO_2 to increase output power. According to the results of simulation, the angle of 5° was chosen as an angle to arrange input and output parts because it obtained the smallest propagation loss. While the best coupling length parameter is 300 μm because it achieved more balanced output power distribution and has the smallest propagation losses. Comparing with previous studies, the proposed waveguide has the best performance in terms of waveguide losses which is 0.53%. The performance can be considered the waveguide to be applied as power splitter for optical telecommunication networking development.

References

[1] Siddiq N A, Khoiro M, Muntini M S, and Pramono Y H 2017 *Int. Sem. Sens. Instrum. Meas. Metrol. (ISSIMM) IEEE Conf. Ser.* **2017** 169
[2] Hatta A M, Farrell G, Semenova Y, and Wang Q 2014 *Opt. Int. J. Light Electron Opt.* **125** 795
[3] Purnamaningsih R W, Poespawati N R, Abuzairi T, and Dogheche E 2019 *Photonics* **6** 63
[4] Ahmadi Tameh T, Isfahani B M, Granpayeh N, and Javan A M 2011 *AEU Int. J. Electron. Commun.* **65** 281
[5] Abdeldayem H A and Frazier D O 2008 *Nonlinear Optics and Applications* (United States: NASA Goddard Space Flight Center)
[6] Rahal A, Benhaoua A, Jlassi M, and Benhaoua B 2015 *Superlattices Microstruct.* **86** 403
[7] Uysal B Ö and Arıer Ü Ö A 2015 *Appl. Surf. Sci.* **350** 74
[8] Khoiro M, Muntini M S, and Pramono Y H 2018 *Int. J. Microw. Opt. Technol.* **13** 462
[9] Khoiro M, Siddiq N A, Annovasho J, Muntini M S and Pramono Y H 2017 *Int. Semin. Sens. Instrum. Meas. Metrol. (ISSIMM) IEEE Conf. Ser.* **2017** 174
[10] Asnawi A, Siddiq N A, Muntini M S, and Pramono Y H 2019 *Int. J. Microw. Opt. Technol.* **14** 468
[11] Rahmati A T and Granpayeh N 2010 *Opt.* **121** 1631
[12] Sudarsono, Gatut Y, Hasto S, Gontjang P, Diky A, Nurrisma P, Susilo I, Iim F, Yunus A R, and Yono H 2019 *J. Phys. Conf. Ser.* **1153** 012100

Getting to Know the Last Five Years Trend on Microcontroller Research: A Bibliometric Analysis

M Yantidewi*, B Jatmiko, I Sucahyo, A Kholiq, N A Lestari, and U A Deta

Department of Physics, Universitas Negeri Surabaya, Indonesia

*Email: metayantidewi@unesa.ac.id

Abstract. A microcontroller is a compressed microcomputer manufactured to control embedded systems in office machines, robots, home appliances, motor vehicles, and several other gadgets. This device has played significant and essential roles in daily human life. Unfortunately, an in-depth analysis related to microcontrollers has not been found. Hence, it is compulsory to conduct a comprehensive analysis to understand better the research themes and trends in a specific scientific scope. This study aims to determine the extent of the research development on microcontrollers carried out throughout 2017-2021 through bibliometric analysis. From the bibliometric analysis that has been performed, we can conclude that research on microcontrollers is dominated by sensors, algorithms, voltage, and power. On the other hand, research that shows the relationship between solar panels and microcontrollers is still lacking. With this finding, it is hoped that researchers will focus their research on potential topics that are rarely looked at.

1. Introduction

Since the early 1980s, a small number of manufacturers began to build microprocessors into their products to reduce the number of chips, with the aim of lowering production costs and thereby reducing end-user prices. While microprocessors greatly improved what they replaced, they were not complete magical components for lowering the cost and complexity of product design. The problem was that in order for a microprocessor to be able to do useful things, it must be surrounded by lots of additional input-output (I/O) chips and other supporting chips. In the 1990s, silicon processing and improved chip manufacturing techniques put more circuits with additional functions and features on a single chip. This chip was then called a microcontroller [1].

The existence of microcontrollers in electronic products is no longer a question and an indisputable fact. Their use is no longer limited to home and office equipment, instruments and machines, but has also penetrated into everyday life [2]. Significant increases in capability, speed and cost reduction have contributed to the widespread use of microcontrollers. Lately, microcontrollers have been embedded in intelligent sensors [3], which are suitable for controlling processes and product functions. The revolution in information technology (IT) has made it possible to have a flexible means of controlling home appliances through microcontroller-based systems. This revolution occurs because microcontrollers bring the flexibility of program control and the computing power of computers to solve any problem. The product with an embedded microcontroller behaves as an intelligent system. These systems use intelligence to automatically control electronic devices like lights, fans, radio sets, television sets, air conditioners, security cameras, etc. They focus more on comfort and efficient utilization of electricity [4]. If necessary, it could be an autonomous user more minor system [5].

However, there has been no bibliometric analysis that has reviewed the trends of microcontroller research over the last five years. The study conducted by Hsiung did review the development of microcontrollers in technology, but the study was conducted in 1992 and did not use bibliometric analysis [6]. There are also the studies by Guven et al. [7] and Shi Y et al. [8], which discusses the concept of a microcontroller-based system and the use of a microcontroller for measurement and control. Nevertheless, both of them did not use bibliometric analysis as a research method. Therefore, it is imperative to know how far the development of microcontroller research during the last five years. Based on these facts, the objective of this study is to determine the extent of research related to microcontrollers by providing bibliometric analysis on the Scopus database so that opportunities for further research can be determined.

2. Method

Bibliometrics is a mathematical and statistical methodology that comprises scientific publications, citations, reports, and patent documents [9]. Bibliometric analysis is a scientific computer-aided review technology that recognizes and identifies core research or writers and their connections by enclosing all publications associated with a particular subject or scope [10]. The bibliometric analysis accommodates an accurate and objective method for measuring the implication of an article to the progress of knowledge and is an often-used tool for scrutinizing trends and performance in a particular subject [11]. The data analyzed in this study was taken from the Scopus database, which is well-known as one of the world's largest literature databases and provides good scientific, academic information [12,13]. The data collection was carried out in August 2021 using the keyword "microcontroller" according to the criteria "titles, keywords, and abstract" and was devoted to 2017-2021. The bibliometric analysis had been employed in this study to map the research trend on the microcontroller.

In this study, VOSviewer was used to visualize and explore the trend. This software was chosen because it provides interactive options and functions that are easy to access and explore bibliometric data networks, such as the number of citations or co-occurrence relationships between key terms and concepts [14]. VOSviewer is capable of constructing publication maps, country maps, journal maps built upon a network (co-citation), forming a keyword map based on shared networks, and establishing maps with a copious amount of objects [15].

3. Results and Discussion

A research paper's keywords are an essential element that contains meaningful information about the paper [9]. Because scientific texts may be reduced to a set of combined appearances between the words they include, keyword analysis has enabled the detection of present or future prominent research subjects based on co-occurrence analysis. If two concepts regularly appear together in one group of papers but rarely do so alone in another, their co-occurrence is exceptionally high [16].

In this study, a search for the keyword "microcontroller" from Scopus yielded 1,662 documents of various types (conference papers, book chapters, reviews, books, notes, and editorials). These 1,662 documents are the result of publications from 2017 to 2021, as seen in Figure 1.

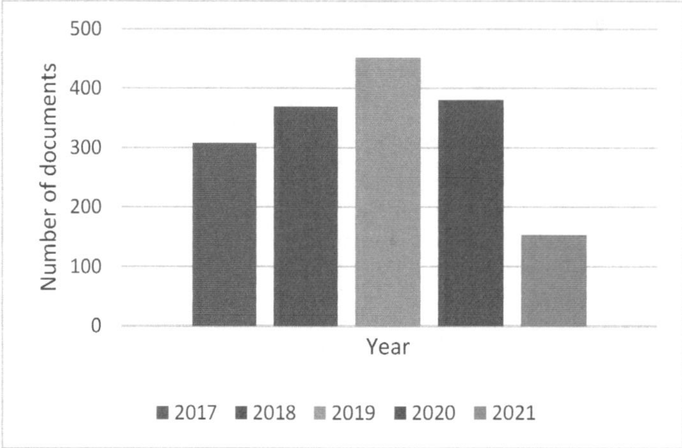

Figure 1. Number of publications on microcontroller topic in 2017-2021 taken from Scopus

We can see in Figure 1, the most microcontroller-topic research documents were produced in 2019. On the other hand, in 2021, the number of publications on this topic reached the lowest number. This number can still increase because the number of documents recorded in Scopus when this data was taken from January to early August 2021.

As previously mentioned, the 1,662 documents consisted of various types of documents with details of 1,131 conference papers, 494 journal articles, 20 book chapters, eight reviews, six books, two notes, and one editorial (Table 1). According to table 1, conference papers dominate research on the microcontroller. This dominance is followed by journal articles, whereas notes and editorial ranked at the bottom.

Table 1. The number of document types

Document type	Number
Conference paper	1,131
Journal article	494
Book chapter	20
Review	8
Book	6
Note	2
Editorial	1

3.1. The overlay visualization of microcontroller

In the overlay visualization, the novelty of the publication year of each published article can be seen based on the keywords/terms described by colour gradations from dark blue to bright yellow, or in other words, showing traces of research history from year to year. Keywords with the dark blue colour indicate that they were included in the research of around 2018. In the meantime, research with the term/keyword "pulse", which is bright yellow concluded that studies related to this keyword were published in 2019. By conducting an in-depth analysis of each yellow keyword/term in the visualization overlay, it will follow new trends in microcontroller research in the world.

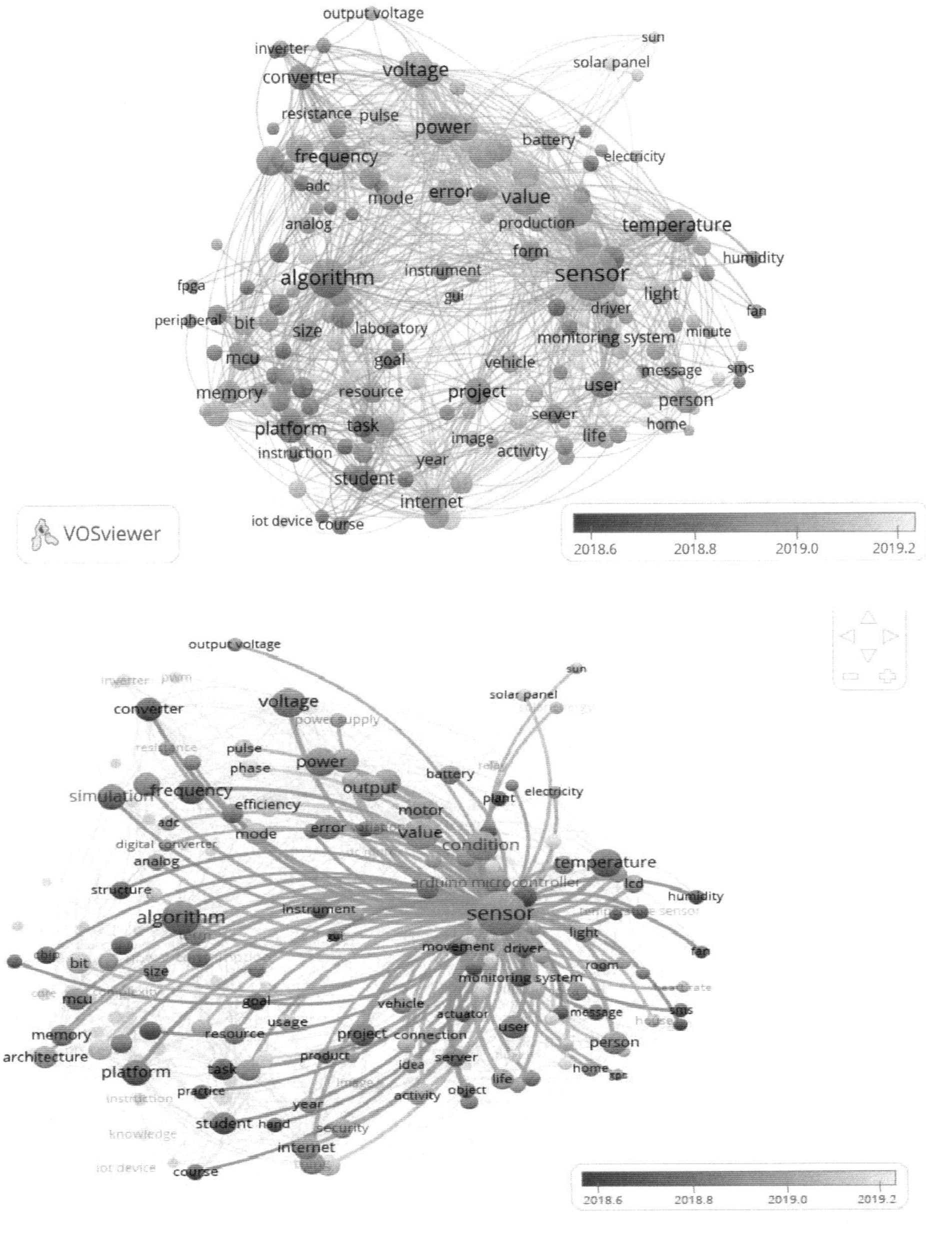

Figure 2. Overlay visualization of "Sensor Keyword"

Judging from the visualization of Figure 2, the lines connecting the points or nodes are pretty many. However, the most prominent is the nodes connecting the temperature and algorithm to the sensor. Both the temperature and algorithm have been around since 2018. Although other nodes showing 2018 also exist, but both of them look the most prominent among the others. This indicates that temperature and algorithms are the most popular topic compared to other topics such as converter, humidity, and platform.

3.2. The density visualization of microcontroller

The density view is beneficial for getting a quick overview of the overall structure of the map and for highlighting the most critical sections on the map [17].

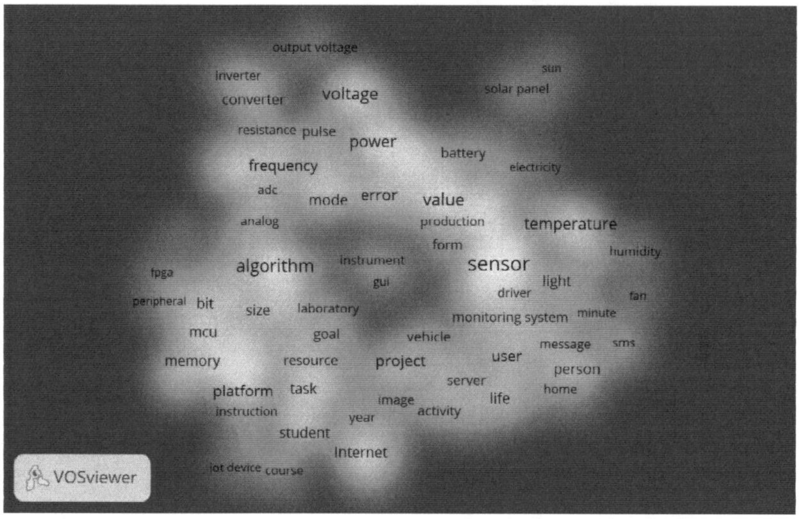

Figure 3. Density visualization of microcontroller topic during 2017-2021

Figure 3 depicts the depth of research related to the topic area. The proportion of two colours mixed depends on the total item density of a point. The closer a point's colour is to the background colour, the lower its overall item density. Also, it is reasonable to assume that higher density means the topic is well-developed to study [18].

In Figure 3, some topics are rarely investigated, for example, FPGA and solar panels. Sensors and algorithms, on the other hand, remain the prima donnas of microcontroller-related research.

3.3. The network visualization of microcontroller

Using VOSviewer, a network visualization was carried out on all these documents (Figure 4). The thickness of the connecting line indicates the strength of the topic field or keyword pair [15]. In the network, the nodes and more prominent words reflect their highest occurrence, the colours indicate clustering, and the lines show the interrelationship of the keywords [19].

There are five clusters (red, green, blue, yellow, and purple) in the mapping of microcontroller research. The cluster represented by different colours shows a relationship between one topic and another, and nodes in the same colour cluster suggested a similar topic among these publications. Of the five clusters, the yellow and purple clusters are topics that are rarely explored. It can also be seen from Figure 4 that sensor, algorithm, voltage, and power are the dominant topic in microcontroller research, and this means that researchers discuss these four topics mainly. Among the four, the sensor is the most popular one because of its largest node size, which indicates that the keyword "sensor" is chosen together in microcontroller-related research.

We can see in the red cluster that temperature and humidity are the most frequently associated topics to the sensor. From this, it can also be known that temperature and humidity sensors are the types of sensors that are developing rapidly in their research when compared to other types of sensors.

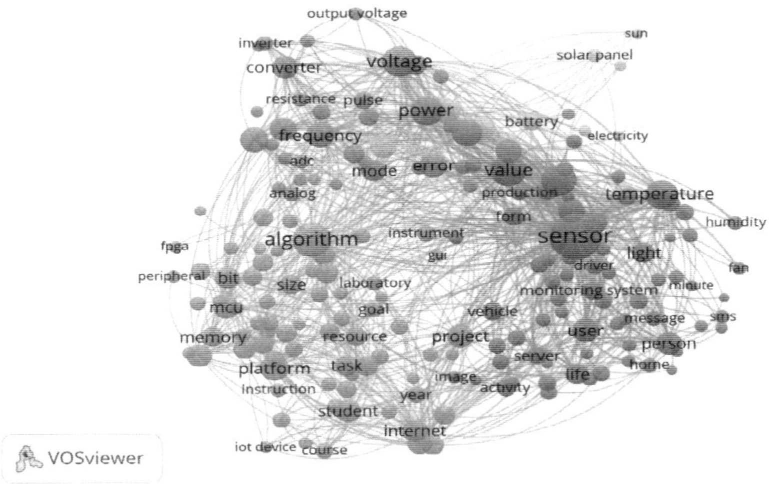

Figure 4. Network visualization of microcontroller topic during 2017-2021

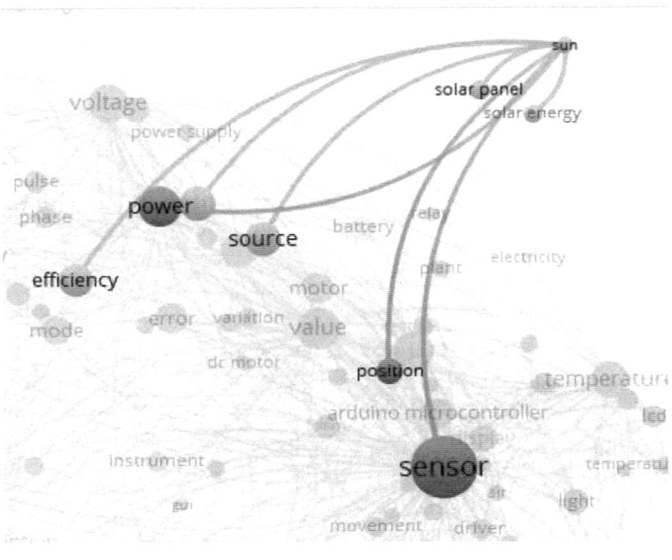

Figure 5. Network visualization of "Sun Keyword"

As we know, one of the nodes in the yellow cluster is the sun. When the sun is chosen as the keyword to visualize the network as pictured in Figure 5, it is clear that there are only seven nodes connected to it. Sensor, power, position, source, efficiency, solar panel, and solar energy are among the seven points. Note that three of the seven points are part of the red and blue clusters, namely sensors and positions that belong to the red cluster and positions that belong to the blue cluster.

Perhaps, not many researchers conduct studies on the sun because it is constrained by the high initial installation cost, which is one of the most significant drawbacks of solar energy systems [20]. This fact corresponds with the findings of David et al. They claimed that articles discussing solar energy from a technical point of view were still few compared to those discussing solar energy management [21].

4. Conclusion

From the bibliometric analysis carried out, the conclusion certified that there were five main clusters in research on microcontrollers. In addition, it was also discovered that sensors, algorithms, voltage, and power continued to dominate research on microcontrollers. On the other hand, research that showed the relationship between sun and microcontrollers was still lacking. Also, through bibliometric analysis, this research is advantageous for depicting a comprehensive research framework and allowing future scientists to focus more on rarely explored specific topics.

Acknowledgment

This study was supported by Penelitian Kebijakan FMIPA Universitas Negeri Surabaya 2021.

References

[1] Trevennor A 2012 *Practical AVR microcontroller* (Berkeley, CA: Apress)
[2] Al Dhaher A H G 2001 *Microprocess. Microsy.* **25** 317
[3] Al Dhaher A H G 2004 *Int. J. Eng. Educ.* **20** 52
[4] Vincent E and Godswill O 2018 *Researchjournali's J. Comput. Sci.* **5** 1
[5] Kelemen M, Prada E, Klemenova T, Mikova L, Virgala I, and Liptak T 2015 *Appl. Mech. Mater.* **816** 248
[6] Hsiung S C 1992 *Research and Development of Microcontroller Experiment Instructional Units and Their Effectiveness with Industrial Technology, Electronic Technology, and Electrical Engineering Technology Majors Retrospective Theses and Dissertation 1003* (USA: Iowa State University)
[7] Guven Y, Cosgun E, Kocaoglu S, Gezici H, and Yilmazlar E 2017 *Res. Inventy Int. J. Eng. Sci.* **6** 38
[8] Shi Y, Qiu G, and Wang N 2020 Basic microcontroller use for measurement and control *Introduction to Biosystem Engineering* eds Holden N M, Wolfe M L, Ogejo J A and Cummins E J (USA: American Society of Agricultural and Biological Engineers and Virginia Tech Publishing)
[9] Shi J, Miao W, and Si H 2019 *Sustain.* **11** 988
[10] Han J, Kang H, Kim M, and Kwon G H 2020 *J. Biomed. Inform.* **109** 103516
[11] Herdianto R, Windyaningrum N, Masruroh B, and Setawan M A 2021 *Belantika Pendidik.* **4** 44
[12] Hamidah I, Sriyana, and Hudha M N 2020 *Indones. J. Sci. Technol.* **5** 209
[13] Klapka O and Slaby A 2018 Visual analysis of search results in Scopus database *Digital Libraries for Open Knowledge TPDL 2018 (Lecture Notes in Computer Sciences* vol 11057*)* eds Mendez E, Crestani F, Ribeiro C, David G and Lopes J (Switzerland: Springer Nature) p 340
[14] Van Eck N J and Waltman L 2011 *ISSI Newsl.* **7** 50
[15] Efendi D N, Irwandani, Anggraini W, Jatmiko A, Rahmayanti H, Ichsan I Z, and Rahman M M 2010 *J. Phys. Conf. Ser.* **1796** 012096
[16] González-Zamar M D, Abad-Segura E, Vázquez-Cano E, and López-Meneses E 2020 *Electron.* **9** 1246
[17] Van Eck N J and Waltman L 2010 *Scientometr.* **84** 523
[18] Chen X, Chen J, Wu D, Xie Y, and Li J 2016 *Procedia Comput. Sci.* **91** 547
[19] De Moraes L L and Kafure I 2020 *RDBCI: Digit. J. Libr. Inf. Sci.* **18** e020016
[20] Kabir E, Kumar P, Kumar S, Adelodun A A, and Kim K-H 2018 *Renew. Sust. Energ. Rev.* **82** 894
[21] David T M, Rizol P M S R, Machado M A G, and Buccieri G P 2020 *Heliyon* **6** e04452

Radiation Profile of 2.45 GHz Bi-Circular Loop Antenna using Parabolic Reflector Made of Frying Pan (Wajanbolic)

J Annovasho[1,2]*, V Rahayu[2], A Ardimas[2,3] and R A Firdaus[4]

[1] Tadris Fisika, IAIN Palangka Raya, Indonesia
[2] Fisika, Universitas Billfath, Kompleks YPP Al-Fattah, Indonesia
[3] Department of Nanoscience and Technology, Chulalongkorn University, Thailand
[4] Fisika, Universitas Negeri Surabaya, Indonesia

*Email: jhelang.annovasho@iain-palangkaraya.ac.id

Abstract. This study aims to determine the dimensions of the antenna and reflector which can optimally work at a frequency of 2.45 GHz. A good antenna is an antenna that has high directive capability, high performance, and inexpensive. In this work, the proposed antenna model was a Bi-Circular Loop (BCL) with a reflector using a frying pan (Wajanbolic). The methods were used in this research for instance computational simulation, fabrication, and characterization. Simulations were carried out using the Finite Different Time Domain (FDTD) technique. The simulation results were compared with the measurement process. In the first simulation, four reflectors sizes could qualify as antennas, namely diameters of 309.00 mm, 335.00 mm, 364.00 mm, and 381.00 mm. The four reflectors sizes were optimized by changing the radius parameter of the BCL antenna. The best results were obtained on the reflector with a diameter of 364.00 mm and a BCL radius of 17.38 mm. The simulation results showed a radiation profile consisting of an RL value of -35.69 dB and a gain value of 16.40 dBi. Based on the fabrication and measurement of the antenna, the RL value was -54.75 dB and the directional antenna gain was 16.00 dBi. An antenna with such performance can be used as a point-to-point Wi-Fi transmitter.

1. Introduction

Utilization of the internet in rural areas will have many positive impacts on community development [1,2]. Internet access is limited in many villages due to the unfavourable geographical location. The difficulty is compounded by the high cost of instruments and required hardware maintenance in order to transmit internet data. The current obstacle is that there are unavailable antenna and working at a frequency of 2.45 GHz with optimal performance requirement.

Many previous studies had been carried out, such as using a step impedance resonator and using a dielectric taper [3,4]. The latest result presented that the gain value of antenna was about 4 dBi and required more costs and long process in terms of characterization and fabrication. A proposed solution for the antenna device to overcome the problematic above was to use a bi-circular loop (BCL) antenna. BCL is a dual-layer Printed Circuit Boards (PCB) based antenna. By using a flat reflector, the BCL simulation obtained a good performance with a return loss (RL) value of -68.009 dB and gain value of 9.04 dBi [5,6]. However, this antenna design did not have a good directivity value yet for long-distance point-to-point communication.

Changing the reflector type on the BCL is remarkably important in order to overcome the directivity problem. An alternative material that can be used is an aluminium frying pan. These materials have a low price, are available in the market, and allow people to carry out self-assembly and maintenance. In addition, an aluminium has constructive corrosion characteristics in consequence the long-term use does not significantly reduce the ability of the material. Several studies on Wajanbolic reflectors had been carried out, but its performance was not optimal. Yurandi et al. found that the gain generated by the Wajanbolic reflector was 17.065 dBi [7]. Moreover, Nafik et al. presented a gain value of 16 dBi on the 3G signal [8]. Kharunnisa and Umben et al. obtained the gain results in the form of signal amplification [9,10]. A gain value of 14 dB and 8 dB were presented by Endri et al. and Handoko et al., respectively [11,12]. However, all of their studies have not discussed the basic radiation profile of the antenna, namely the Return Loss (RL) parameter.

Therefore, the purpose of this study is to design antenna and reflector devices that work at a frequency of 2.45 GHz. This study also discusses the characteristics of the RL in order to ensure the antenna having optimal performance. Antenna parameters were optimized using the Finite Different Time Domain (FDTD) technique. Then, the antenna was fabricated and measured to get the value of RL, gain, and radiation pattern. This research is quite important to produce antennas that allows rural communities getting easier internet access.

2. Method

2.1. Circular loop antenna

In the far-field region (kr>>1), the magnetic and electric fields in the circular loop are expressed by [13].

$$H_\theta = -\frac{k^2a^2I_0e^{-jkr}}{4r}\sin\theta = -\frac{\pi SI_0e^{-jkr}}{\lambda^2 r}\sin\theta \tag{1}$$

$$E_\phi \simeq \eta\frac{k^2a^2I_0e^{-jkr}}{4r}\sin\theta = \eta\frac{\pi SI_0e^{-jkr}}{\lambda^2 r}\sin\theta \tag{2}$$

$$H_r \simeq H_\phi = E_r = E_\theta \tag{3}$$

The ratio between $-E_\phi/H_\theta$ is the wave impedance which can be written in Equation (4).

$$Z_w = -\frac{E_\phi}{H_\theta} \simeq \eta \tag{4}$$

where,
Z_w= wave impedance
η = characteristic impedance

The radiation intensity U is expressed in Equation (5).

$$U = r^2W_r = \frac{\eta}{2}\left(\frac{k^2a^2}{4}\right)^2 |I_0|^2 \sin^2\theta \tag{5}$$

U_{max} occurs when phase $\theta=\pi/2$, then U_{max} is presented in Equation (6).

$$U_{\max} = U\Big|_{\theta=\frac{\pi}{2}} = \frac{\eta}{2}\left(\frac{k^2a^2}{4}\right)^2 |I_0|^2 \tag{6}$$

Loop directivity is calculated in Equation (7).

$$D_0 = 4\pi\frac{U_{maks}}{P_{rad}} \tag{7}$$

The input impedance and characteristic-impedance in a circular loop are expressed by Equation (8) [14].

$$Z_i = Z_o \cdot \frac{e^{2\pi b\gamma}-1}{e^{2\pi b\gamma}+1} \tag{8}$$

Z_0 and γ are expressed in Equations (9) and (10).

$$Z_o \approx \sqrt{\frac{R + j\omega L}{j\omega C}} \tag{9}$$

$$\gamma \approx \sqrt{-\omega 2LC + j\omega CR} \tag{10}$$

where R is the resistance, L is the inductance, and C is the capacitance of the circular loop.

The Return Loss (RL) is the basic radiation profile of the antenna. The RL is expressed by Equation (11).

$$RL = -10\log\frac{P_r}{P_i} \tag{11}$$

where $P_i = \frac{V_i^2}{R}$ and $P_r = \frac{V_r^2}{R}$, then the RL can be obtained in Equation (12).

$$\begin{aligned} RL &= -10\log\left|\frac{V_r^2}{V_i^2}\right| \\ &= -20\log\left|\frac{V_r}{V_i}\right| \\ &= -20\log|\Gamma| \end{aligned} \tag{12}$$

where,

P_i = input power

V_r = reflected voltage

P_r = reflected power

Γ = reflection coefficient

V_i = input voltage

2.2. Parabolic Reflector

By using the method of images as depicted in Figure 1, where the directrix axis is the symmetry axis on the parabola from the focal point. Thus, the distance from the focal point F(0, f) to M(x, y) is the same as the distance from M(x, y) to the directrix axis or point D(x, -f) [8]. So, it can be written in Equation (13).

$$\sqrt{(x-0)^2 + (y-f)^2} = \sqrt{(x-x)^2 + (y-(-f))^2} \tag{13}$$

therefore,

$$\begin{aligned} x^2 &= 4yf \\ y &= \frac{x^2}{4f} \end{aligned} \tag{14}$$

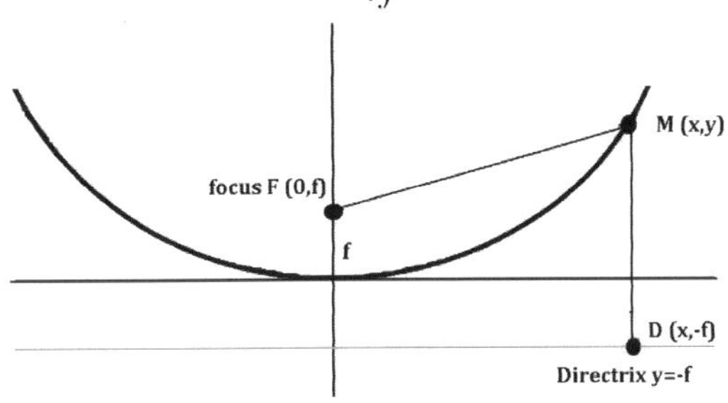

Figure 1. Illustration of parabolic reflector

Thus, the focal distance *f* as the reflector distance to the antenna can be expressed in Equation (17).

$$f = \frac{D^2}{16h} \tag{17}$$

If the parabolic antenna is a circular aperture, the equation to determine the approximate gain value can be written as [8].

$$G \approx \frac{(\pi^2 D^2)}{\lambda} \tag{18}$$

where,
G = isotropic gain
D = diameter of the reflector
λ = wavelength

2.3. Antenna design
Figures 2 and 3 show the schematic of the installation of the antenna and reflector in the simulation program. Parameters D and h were obtained from the measurement results of the reflector dimensions. The reflector was mounted on the back of the antenna. Thus, the reflector will face the direction of the wave propagation emitted by the antenna. The distance between the axis of the reflector (vertex) and the antenna are referred as well as the focal distance.

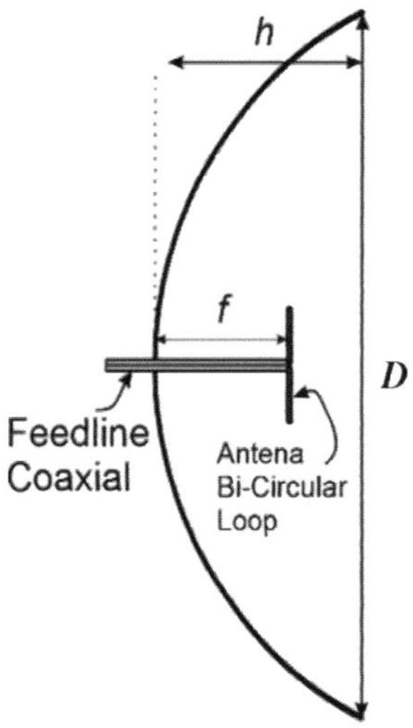

Figure 2. Bi-Circular loop antenna design with a wajanbolic reflector, two-dimensional schematic seen from the side

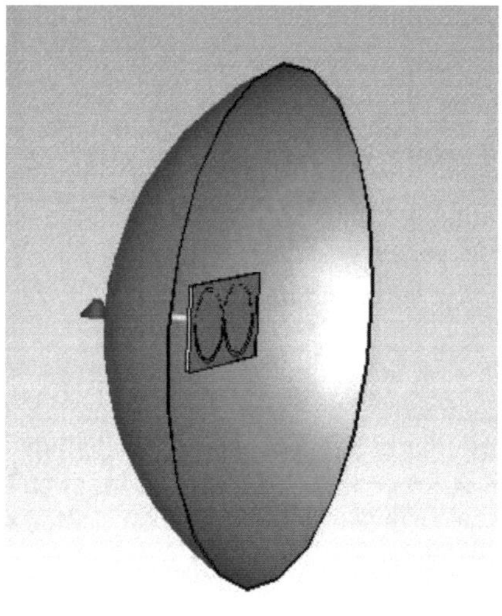

Figure 3. Bi-Circular loop antenna design with a wajanbolic reflector, 3-dimensional image on the simulation program

The focal distance *f* is calculated based on Equation (17) [8,9]. The calculation results of the focal distance are presented in the Table 1.

Table 1. Parameters of frying pan dimension

Frying Pan No.	*D* (mm)	*h* (mm)	*f* (mm)
1	309	96.20	62.03
2	335	101.50	69.10
3	364	105.00	78.87
4	381	111.80	81.15

The prior simulation, the used of Bi-Circular Loop antenna with parameters from the simulations have been carried out by Annovasho et al. [5] and Bustomi et al. [6]. These parameters are the loop radius of 14.95 mm and the link radius of 1.15 mm. Other parameters are presented in Table 2.

Table 2. Parameter of Bi-Circular loop antenna

Parameter	Size (mm)
W	86.11
H	66.98
substrate thickness	1.54
w	2.00
copper thickness	0.03

2.4. Simulation method
Simulation was used to predict the antenna properties that was properly designed, so it is unnecessary to fabricate at first. It is believed that the simulation can save research material and can easily to understand various characteristics of the antenna. In this study, the Finite Integration Technique (FIT) analysis was used for characterizing the antenna which is relevant to FDTD for the transient solver [15].

2.5. Fabrication and measurement method
Installation of the antenna was applied on the reflector of the frying pan. PVC pipe can be used as testing purpose. PVC pipe is an insulating material which is lightweight and malleable. The PVC pipe was firstly measured, then in some parts it was shaped like a flattened T. To combine with other parts, shock/pipe connections were used on either straight or turn connections. The position of the antenna mount to the reflector and the reflector to the mast can be seen in Figures 4 and 5.

Figure 4. Mounting of the antenna on the reflector (front side)

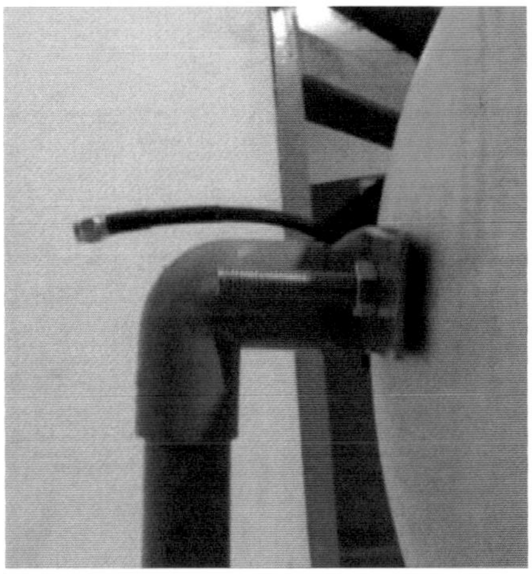

Figure 5. Mounting the reflector on the support post (back side)

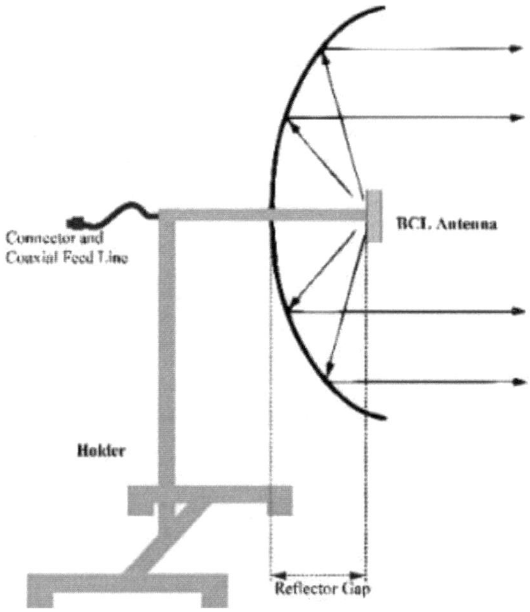

Figure 6. Antenna and reflector installation

The measurement process was carried out with the NanoVNA V2 device [16]. For the measurement of radiation pattern, it can be applied on one receiving signal unit, one signal generation unit, and and as well a directional transmitting antenna.

3. Results and Discussion

By using the BCL parameter in the previous explanation, the results were obtained in the following table.

Table 3. Minimum peak in first simulation

Frying Pan No.	Minimum Peak	
	RL (dB)	Frequency (GHz)
1	-14.80	2.99
2	-11.14	2.87
3	-8.27	2.69
4	-7.88	2.66

The next simulation step has been carried out to make the RL minimum peak shiftable at a frequency of 2.45 GHz by changing the loop radius. In the loop radius, if the value is increased, RL minimum peak will shift to the left. In contrast, if it is reduced, it will make the RL minimum peak shifts to the right [5]. From the simulation results of this parameter optimization, the following results are obtained.

Table 4. Optimization simulation results at a frequency of 2.45 GHz

Frying Pan No.	Loop Radius (mm)	RL (dB)	Gain (dBi)
1	21.09	-21.28	11.90
2	18.91	-18.14	13.90
3	17.38	-35.69	16.40
4	17.05	-19.00	16.80

Table 4 shows that the larger the reflector pan diameter gives the greater gain value. The antennas and reflectors as the researchers fabricated are those based on optimization simulations that have a minimum RL value, namely the combination of antenna and reflector or frying pan number 3. The focal distance affects the antenna impedance because the antenna polarization is parallel to the reflector plane [13]. In frying pan number 3, the focal distance provides optimal impedance to the antenna so that the antenna has a radiation profile in the form of a minimum RL value. In this condition, there is an impedance matching due to the characteristic-impedance as in Equation (9). The antenna and reflector system were equal to the input impedance of 50 Ω. The amount of radiation profile in the form of gain was influenced by the reflector diameter, as shown in Equation (18). The bigger diameter of the pan, the better the gain will be. In this condition, impedance matching has not achieved yet since the input impedance used is 50 Ω. However, it is important to pay attention to the RL value so that the antenna can radiate electromagnetic waves maximally to avoid power losses.

After the fabricated antenna has been measured, the results were obtained in frying pan no.3, loop radius BCL 17.38 mm, RL -54.75 dB, and Gain 16.00 dBi. From the radiation pattern graph (Figure 8), the antenna has the highest gain at an angle of 0°. It can be said that the antenna has a directional radiation pattern. This shows that the simulation results are valid. However, the antenna can be used as a medium-distance point-to-point network link (1-10 km). The results of RL and radiation pattern can be seen in the Figures 7 and 8.

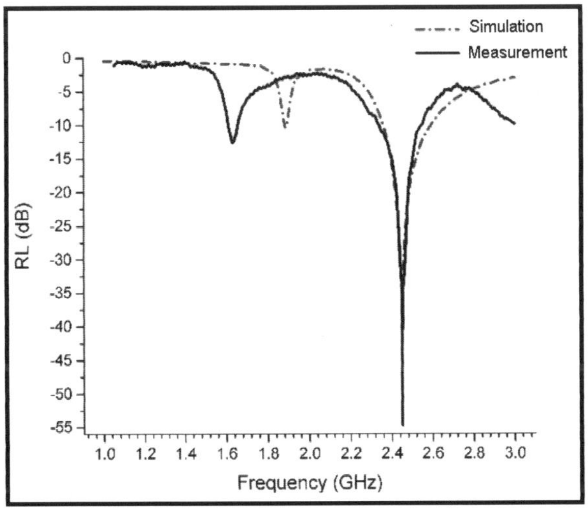

Figure 7. RL vs frequency graph

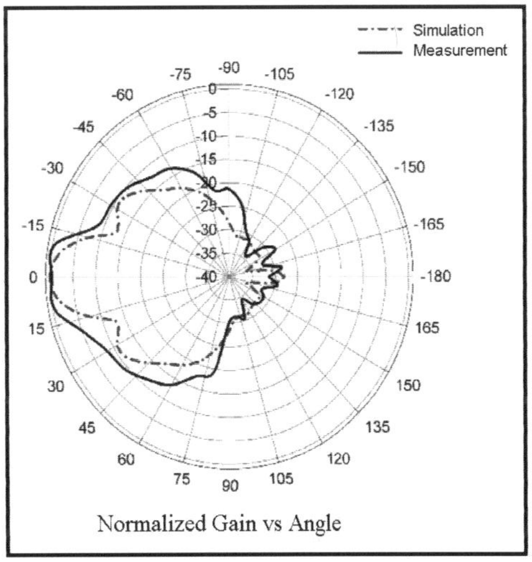

Figure 8. Radiation pattern at 2.45 GHz

Compared to previous studies, the obtained result in this study provides a complete explanation of the radiation profile of the Wajanbolic reflector. The RL value indicates that the antenna works optimally with small power losses. Gain value indicates that the measured value is in accordance with the theory. This study can be used as reference in research especially on Wajanbolic reflectors. This research also opens the opportunity to investigate other characteristics of Wajanbolic antennas and reflectors in the future. The investigations that can be carried out later on in term of the combinations of other types of antennas, optimization of reflectors with smaller sizes, and trials of their use in real conditions.

4. Conclusion

From the description above, it can be concluded that the best results were obtained on the reflector with a diameter of 364.00 mm and a BCL radius of 17.38 mm. Based on the simulation, these parameters produced a radiation profile in the form of RL value is -35.69 dB and a gain of 16.40 dBi. Based on the fabrication and measurement, the antenna had a radiation profile of the RL value of

-54.75 dB and the gain of 16.00 dBi on directional antenna. An antenna with such performance can be used as a point-to-point Wi-Fi transmitter.

Acknowledgment
The authors would like to thank The Ministry of Research and Technology/National Research and Innovation Agency (Kemenristek/BRIN) of Republic Indonesia who have funded this research in 2020.

References
[1] Badri M 2016 *J. Risal.* **27** 62
[2] Subiakto H 2013 *Masy. Kebud. Politik* **26** 243
[3] Firdaus R A, Pramono Y P, Yahya E, Endarko, and Gunawan W H *Proc. Int. Conf. Radar Antenna Microw. Electron. Telecommun. (ICRAMET)* p 24
[4] Firdaus R A,Pramono Y P, and Endarko 2019 *Int. J. Microw. Opt. Technol. (IJMOT)* **14** 4
[5] Annovasho J, Buka O, Khoiro M, and Pramono Y H 2017 *Int. Semin. Sens. Instrum. Meas. Metrol. (ISSIMM)* p 154
[6] Bustomi M A, Annovasho J, and Pramono Y H 2021 *Int. J. Adv. Trends Comput. Sci. Eng.* **10** 800
[7] Yurandi N, Jambola L, and Darlis A R 2013 *J. Reka Elkomika* **1** 233
[8] Nafik I and Yuniarto Y 2012 *Gema Teknol.* **17** 1
[9] Khairunnisa K 2017 *Poros Teknik* **8** 25
[10] Umben A, Elbani A, and Marpaung J 2021 *J. Teknik Elektro Univ. Tanjungpura* **2** 48235
[11] Endri J, Handayani A S, Jannah R, and Al-Kausar J 2019 *J. Phys. Conf. Ser.* **1167** 012020
[12] Handoko F 2018 *J. Tek. Elektro Univ. Tanjungpura* **2** 27320
[13] Balanis C A 2016 *Antenna Theory: Analysis and Design, Fourth Edition* (West Sussex: John Wiley & Sons Ltd)
[14] Yang Y, Zeng Y, and Chen Q 2017 *Telecommun. Syst.* **65** 331
[15] Weiland T, Timm M, and Munteanu I 2008 *IEEE Microw. Mag.* **9** 62
[16] Derickson D, Jin X, and Bland C C 2021 *ASEE Pac. Southwest Conf.* 35185

Journal of Physics: Conference Series **2110** (2021) 012008 doi:10.1088/1742-6596/2110/1/012008

Determination of Resonance Parameters of The PANI Thin Film Fabricated using Spin Coating Method

N P Putri[1,*], F Al Azis [2], D D Kamasi [2], and S P Sakti[2]

[1]Physics Department, Faculty of Mathematics and Natural Science, Universitas Negeri Surabaya, Indonesia
[2]Physics Department, Faculty of Mathematics and Natural Science, Brawijaya University, Indonesia

*Email: nugrahaniprimary@unesa.ac.id

Abstract. This study aims to determine the resonance parameters of polyaniline thin films to better understand the viscoelasticity properties of polyaniline films. The spin coating method was used with varying solvent concentrations and rotating speed of spin coater during the deposition Polyaniline (PANI) thin film on a quartz crystal microbalance (QCM). To determine the resonance parameters of the PANI thin film, the impedance and QCM frequency measurements were first carried out before and after coated with PANI. The modelling used is a modified BVD model, and the determined resonance parameters are C_0, C_1, L_1, R_1, L_2 and R_2. From the results of the analysis using the GRG Nonlinear solver program, it was found that solution concentration and rotational speed in the spin coating process has a significant effect on resonator parameters of PANI thin film. The best solution concentration from this study was 2% DMF with a rotational speed of 2,000 rpm. This is because in these conditions it provides a minimal damping effect on QCM.

1. Introduction

Quartz Crystal Microbalance or commonly known as QCM is a resonator that has piezoelectric properties. When a potential difference is operated on a quartz crystal, the quartz oscillates at certain frequency. The change in mass that occurs on the quartz surface is directly correlated with the change in the crystal oscillation frequency. The effect of adding mass on the QCM surface is in accordance with the following Saubrey equation

$$\Delta m = \frac{A\sqrt{\rho_q \mu_q}}{2f_0^2}.\Delta f \quad (1)$$

With Δf is frequency changes (Hz), f_0 is the fundamental resonant frequency of QCM, A is the area of the deposited quartz (m^2), ρ_q is quartz density (2,684 g/cm^3) and μ_q is quartz modulus (2,947 x 10^{11} g/cm s^2) [1].

To better understand the resonance behaviour of QCM, several mathematical models have been proposed, one of which is the Butterworth Van Dyke (BVD) model [2]. Based on the physical properties of QCM, the BVD design uses four parameters, namely two capacitors, one inductor and one resistor (Figure 1a), which are divided into two branch circuits. One of the branches consists of only one capacitor, C_0, which is known as the electrical branch. The other branch, the acoustic branch, consists of a capacitor C_1, an inductor L_1 and a resistor R_1 connected in series. A capacitor C_0 is the electrical capacitance across the electrodes, an inductor L_1 is the motional inductance proportional to the mass, and a capacitor C_1 is the motional capacitance inversely proportional to the stiffness. A resistor R_1 is the motional resistance that quantifies dissipative losses. QCM that has been coated can be modelled with a modified BVD design (Figure 1. b**)**. There is an additional resistor R_2 and an inductor L_2 connected in series on the branch of the motional arm. R_2 and L_2 can be described as the damping parameter of the QCM caused by the layer [3].

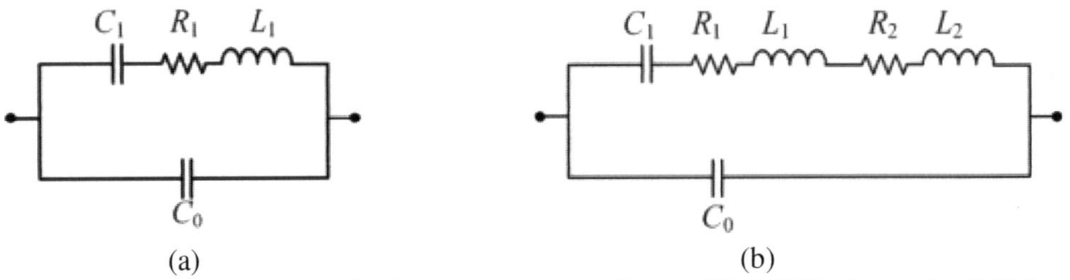

Figure 1. (a) BVD design for QCM without coating and (b) modified BVD design for QCM + PANI thin film

In order to function as a chemical sensor or biosensor, the QCM must be given an appropriate, sensitive coating. Polyaniline (PANI) is a conductive polymer with many applications, one of which is an active sensor material. Polyaniline coated on the QCM surface has been reported to detect Volatile Organic Compounds (VOCs), especially alcohol, ether, carbon tetrachloride, chloroform and ethyl acetate [4-6]. Several methods can carry out the deposition of a PANI thin layer on the QCM surface. The simplest deposition method is spin coating [7]. The spin coating method requires a sample in a solution, so the PANI powder must be dissolved first using the suitable solvent. There are several parameters of spin coating deposition that can affect both the quality of the thin film and the viscoelasticity of the PANI thin film. These parameters include the concentration of the solution and the rotational speed of the spin coater [8].

The results of a thin film of PANI deposition on the QCM surface using the spin coating method were mentioned in the previous article [9]. In this study, the discussion will be devoted to the effect of PANI solution concentration and spin coater rotational speed on changes in the resonance parameters of the deposited PANI thin film. The resonance parameter's determination is determined by inputting the impedance measurement data using the impedance analyzer in the Microsoft Excel program with the Nonlinear Generalized Reduced Gradient (GRG) method.

2. Method

The oxidation method was used in the synthesis of PANI powder, as demonstrated in a preceding study [10]. The synthesized PANI powder was then dissolved in Dimethyl Formamide (DMF) solvent with a 2 and 4% solution concentration. The solution is then stirred with a magnetic stirrer to obtain a homogeneous solution. The PANI solution was dripped on the QCM surface and rotated with a spin coater with a 2,000 and 4,000 rpm rotational speed. After that, the QCM was dried at 100° for 10 minutes.

The QCM used in this study was the AT-cut quartz crystal in a standard package obtained from Great Microtama Surabaya Industry. It has a fundamental resonance frequency of 10MHz. Omicron Bode 100 Impedance Analyser was used to measure the impedance on the QCM before and after coating with PANI. Impedance measurements were carried out in the frequency range 9,960,000 to 10,070,000 Hz. Figure 2 shows the impedance spectrum of the QCM without coating. The impedance data is then input into the Microsoft Excel program using the nonlinear GRG method to obtain the resonance parameters of QCM and PANI thin film according to the modified BVD design.

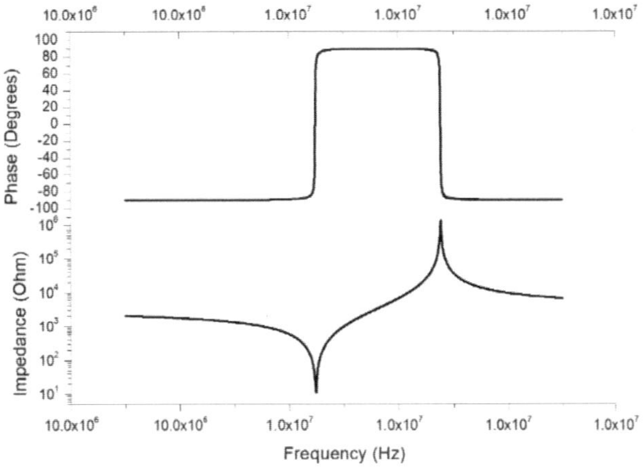

Figure 2. The impedance graph of QCM without coating

3. Results and Discussion

Determination of the value of the resonance parameter of the PANI thin film begins with inputting impedance data from the Omicron Bode 100 Impedance Analyser into the Microsoft Excel program with the .xlsm file type. The input data is the impedance data from the QCM before and after being coated. According to the BVD circuit model, impedance data from uncoated QCM is used to determine the parameter values C_0, C_1, R_1 and L_1. At the same time, the QCM impedance data after being coated is used to determine the parameter values of R_2 and L_2 according to the modified BVD model. Determination of resonance parameters using the Non Linear GRG method has been successfully carried out by several previous studies [1,2,12]. The resonance parameter values for variations in solution concentration and rotational speed can be seen in Table 1. It can be seen that the values of C_0, C_1, R_1 and L_1 for each sample are varied because the QCM used is different.

Table 1. Parameters resonance of PANI thin films

Sample	Resonance Frequency (Hz)	Resonator Parameters					
		C_0 (pF)	C_1 (pF)	R_1 (Ω)	L_1 (mH)	R_2 (Ω)	L_2 (mH)
dmf1_2000	10009020	4.71370	0.02219	9.48530	11.39336	79.01295	0.00144
dmf1_4000	9994879	4.69667	0.02227	7.28714	11.34756	45.45017	0.03891
dmf2_2000	10007311	4.82962	0.02309	6.62864	10.94940	25.92537	0.00662
dmf2_4000	10007304	4.84101	0.02210	10.26251	11.43310	18.10632	0.00961
dmf3_2000	10009145	4.79960	0.02220	6.98080	11.38926	105.922038	0.00076
dmf3_4000	10009042	4.83777	0.02264	6.82825	11.16738	58.19544	0.00106
dmf4_2000	10009124	4.80761	0.02264	7.52692	11.16504	118.88732	0.00278
dmf4_4000	10004812	4.81013	0.02193	11.18735	11.53692	66.00451	0.00970

The value of the resonance parameter of the PANI thin film can be seen from the values of R_2 and L_2. It can be seen that the variation of the solution concentration (1 – 4%) and the variation of the spin coater rotational speed (2,000 and 4,000 rpm) can affect the value of the resonance parameter of the PANI layer deposited above the QCM. The R_2 parameter represents the dissipative properties of the BVD circuit, while the L_2 parameter represents the inertial mass which gives a loading effect on the QCM [3]. The smallest R_2 value was obtained in the dmf2_4000 sample, namely the PANI layer deposited from a 2% PANI/DMF solution with a spin coater rotational speed of 4000 rpm. On the contrary, the most significant R_2 value was obtained from the sample dmf4_2000, namely the PANI layer deposited from a 4% PANI/DMF solution with a spin coater rotational speed of 2,000 rpm. This implies that the energy dissipation process that occurs is not the same in each sample. The most considerable energy dissipation process occurs in the dmf4_2000 sample. It can be interpreted that there is a substantial energy loss in the sample, much more remarkable when compared to other samples.

The L_2 parameter values are also different for all samples. The smallest L_2 value is obtained in the dmf3_2000 sample, while the most significant L_2 value is obtained in the dmf1_4000 sample. As mentioned above, the L_2 value represents the inertial mass deposited above QCM, so it can be interpreted that the enormous mass deposited above QCM is in the sample dmf1_4000. This result is supported by calculating the amount of mass deposited using equation (1), and the results are given in Table 2. The deposited mass is calculated from the change in the resonant frequency. The significance of the mass deposited above QCM was found in the dmf1_4000 sample. The lowest mass is in the sample dmf3_2000. This supports the results of determining the value of L_2 in Table 1 above. The deposited mass came from PANI powder and DMF solvent. DMF solvent does not evaporate when heated because it is trapped on the surface of the QCM.

Table 2. PANI mass deposited on the QCM surface

Sample	**Resonance Frequency (Hz)**		**Calculated deposited mass (g)**
	Before coating	**After coating**	
dmf1_2000	10009652	10009020	5.64589×10^{-7}
dmf1_4000	10011980	9994879	1.52699×10^{-5}
dmf2_2000	10010349	10007311	2.71358×10^{-6}
dmf2_4000	10011484	10007304	3.73278×10^{-6}
dmf3_2000	10009478	10009145	2.97492×10^{-7}
dmf3_4000	10009539	10009042	4.43999×10^{-7}
dmf4_2000	10010363	10009124	1.10669×10^{-6}
dmf4_4000	10009019	10004812	3.75875×10^{-6}

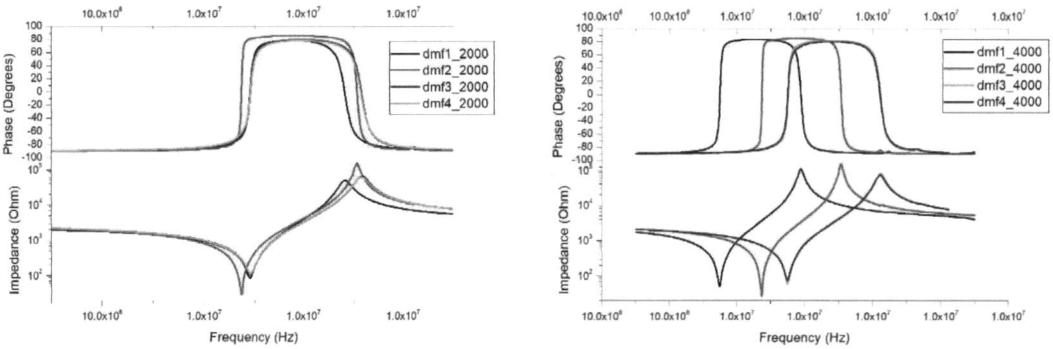

(a) (b)

Figure 3. The impedance graph of QCM coated with PANI thin film at (a) 2000 rpm; and (b) 4000 rpm rotating speed

Figure 3 shows the impedance graph of the PANI film coated on the QCM. The top graph is the phase spectrum, while the bottom graph is the impedance spectrum of the PANI film. An abrupt phase can be observed occurring simultaneously with the minimum and maximum impedance [11]. It can be seen that there are differences in the shape of the spectrum, both the phase spectrum and the impedance spectrum. Figure 3(a) shows that the red phase spectrum (sample dmf2_2000) has a symmetrical shape. The shape of the impedance spectrum is also symmetrical, where the shape of the two peaks is tapered. From the spectrum's shape, it can be seen that the deposited PANI film has no loading effect on QCM. In Figure 3(b), the red spectrum for the dmf2_4000 sample has almost the same shape as the spectrum for the dmf2_2000 sample. The sharp peak of the impedance spectrum indicates that the PANI film is included in the glassy coating material [12]. When related to the results of determining the value of the R_2 parameter in Table 1, the two samples have a fairly small R_2 value compared to other samples. Meanwhile, when viewed from the mass deposited above the QCM, the two samples have a less significant mass. There are several impedance spectrums with peaks that are not sharp. A non-tapered peak indicates that the coating includes rubbery coating and will have a loading effect on QCM. These elastic properties can occur as a result of the interaction between QCM and DMF solvent trapped on QCM's surface.

From the results of the determination of the resonance parameters of the PANI thin film deposited on the QCM using the spin coating method, it can be seen that the concentration of the solution and the spin coater rotational speed significantly affect the value of the resonance parameter. PANI/DMF solution with a solution concentration of 2% will give the lowest dissipative parameter value (R_2). The effect of rotational speed on the amount of mass deposited can be seen in the sample with a solution concentration of 1% (Table 3). The difference in the amount of mass deposited is quite significant. A rotating speed of 2,000 rpm provides significantly less mass amount when compared to 4,000 rpm. The lower rotational speed prevents the DMF solvent from being trapped on the QCM surface, and when heated, the DMF can evaporate [13].

4. Conclusion

From the research that has been done, it can be concluded that the resonance parameters of the PANI thin film prepared by the spin coating method can be determined using a modified BVD model. The solution's concentration and the spin coater's rotational speed have a significant impact on the value of the resonance parameter and the amount of mass deposited. When using DMF solvent, the best parameter values will be obtained with a solution concentration of 2% and a rotational speed of 2000 rpm because it has small R_2 value and small deposited mass.

References

[1] Sakti S P, Masruroh, Kamasi D D, and Khusnah N F 2019 *Mater. Today Proc.* **13** 53
[2] Sakti S P 2014 *Natural-A: J. Sci. Model. Comput.* **1** 16
[3] Mohamoud M A, Hillman A R, and Efimov I 2008 *Electrochimica Acta* **53** 6235
[4] Ayad M M, El-Hefnawey G, and Torad N L 2009 *J. Hazard Mater.* **168** 85
[5] Shinen M H, Essa F O, and Naji A S 2014 *Chem. Mater. Res.* **6** 7
[6] Banerjee S, Konwar D, and Kumar A 2014 *Sens. Actuators B Chem.* **190** 199
[7] Banerjee S, Konwar D, and Kumar A 2012 *Sens. Actuators B. Chem.* **171** 924
[8] Sahu N, Parija B, and Panigrahi S 2009 *Indian J. Phys.* **83** 493
[9] Putri N P, Pravitasari D W, Al Aziz F, Maulidiah, Santjojo D J D H, Masruroh, and Sakti S P 2019 *J. Phys. Conf. Ser.* **1417** 012002
[10] Putri N P, Kusumawati D H, Widiyanti N, and Munasir 2018 *J. Phys. Conf. Ser.* **997** 012009
[11] Sakti S P, Kamasi D D, Nuha S A, and Setiana M 2020 *J. Phys. Conf. Ser.* **1428** 012060
[12] Sakti S P, Masruroh, Istiroyah, Kamasi D D, and Tyas N Z 2021 *Mater. Today Proc.* **44** 3217
[13] Radacsi N, Campos F D, Chisholm C R I, and Giapis K P 2018 *Nat. Commun.* **9** 4740

Journal of Physics: Conference Series 2110 (2021) 012009 doi:10.1088/1742-6596/2110/1/012009

XRD and EDX Analyses on the Formation of $MgTiO_3$ Phase in $(Mg_{0.6}Zn_{0.4})(Ti_{0.99}Sn_{0.01})O_3$ Powders Due to Calcination Temperature Variations

F U Ermawati

Physics Department, Faculty of Mathematics and Natural Sciences, Universitas Negeri Surabaya, Indonesia

Email: frida.ermawati@unesa.ac.id

Abstract. $MgTiO_3$-based dielectric ceramics have been recognized as functional materials in the microwave telecommunications industry. Research and development on $MgTiO_3$ dielectric ceramics has therefore developed rapidly. This paper reports x-ray diffraction (XRD) and energy dispersive x-ray (EDX) analyses on the formation of $MgTiO_3$ phase in $(Mg_{0.6}Zn_{0.4})(Ti_{0.99}Sn_{0.01})O_3$ powder due to variations in calcination temperature from 550 to 700°C for 2 h. The powder was synthesized via the dissolved metal mixing course using magnesium, zinc, titanium and tin metal powders (Merck) as starting materials. The Rietveld refinement on the XRD patterns of the samples revealed that increasing the calcination temperature reduces the molar% content of $MgTiO_3$ phase of from (97.91±1.51) at 550°C to (87.81±1.29) at 700 °C and causes a decrease in the diffraction peak intensity. The remaining % belongs to TiO_2 rutile. The calcination temperature also enlarged the size of $MgTiO_3$ unit cell volume. The EDX data on the atomic% ratio of the elements confirmed the presence of the phases. Discussion of these results is presented in detail in this paper.

1. Introduction

Geikielite $MgTiO_3$ based dielectric ceramics have been recognized as their main functional material for various applications in the satellite telecommunications industry, including global positioning systems, resonator dielectrics and high frequency capacitors [1–8]. Fabrication of Geikielite $MgTiO_3$-based dielectric ceramics has developed rapidly by creating various compositions and designs with the aim of obtaining superior dielectric characteristics that can meet industrial needs [1–3, 9,10]. In each of these compositions, the presence of the $MgTiO_3$ as the main phase is very important to ensure these superior properties. This paper reports the analysis of the formation of the $MgTiO_3$ phase in the new composition, namely $(Mg_{0.6}Zn_{0.4})(Ti_{0.99}Sn_{0.01})O_3$ powder prepared by dissolved metal mixing method as a function of the calcination temperature based on the XRD and EDX measurements. The aim is to ensure that in the new composition, whether the $MgTiO_3$ phase can be formed as the main phase, and at what calcination temperature the content of the $MgTiO_3$ phase reaches its maximum value.

2. Method

The new composition of $(Mg_{0.6}Zn_{0.4})(Ti_{0.99}Sn_{0.01})O_3$ powders (abbreviated to MZTS) were prepared from magnesium, zinc, titanium and tin high purity metal powders (Merck). Each metal powder was weighed according to the expected stoichiometry and dissolved homogeneously in 12 molar hydrochloric acid using a magnetic stirring hotplate. The four liquid solutions were then mixed consistently, dried and crashed to obtain an amorphous powder with a uniform particle size. The resulting powder was then calcined at temperatures of 550, 600, 650 and 700 °C for 2 h using a BWS furnace with a heating step of 10 °C/min. The temperature of 550 °C was chosen because the thermo gravimetric analysis/differential thermal analysis (TGA/DTA) on the $(Mg_{0.8}Zn_{0.2})TiO_3$ powder reported in [11] revealed that the diffraction peaks belonging to the $MgTiO_3$ phase were formed at 500 °C and the powder was calcined at 550 °C. Ermawati et. al in [3] also calcined $(Mg_{1-x}Zn_x)TiO_3$ powders for x = 0.1 – 0.5 at 550 °C. In this study, as already mentioned, the powder was calcined at temperatures of 550, 600, 650 and 700 C in order to study the effect of variations in calcination temperature on the formation of the $MgTiO_3$ phase in the new composition, MZTS powders.

Identification of the crystalline phases formed in the MZTS powders was carried out using a *Match!* Software on the MZTS's x-ray diffraction (XRD) pattern. The XRD measurement was accomplished using XPert MPD (Philips) with Cu K_α radiation at a diffraction angle (2θ) range of 15-70° and at the step size of 0.017 °/min. Analysis on the composition of the identified phases was performed using the Rietveld method implemented in *Rietica* software. Elemental composition of the MZTS calcined powders was examined via energy dispersive x-ray (EDX) technique using $EDAX^{TSL}$ (AMETEX). Figure 1 shows the flowchart of all the experimental stages above.

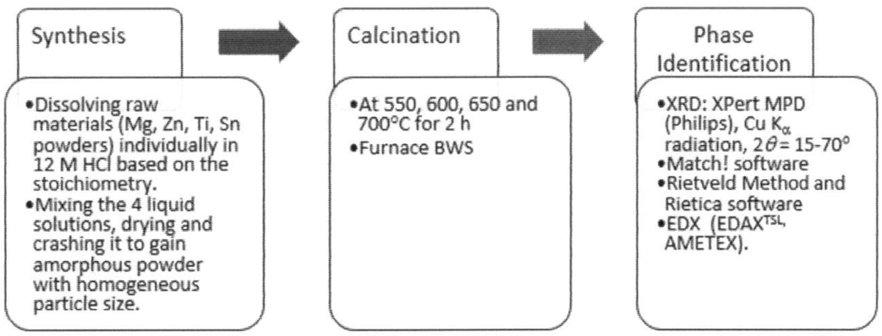

Figure 1. The experimental stages carried out in this work

3. Results and Discussion

3.1. XRD Analysis

Figure 2 shows the XRD patterns of the MZTS powders calcined at 550, 600, 650 and 700 °C for 2 h.

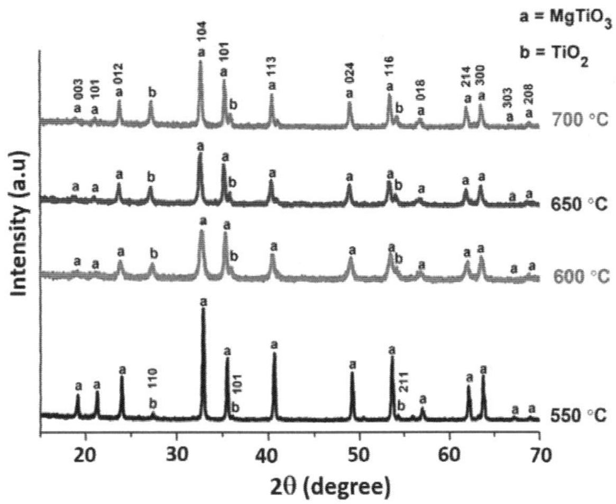

Figure 2. XRD patterns of MZTS powders calcined at 550, 600, 650 and 700 °C for 2 h. a = $MgTiO_3$ (PDF #06-0494), b = TiO_2 rutile (PDF #21-1276)

As shown in Figure 2, the diffraction patterns of the four samples were identified as belonging to the $MgTiO_3$ phase (PDF #06-0494), namely peaks with the label "a" above them. However, three extra peaks at $2\theta = 27.44°$, 36.08° and 54.32° belonging to TiO_2 rutile phase (PDF #21-1276), i.e. the peaks with the label "b", were also identified. The identification of these phases was completed using the *Match!* Software. As the calcination temperature increases from 550 to 700 °C, the intensity of the highest diffraction peak belonging to the $MgTiO_3$ phase at $2\theta = 32.87°$ (i.e. the peak with hkl = 104) decreases. At the same time, the intensity of the diffraction peak belonging to the TiO_2 rutile phase at $2\theta = 27.44°$ (i.e. the peak with hkl = 110) is increasing. The decrease in the intensity of the (104) peak might be closely related to the vibration of the atoms in the sample as the temperature increases. In addition to the decrease in the intensity of the (104) peak followed by peak broadening, it turns out that the 2θ position of the peak also shifts towards a smaller 2θ value as the calcination temperature increases, as shown in Figure 3. The facts in Figures 2 and 3 will then be discussed in the following paragraphs based on the theoretical review and confronted with the XRD quantitative data obtained in this work and that was presented in Figures 4-9. So far, the ordered expanses of atoms in a crystal have often been thought to lie at fixed points in the lattice. The truth is that the atoms undergo thermal vibrations at their average position, even at absolute zero temperature. The amplitude of the thermal vibration will increase along with the increase in temperature, resulting in a shift in the position of the atoms from their average position. Theoretically, Cullity [12] explained that the increase in atomic thermal vibration due to an increase in temperature can cause the following, namely: a) the size of the unit cell volume becomes larger so that the periodicity of the distance between the closest crystal planes (d) changes, and therefore the 2θ position of the diffraction peak also shifts from its original position; b) the intensity of the diffraction peak is reduced, and c) the intensity of the background scattering between the diffraction peaks increases.

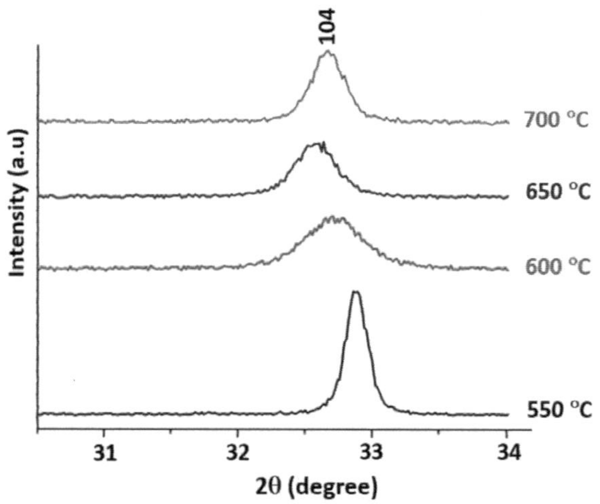

Figure 3. The shift in the 2θ position of the (104) peak belonging to the $MgTiO_3$ phase as a function of the calcination temperature. The data is taken from Figure 2

Figures 4, 5, 6, and 7 provide the results of the refinement of the x-ray diffraction pattern of the four samples in Figure 2 using the Rietveld method and *Rietica* software. Figures of merit (*FoM*) values of the refinement that consist of the goodness of fit (*GoF*), the profile factor (R_p) and the weighted profile factor (R_{wp}) are also given for each refinement. In Figures 4-7, the +++ spectrum represents the experimental pattern, the red line is the model pattern generated based on the crystallographic data of the identified $MgTiO_3$ and TO_2 rutile phases (i.e. PDF #06-0494, PDF #21-1276, ICSD #65794 and ICSD #64987). The green line is the difference in intensity height between the experimental pattern and the model pattern. The small vertical blue lines are the Bragg peaks positions of the identified $MgTiO_3$

and TiO_2 rutile phases. Figures 8-9 show the variation of unit cell volume of $MgTiO_3$ phase and the Molar% of $MgTiO_3$ and TiO_2 phases as a function of calcination temperature.

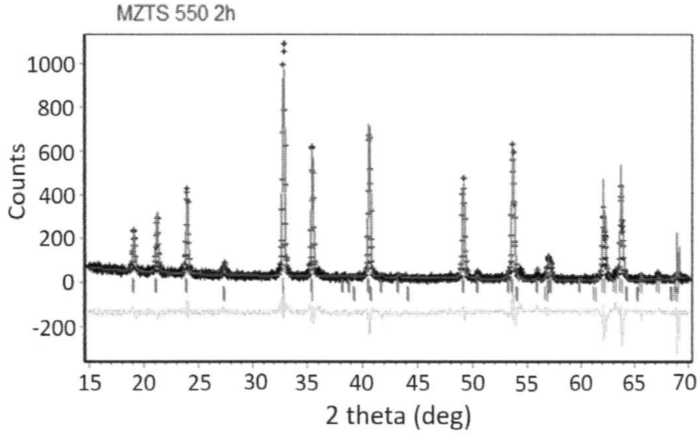

Figure 4. The Rietveld refinement on the XRD pattern of MZTS powder calcined at 550°C for 2 h. *GoF* = 1.62, R_p = 10.81 and R_{wp} = 12.78

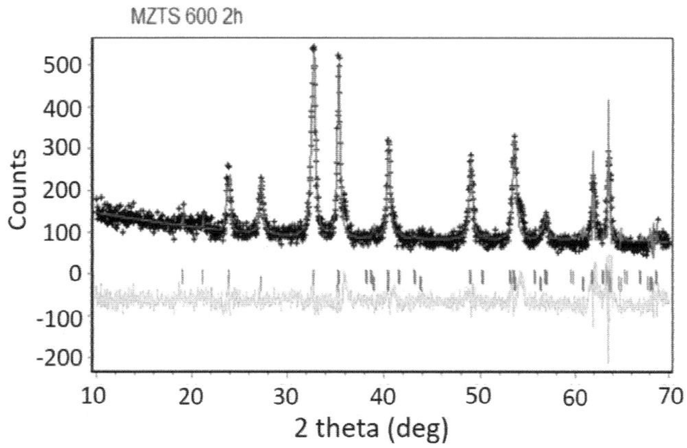

Figure 5. The similar Rietveld refinement result as in Figure 4, but this is for MZTS powder calcined at 600°C for 2 h. *GoF*=1.76, R_p=10.73 and R_{wp}=10.56

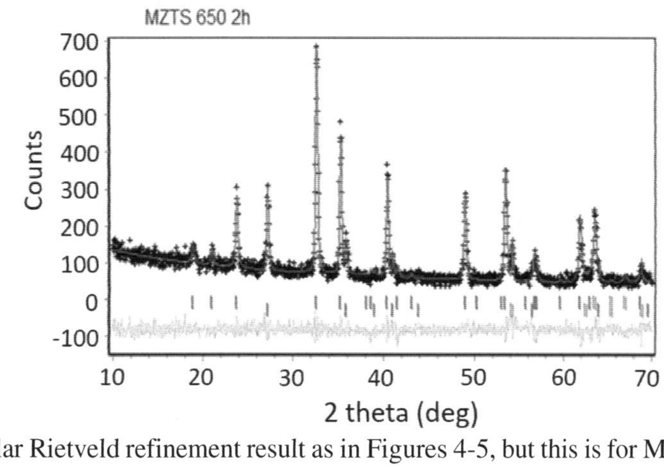

Figure 6. The similar Rietveld refinement result as in Figures 4-5, but this is for MZTS powder calcined at 650°C for 2 h. *GoF*=1.99, R_p=9.68 and R_{wp}=9.92

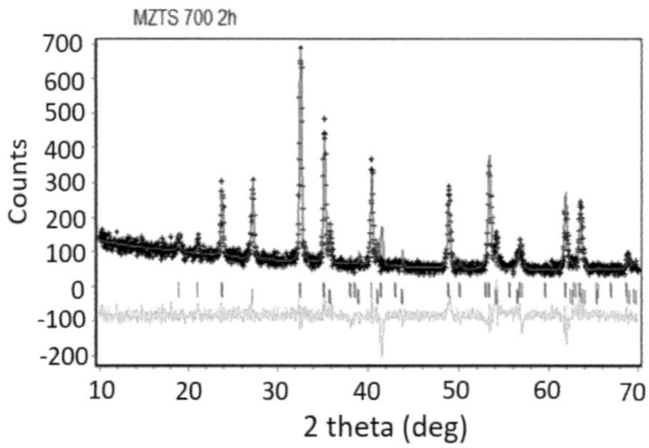

Figure 7. The similar Rietveld refinement result as in Figures 4-6, but this is for MZTS powder calcined at 700°C for 2 h. *GoF*=1.50, R_p=11.37 and R_{wp}=11.11

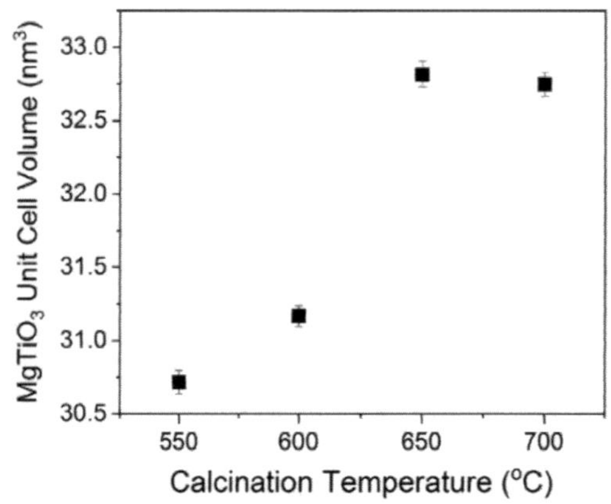

Figure 8. Variation of the unit cell volume of $MgTiO_3$ phase in MZTS powders as a function of calcination temperature. The data was the output of the Rietveld refinement in Figures 4-7

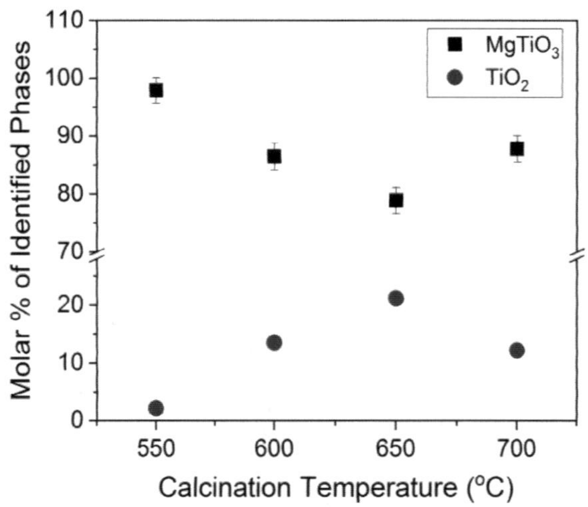

Figure 9. Variation of Molar % of $MgTiO_3$ and TiO_2 rutile phases in MZTS powders as a function of calcination temperature. The data was also the output of the Rietveld refinement in Figures 4-7

As shown in Figures 4-7, all the diffraction peaks belonging to the $MgTiO_3$ and TiO_2 phases can be modelled and their positions coincide with the positions of the Bragg peaks. In addition, the difference in intensity between the experimental peaks and the model peaks is almost all low, as evidenced by the GoF values which are less than 2.00 [3,11,13,14]. As a result, the output of the Rietveld refinement, in this case the unit cell volume of the $MgTiO_3$ phase in Figure 8 and the molar % of the two phases in Figure 9, can be used for further analysis. In Figure 8, when the calcination temperature was increased from 550 °C to 650 °C, the unit cell volume of the $MgTiO_3$ phase increased significantly from (30.72±0.08) nm^3 to (32.81±0.09) nm^3. When the temperature was raised to 700 °C, however, the volume of the unit cell actually decreases slightly to (32.75±0.08) nm^3. The decrease is still well above the value at 550 °C. The fact in Figure 8 is consistent with the shift of the (104) peak position in Figure 3, i.e. the increase in temperature from 550 °C to 650 °C causes the peak to shift towards the smaller 2θ value, but at 700 °C the peak shifts slightly in the opposite direction of 2θ. The reason why the phenomenon of 2θ shift direction changes between 650 - 700 °C occurs, unfortunately cannot be explained yet. Perhaps further calcination of the sample at 800 °C could provide an answer to this phenomenon. Nevertheless, the facts in Figs. 3 and 8 are all in accordance with the theory, especially at the points a) and b). The point c) seems also to appear, in Figures 4-7.

In Figure 9, an increase in the calcination temperature from 550 °C to 700 °C also resulted in a molar % change of the $MgTiO_3$ phase in the sample, but with the opposite direction of the change in the unit cell volume in Figure 8. At 550 °C, the molar % of the $MgTiO_3$ phase is (97.91±2.21). The value then dropped to (78.81±2.28) at 650 °C before rising to (87.81±2.29) at 700 °C. The remaining % molar belongs to the rutile TiO_2 phase. As far as the researcher is concerned, there are no publications reporting on this composition so no comparisons can be made. However, the closest composition ever reported was $Mg(Ti_{1-x}Sn_x)O_3$ with a variation of $x = 0$, 0.01, 0.02, 0.03 and 0.04 but with a calcination temperature of 800 °C [15]. It was reported that for all these x variations, the $MgTiO_3$ phase was detected as the main phase accompanied by TiO_2 rutile as the impurity phase. The composition with $x = 0.01$, i.e. $Mg(Ti_{0.99}Sn_{0.01})O_3$ contained (97.03±2.52) molar % of $MgTiO_3$ phase and the unit cell volume of (30.72±1.13) nm^3.

3.2. EDX Analysis

Figures 10-13 depict the EDX analysis on the elemental compositions of the MZTS powders calcined at 550, 600, 650 and 700 °C for 2h using EDAX^{TSL} (AMETEX).

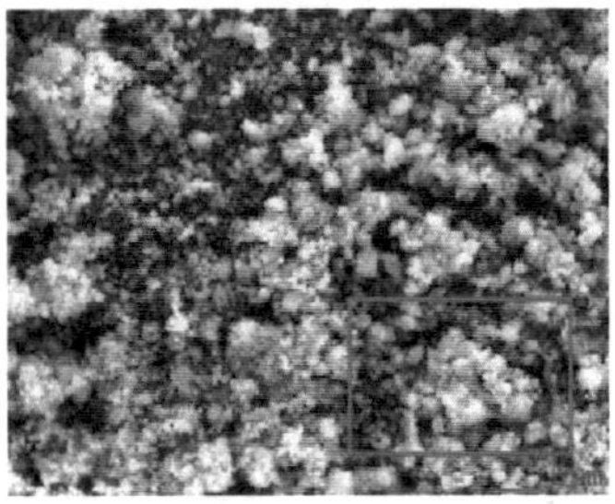

Element	Wt.%	At.%
OK	38.88	57.97
MgK	24.49	24.03
SnL	00.51	00.10
TiK	35.45	17.65
ZnK	00.67	00.25
Matrix	Correction	ZAF

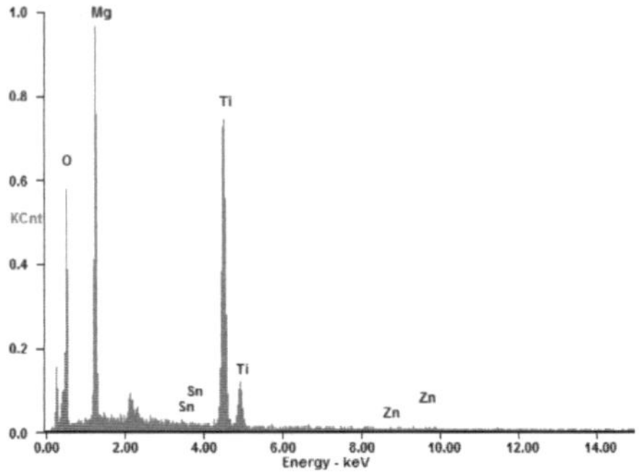

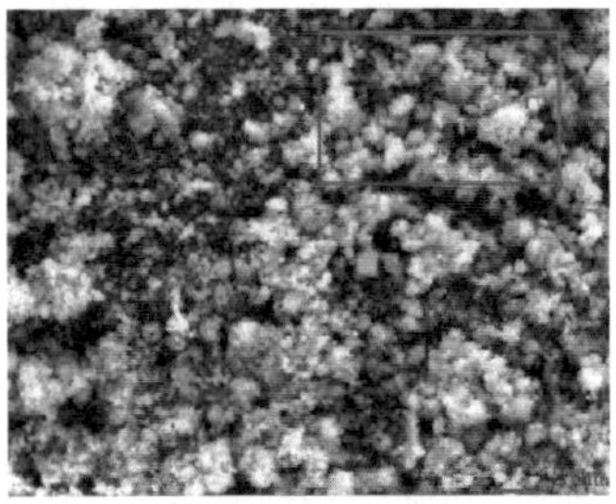

Element	Wt.%	At.%
OK	36.55	56.30
MgK	22.83	23.14
SnL	00.73	00.15
TiK	39.09	20.11
ZnK	00.80	00.30
Matrix	Correction	ZAF

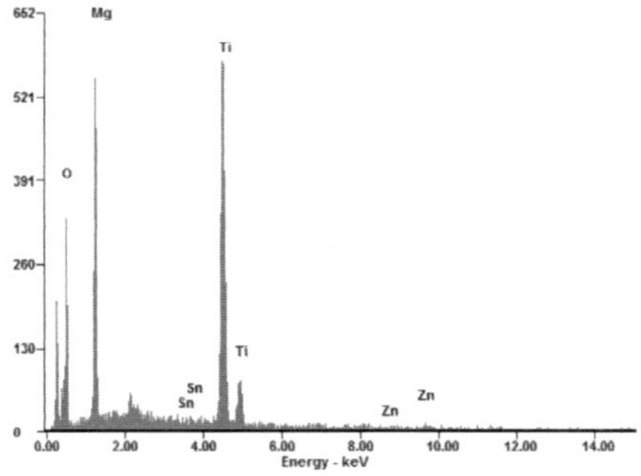

Element	Wt.%	At.%
OK	38.58	57.54
MgK	24.75	24.30
SnL	00.00	00.00
TiK	35.88	17.88
ZnK	00.78	00.29
Matrix	Correction	ZAF

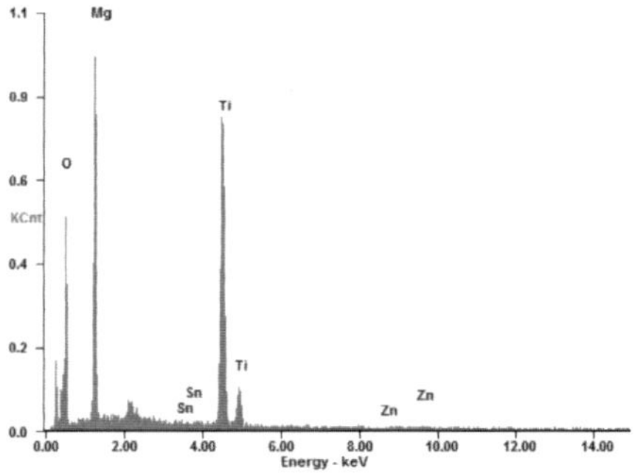

Figure 10. EDX elemental analysis on the MZTS powder calcined at 550 °C in the red square areas

Element	Wt.%	At.%
OK	30.08	52.51
MgK	15.65	17.98
SnL	00.00	00.00
TiK	40.62	23.68
ZnK	13.64	05.83
Matrix	Correction	ZAF

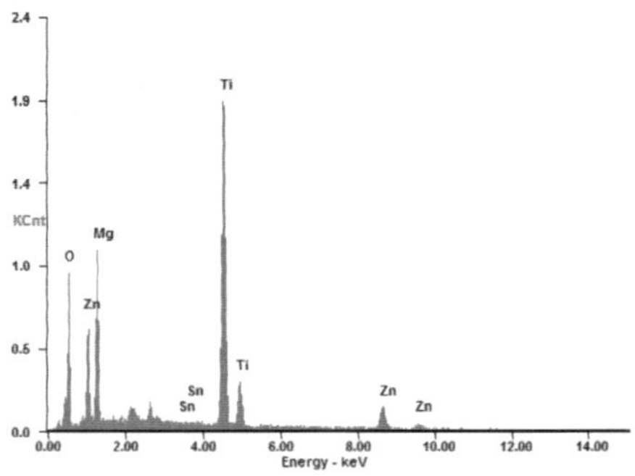

Element	Wt.%	At.%
OK	22.70	44.86
MgK	12.08	15.72
SnL	00.59	00.16
TiK	45.35	29.94
ZnK	19.29	09.33
Matrix	Correction	ZAF

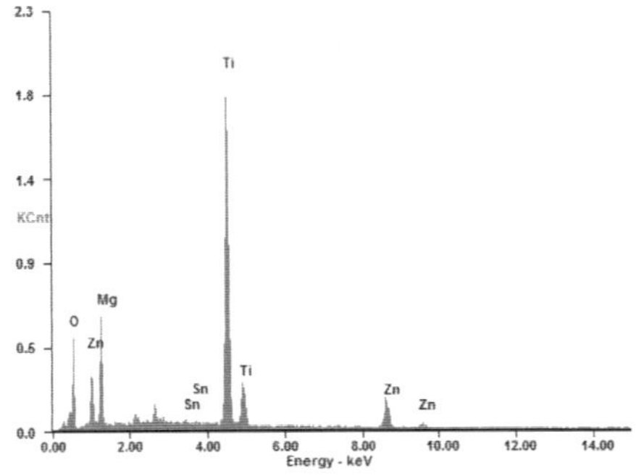

Element	Wt.%	At.%
OK	35.35	58.96
MgK	12.71	13.95
SnL	00.53	00.12
TiK	40.23	22.41
ZnK	11.18	04.56
Matrix	Correction	ZAF

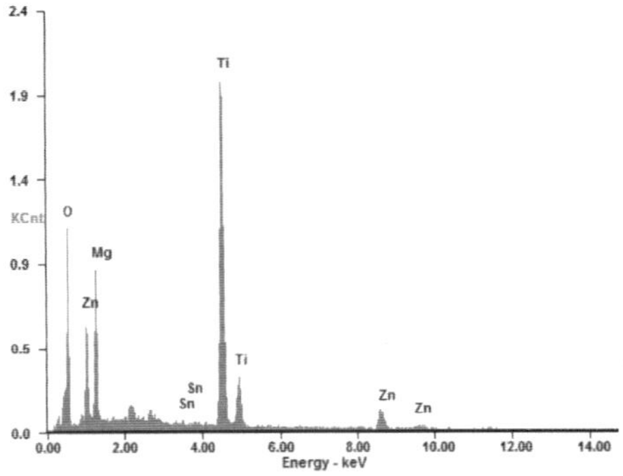

Figure 11. The similar EDX elemental analysis as in Figure 10, but this is for the powder calcined at 600°C

In Figures 10-13, the EDX measurements were performed at three different spots in the red boxes. The weight % and atomic % data along with the spectra for each element are also shown. Overall, the atomic % data of each element is relatively consistent between the three spots. By taking the atomic % data in one of spots, in Figure 10, the ratio of

$$(Mg_{0.6}Zn_{0.4}):(Ti_{0.99}Sn_{0.01}):O=(24.03+0.25):(17.65+0.10):57.97=24.28:17.75:57.97 \quad (1)$$

The ideal atomic % ratio is $\approx 19.33:19.33\ 57.97=1:1:3$ (2)

By comparing the data in Equations (1) and (2), it is seen that the % of Ti atoms in Figure 10 is less than it should be. In other words, the % of Ti atoms in Equation (1) is mainly for the $MgTiO_3$ phase. This hypothesis is in accordance with the data presented in Figures 2 and 9 for the 550 °C where the TiO_2 phase is trivial, i.e. only (2.09±1.77) molar %. With the same analysis, in Figure 11, the ratio of

$$(Mg_{0.6}Zn_{0.4}):(Ti_{0.99}Sn_{0.01}):O=(15.72+9.33):(29.94+0.16):44.86=25.05:30.01:44.86 \quad (3)$$

The ideal atomic % ratio is $\approx 14.95:14.95:44.86=1:1:3$ (4)

Therefore the % of Ti atoms in Equation (3) is excessive. This means that apart from the $MgTiO_3$ phase, some of the Ti belongs to another phase, presumably TiO_2 rutile. The hypothesis also fits the data in Figures 2 and 9 for 600 °C.

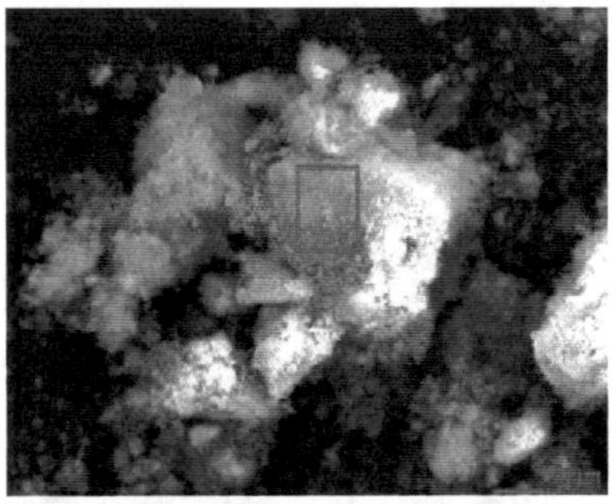

Element	Wt.%	At.%
OK	27.40	47.97
MgK	19.94	22.98
SnL	01.42	00.33
TiK	43.29	25.31
ZnK	07.95	03.41
Matrix	Correction	ZAF

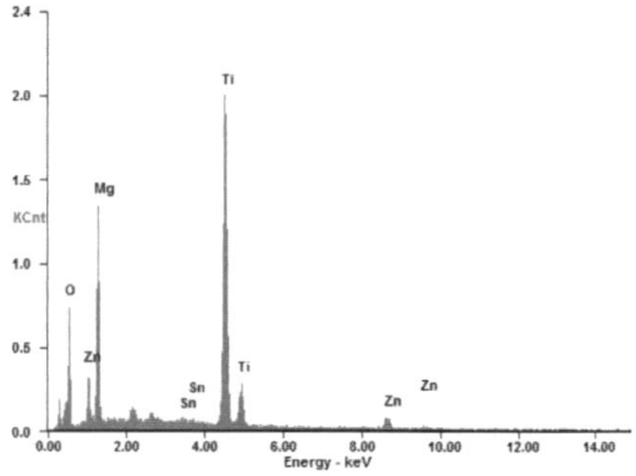

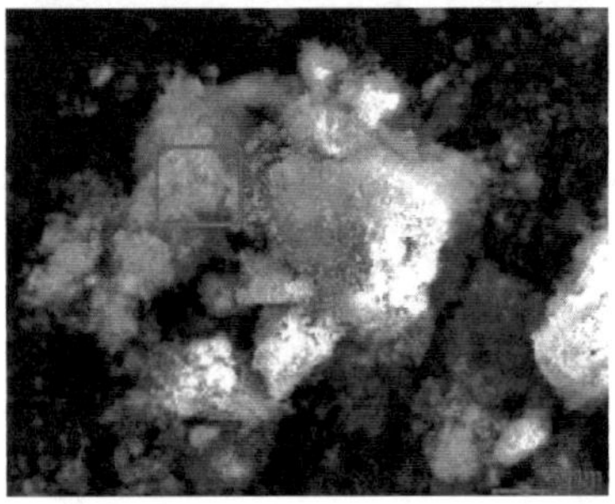

Element	Wt.%	At.%
OK	28.96	50.80
MgK	17.42	20.11
SnL	00.59	00.14
TiK	39.50	23.14
ZnK	13.53	05.81
Matrix	Correction	ZAF

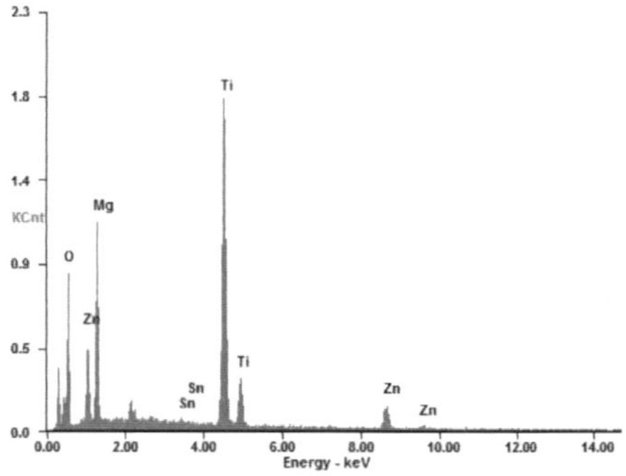

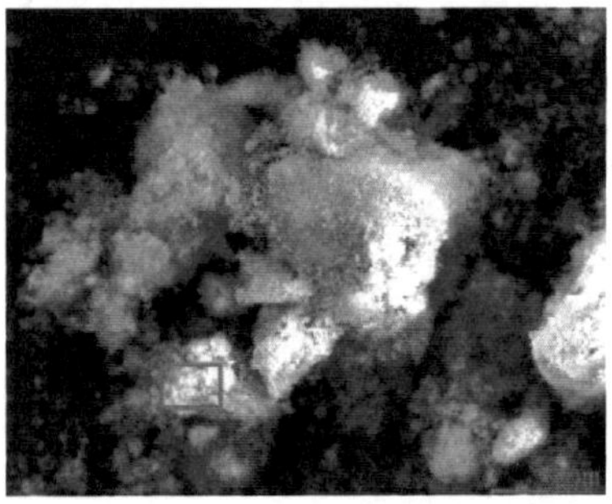

Element	Wt.%	At.%
OK	24.86	47.46
MgK	11.33	14.23
SnL	00.88	00.23
TiK	50.95	32.48
ZnK	11.98	05.60
Matrix	Correction	ZAF

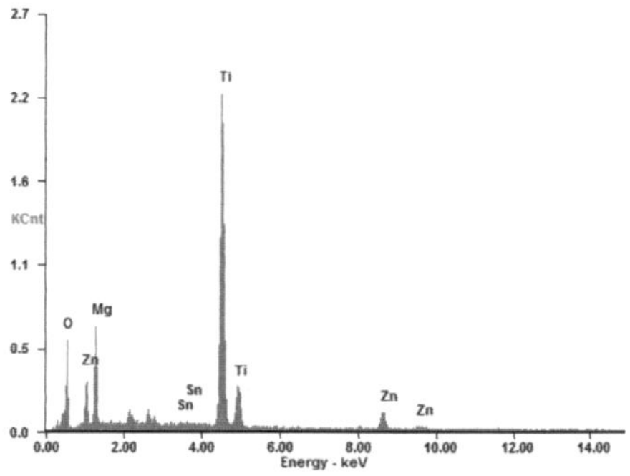

Figure 12. The similar EDX elemental analysis as in Figures 10-11, but this is for the powder calcined at 650°C

With the same analysis, in Figures 12 and 13, the ratio of ($Mg_{0.6}Zn_{0.4}$): ($Ti_{0.99}Sn_{0.01}$): O are given in Equations (5) and (7), while the ideal atomic % ratio are given in Equations (6) and (8).

$$(Mg_{0.6}Zn_{0.4}):(Ti_{0.99}Sn_{0.01}):O=(14.23+5.60):(32.48+0.23):47.46=19.83:23.71:47.46 \quad (5)$$

$$\text{The ideal \% atomic ratio is } \approx 15.82:15.82:47.46=1:1:3 \quad (6)$$

$$(Mg_{0.6}Zn_{0.4}):(Ti_{0.99}Sn_{0.01}):O=(15.64+6.16):(27.52+0.32):50.36=21.80:27.84:50.36 \quad (7)$$

$$\text{The ideal \% atomic ratio is } \approx 16.79:16.79:50.36=1:1:3 \quad (8)$$

Based on the % of Ti atoms data in Equations (5) and (7), it turns out that the % of Ti atoms at 650 and 700 °C is excessive so that the presence of other phases containing Ti besides $MgTiO_3$ becomes clear. As is the case at 550 and 600 °C, the intended phase may be TiO_2 rutile. Again, this hypothesis corresponds to data in Figure 9 for 650 and 700 °C.

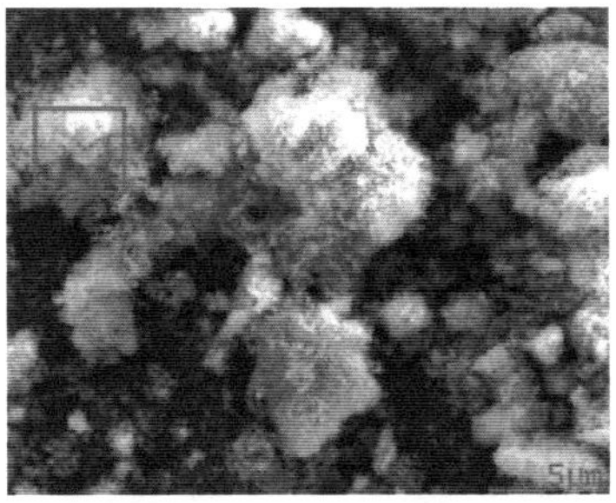

Element	Wt.%	At.%
OK	30.15	53.51
MgK	12.74	14.88
SnL	01.16	00.28
TiK	44.37	26.30
ZnK	11.59	05.03
Matrix	Correction	ZAF

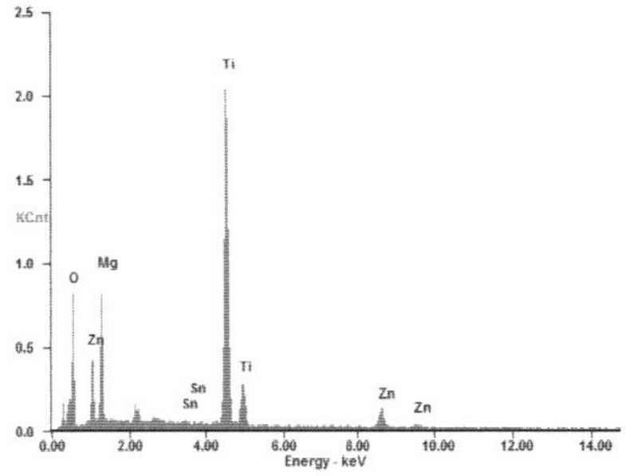

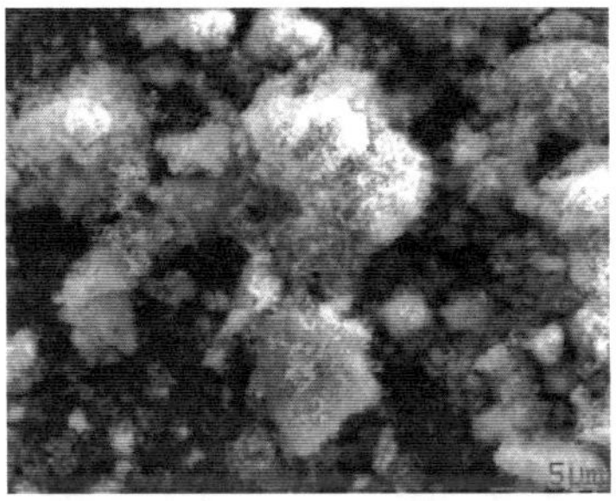

Element	Wt.%	At.%
OK	27.36	50.36
MgK	12.91	15.64
SnL	01.30	00.32
TiK	44.76	27.52
ZnK	13.67	06.16
Matrix	Correction	ZAF

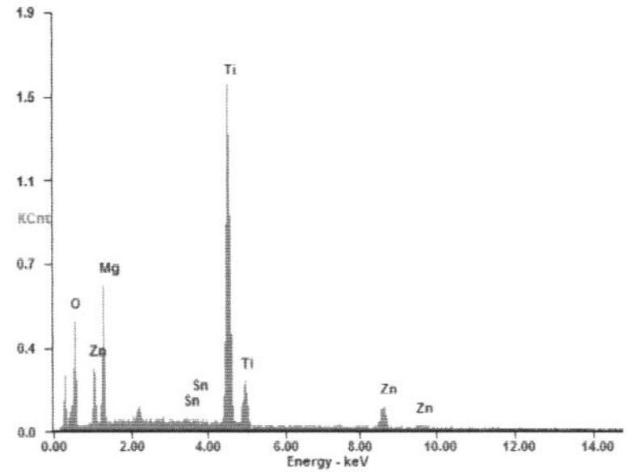

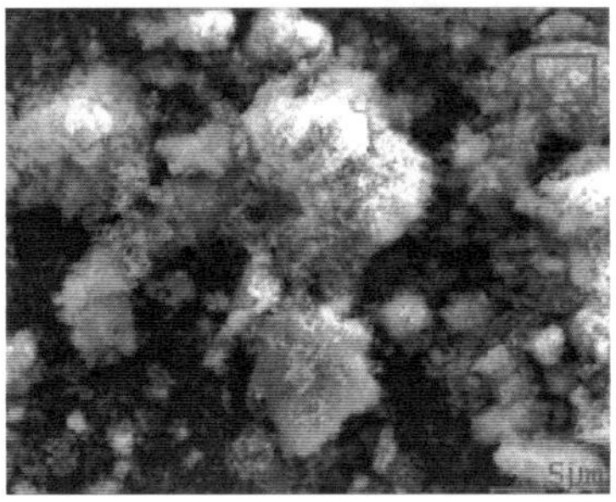

Element	Wt.%	At.%	
OK	32.50	54.56	
MgK	17.86	19.73	
SnL	00.87	00.20	
TiK	36.61	20.53	
ZnK	12.16	04.99	
Matrix	Correction	ZAF	

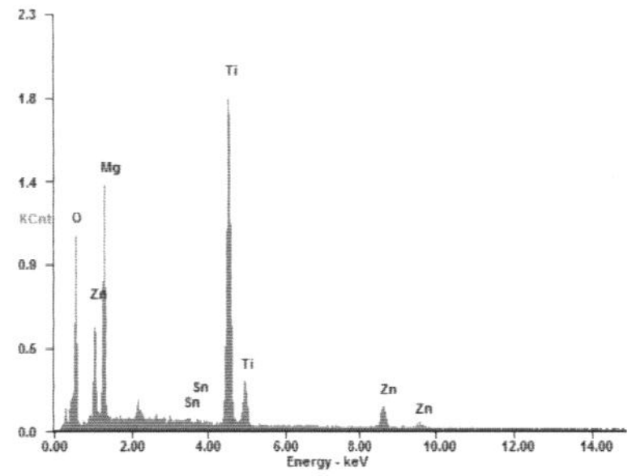

Figure 13. The similar EDX elemental analysis as in Figures 10-12, but this is for the powder calcined at 700°C

4. Conclusion

A series of analyses on the formation of the $MgTiO_3$ phase in $(Mg_{0.6}Zn_{0.4})(Ti_{0.99}Sn_{0.01})O_3$ powders with a calcination temperature of 550-700°C using XRD and EDX data have been completed. The two types of data support each other, i.e. the $MgTiO_3$ phase was formed as the main phase accompanied by TiO_2 rutile as the impurity phase. The increase in calcination temperature causes the $MgTiO_3$ content to decrease. The temperature of 550°C gives the most optimum $MgTiO_3$ molar %.

References

[1] Zhang J, Yue Z, Luo Y, and Li L 2018 *Ceram. Int.* **44** 21000
[2] Wu H T, Jiang Y S, Cui Y J, Zhang X H, Jia X, and Yue Y L 2012 *J. Electron. Mater.* **42** 445
[3] Ermawati F U, Pratapa S, Suasmoro S, Hübert T, and Banach U 2016 *J. Mater. Sci. Mater. Electron.* **27** 6637
[4] Kadarosman A and Ermawati F U 2021 *J. Neutrino* **13** 67
[5] Ermawati F U 2021 *J. Phys. Conf. Ser.* **1805** 012039
[6] Izza L and Ermawati F U 2021 *J. Ilmu Fis.* **13** 62
[7] Samyuktha V S, Subbarao T, Suvarna R P, and Kumar A G 2017 *J. Ovonic Res.* **13** 33
[8] Hsu C H and Chang C H 2015 *Ceram. Int.* **41** 6965
[9] Zhang M, Li L, Xia W, and Liao Q 2012 *J. Alloys Compd.* **537** 76
[10] Xue X, Yu T, and Xu G 2013 *J. Mater. Sci. Mater. Electron.* **24** 1287

[11] Ermawati F U, Suasmoro S, and Pratapa S 2015 *Adv. Mater. Res.* **1112** 47
[12] Cullity D and Stock S R 2014 *Elements of X-Ray Diffraction 3rd Edition* (New York: Pearson)
[13] Rietveld H M 1969 *J. Appl. Crystallogr.* **2** 65
[14] Post J E and Bish D L 2018 *Rietveld Refinement of Crystal Structures Using Powder X-Ray Difraction Data in: Modern Powder Diffraction* (USA: De Gruyter) p 277
[15] Ermawati F U, Ahfa M H, and Supardiyono 2019 *J. Phys. Conf. Ser.* **1171** 012038

Synthesis of Nanofiber Polyvinyl Alcohol (PVA) with Electrospinning Method

D H Kusumawati*, K V N Istiqomah, I Husnia, and N Fathurin

Department of Physics, Universitas Negeri Surabaya, Indonesia

*Email: diahkusumawati@unesa.ac.id

Abstract. PVA nanofibers are PVA solution-based nanofibers. PVA solution is a synthetic polymer that is non-toxic, soluble in air, and has excellent thermal, gas permeability, and chemical resistance qualities. Nanofibers were synthesized using PVA (Polyvinyl alcohol) solutions at concentrations of 8%, 10%, and 12%. The electrospinning method is used in this investigation, with input voltages of 17 kV and 20 kV, needle distances of 7 cm, 10 cm, and 15 cm from the collector, and a flow rate of 5 ml/hour for each concentration. An optical microscope was used to examine the nanofiber synthesis results in order to assess the morphology and diameter of the fiber. PVA nanofiber with the best fiber from electrospinning has a diameter of 1.06 μm and homogeneous fiber without beads is a synthesized nanofiber with a variation of 10% PVA solution concentration, 15 cm distance from the syringe needle to drum collector, and 20 kV voltage.

1. Introduction

Nanotechnology is one of the most important and interesting fields of physics, chemistry, biology and engineering in recent years. One of the nanotechnology developments is nanoparticles. Nanoparticle research is growing rapidly because it can be widely applied in the fields of environment, electronics, optics and biomedicine [1]. Nanotechnology is a scope for study of nanosized materials ranging in size from 1–100 nm. These nanosized materials include nanofibers, nanowires, nanorods, nanocrystals, nanofoams, and nanotubes [2]. Currently, nanofibers are becoming the most desirable material because their applications in various fields such as drug delivery, catalysis, textiles, tissue engineering, filter media, cosmetics, and sensors [3].

Fabrication methods for nanofibers are various, such as electrospinning, centrifugal spinning, melt-blown technology, phase separation, and template synthesis [2]. But the electrospinning method is most extensively used and most widely accepted [4]. This method has the potential to create infinite and continuous nanofibers [5]. This method can also make nanofibers with various surface topography and morphology from solution parameters, operational parameters, and environmental conditions [6]. The solution parameters consist of concentration, molecular weight, elasticity, viscosity, surface tension, and conductivity. The operational parameters are the pressure of the syringe pump, high electric voltage, and the spinning distance between the needle tip of syringe and collector drum. Parameters of environmental conditions are temperature and humidity [7].

The electrospinning process in which a polymer solution is charged by an electric field. The polymer solution is flowed at high voltage through a syringe. The polymer solution is removed from the tip of the needle to form nanofibers continuously due to the high voltage. At that point, the droplets of the polymer solution at the tip of the needle form a cone called a Taylor cone [8].

Various synthetic polymers are used to produce nanofibers including poly glycolic acid (PGA), polyvinyl alcohol (PVA), polyurethane (PU), poly lactic acid (PLA), pollactic-co-glycolic acid (PLGA), and polyvinyl pyrrolidone (PVP), as well as natural biopolymers such as chitosan, collagen, gelatin and cellulose [9]. Of the many synthetic polymers, in this research using PVA due to the characteristics of PVA which is very soluble in liquids, in this case distilled water, non-toxic, has good thermal characteristics and chemical stability. PVA is also easy to obtain and can form fibers with small diameters, is homogeneous and can be used in various applications. [10]. Nanofibers with small and homogeneous diameters can be obtained by varying the synthesis process (electrospinning). According to research by Ewa et al. [11], the parameters that affect the fiber formed from the electrospinning process are voltage, flow rate, needle distance to collector, concentration. and solution viscosity, conductivity, temperature and humidity. To obtain the desired fiber, it can be done by adjusting the variation during the electrospinning process. Variations made in the manufacture of PVA fiber by electrospinning method include solution concentration, voltage, flow rate, and the distance of the needle tip to the collector drum.

2. Method

The research was initiated by making PVA solutions in various concentrations which can form fibers (8%, 10%, and 12%). The PVA solution was put into a syringe and put into an electrospinning device. At the time of fiber manufacture, variations are also made to the tension and distance of the needle to the collector drum. The diameter of the resulting nanofiber was measured and its morphology and homogeneity were observed with an Optical Microscope (Microscope Metallurgy BX53M Olympus).

2.1. Materials and Tools

The tools used in this study were 250 ml pyrex beaker, 5 ml pyrex measuring cup, dropper pipette, spatula, thermometer, aluminum foil, digital balance, magnetic stirrer, glass slide, 10 ml syringe, and Nachriebe 601 Electrospinning and for characterization using Microscope Metallurgy BX53M Olympus. While the materials used in this study were polyvinyl alcohol 72,000 g/mol (Merck) and distilled water.

2.2. Electrospinning Method

The electrospinning method is a technique for creating fiber used electric field. The electrospinning process is broken down into three steps, each of which is described as a jet model. The first stage is the initiation phase, which involves the creation of solution droplets at the needle end until the solution radiates towards the collector, causing the fiber's diameter to shrink. The second stage involves thinning the fiber diameter. The diameter of the fiber decreases as the distance between the tip of needle and the collector increases. Meanwhile, the third stage is the process of solvent evaporation and fiber freezing (jet solidification) [12]. The formed fiber is placed on a glass slide attached to the collector drum, so that it is easier to pick up and characterize it with an optical microscope to determine the diameter of fiber and its morphology formed.

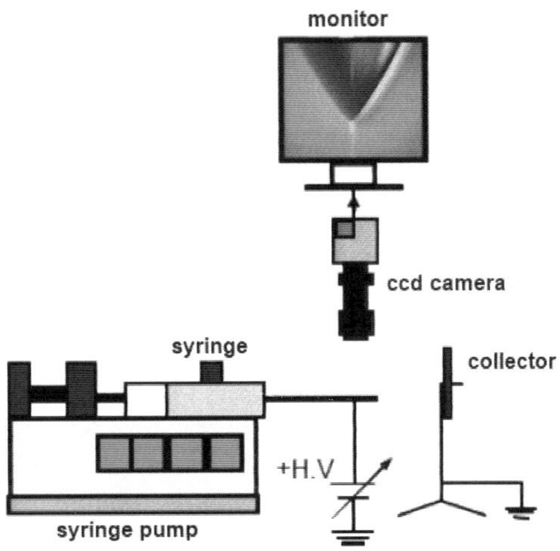

Figure 1. Schematic of the electrospinning equipment system [13]

2.3. Preparation of PVA Nanofibers

The basis material is PVA, which has a molecular weight of 72,000 g/mol. The PVA was used in the form of a PVA solution with concentrations of 8%, 10%, and 12%. Eight percent PVA solution was made by adding 8 grams of PVA into 100 ml of distilled water. Eight grams of PVA were mixed with 100 ml of distilled water to get an eight percent PVA solution. Ten grams of PVA were mixed with 100 ml of distilled water to make a 10% PVA solution. Twelve grams of PVA were mixed with 100 ml of distilled water to get a 12 percent PVA solution. The mixture was then agitated at 100°C until it was totally homogenous. Because PVA is hydrophilic, it may dissolve in water, adding aquadest to a PVA solution tries to dissolve it.

The resultant solution was then electrospinning at 17 kV and 20 kV, with a distance of 7 cm, 10 cm, and 15 cm between the needle tip of syringe and the collector drum, and a flow rate of 5 ml/hour and the electrospinning process time is 1 hour.

2.4. Nanofibers characterization method

The fibers generated as a result of the electrospinning process was characterized using an optical microscope (Microscope Metallurgy BX53M Olympus). It was used to examine the morphology of samples and determine the fibers diameter. It will be able to determine the best variation of voltage, distance between the needle tip and collector, and concentration of solution.

3. Results and Discussion

PVA solution with three different concentrations can produce fiber after electrospinning process. Overall, the resulting fiber has a diameter of less than 2 μm, where the largest fiber has a diameter of 1.45 μm and the smallest fiber is 0.42 μm. The diameter of the fiber produced with variations in voltage, concentration of PVA solution and the needle distance to the collector as shown in Table 1. Fiber diameter and fiber morphology were observed using an optical microscope, so that the accuracy of the diameter measurement results is still in micro units, so that higher accuracy needs to be measured with another characterization tool.

PVA nanofibers with a voltage variation of 17 kV mostly have smaller diameters than nanofibers with a voltage of 20 kV, this is contrary to research from Ewa et al [11], where the higher the voltage, the smaller the fiber diameter. However, the sample with a voltage of 20 kV is much more homogeneous than the sample with a voltage variation of 17 kV.

PVA nanofiber with a voltage of 20 kV has a fiber shape that looks neater, tighter, and stronger (Figures 2, 3, and 4). The nanofibers formed at a voltage variation of 20 kV also did not appear to have beads at all, unlike the nanofibers that used a voltage of 17 kV, where beads still formed on some parts of the nanofibers. In addition, although the diameter of the fiber in the 20 kV PVA nanofiber is larger than that of the 17 kV PVA nanofiber, the fiber with a strong morphology is more likely to be applied according to the PVA function in the process of forming fibers using the electrospinning method, namely as a polymer carrier material to be formed into nanofibers. Thus, the voltage variation that produces PVA nanofibers with the best morphological properties is at a voltage of 20 kV.

Table 1. Variations in solution concentration, needle distance to the collector, and voltage were used to test the synthesis of PVA nanofibers

Concentration of PVA solution (%)	Needle distance to collector(cm)	Nanofiber diameter on 17 kV (μm)	Nanofiber diameter on 20 kV (μm)
8	7	0.89	0.90
	10	1.13	1.45
	15	1.26	1.11
10	7	0.57	0.42
	10	0.95	1.17
	15	1.07	1.06
12	7	0.63	0.83
	10	1.15	1.35
	15	0.75	1.04

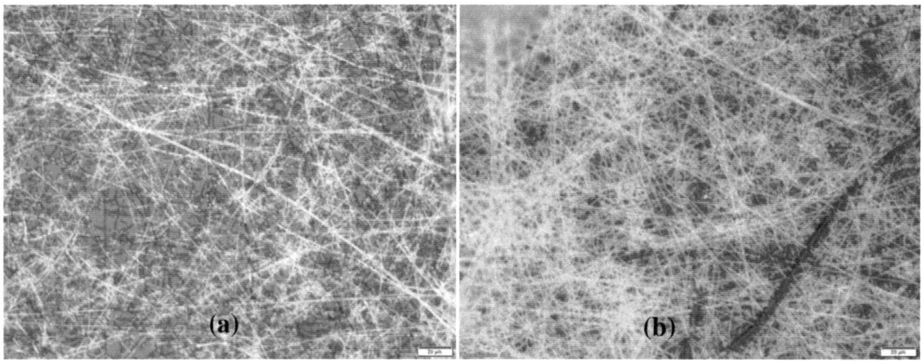

Figure 2. 10% PVA nanofibers with 7 cm needle distance to collector, (a) 17 kV voltage and (b) 20 kV voltage

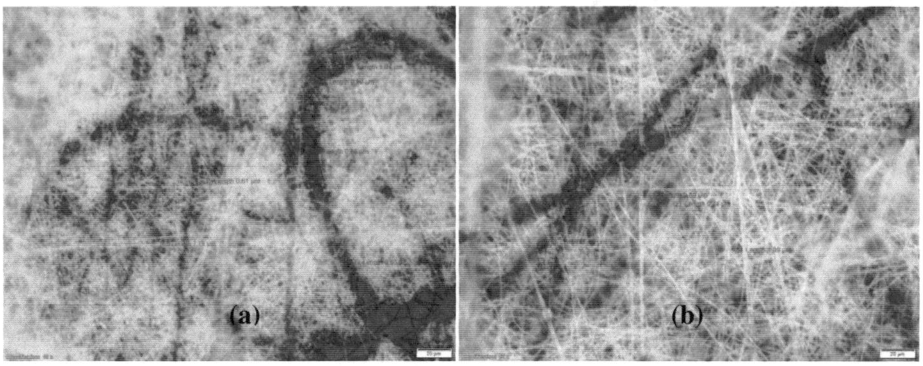

Figure 3. 10% PVA nanofibers with 10 cm needle distance to collector, (a) 17 kV voltage and (b) 20 kV voltage

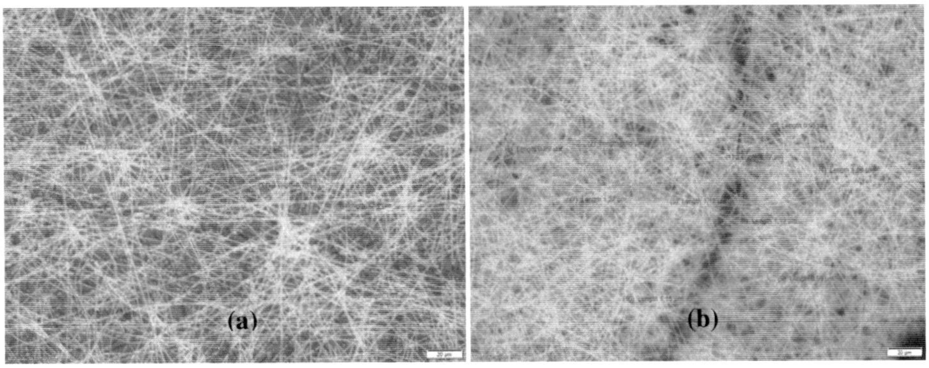

Figure 4. 10% PVA nanofibers with 15 cm needle distance to collector, (a) 17 kV voltage and (b) 20 kV voltage

According to research, homogeneity of nanofibers depends on the concentration percentage of PVA solution, the higher concentration will produce more homogenous PVA nanofiber. However, the diameter of the nanofiber was bigger at the greatest solution concentration, 12 %, than it was at a solution concentration of 10%. (Figure 5). The concentration of 10% nanofiber generated at a voltage of 20 kV was quite homogeneous (Figures 6, 7, and 8). So, in terms of solution concentration variations, a concentration of 10% is the best nanofiber.

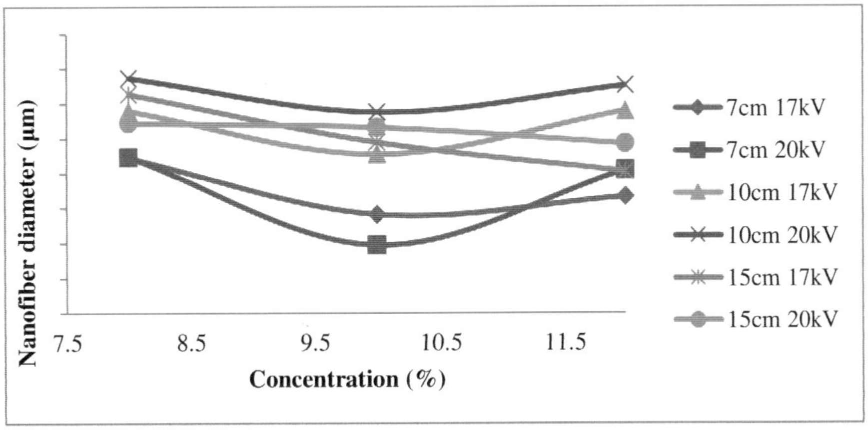

Figure 5. Effect of solution concentration variation on PVA nanofiber diameter

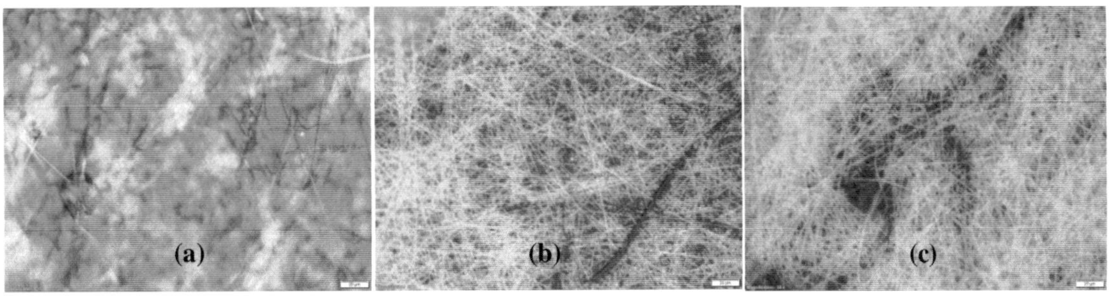

Figure 6. 20 kV PVA nanofiber with 7 cm needle distance to collector, (a) 8 % concentration, (b) 10 % concentration, and (c) 12 % concentration

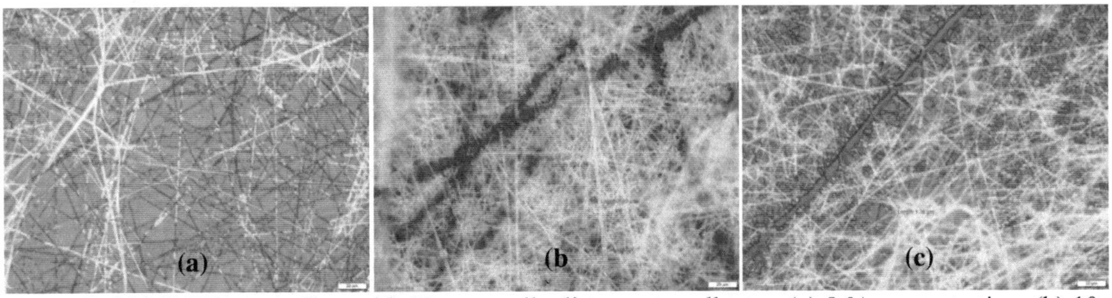

Figure 7. 20 kV PVA nanofiber with 10 cm needle distance to collector, (a) 8 % concentration, (b) 10 % concentration, and (c) 12 % concentration

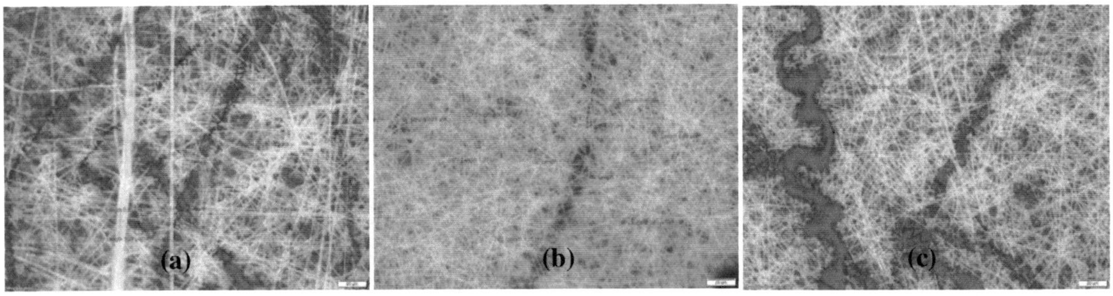

Figure 8. 20 kV PVA nanofiber with15 cm needle distance to collector, (a) 8 % concentration, (b) 10 % concentration, and (c) 12 % concentration

PVA nanofiber diameter with 7 cm needle to collector distance creates nanofibers with the smallest diameter than 10 cm and 15 cm distance, but 7 cm needle to collector distance produces inhomogeneous nanofibers, according to the modification of needle to collector distance (Figure 9). In this investigation, the best fiber morphology with variations in needle distance occurred when the distance was set to 15 cm. This is because the diameter of the nanofiber is less at a distance of 15 cm than it is at the distance of 10 cm. Furthermore, fiber morphology is more homogenous, stronger, and neat at a distance of 15 cm than at a distance of 10 cm (Figure 10).

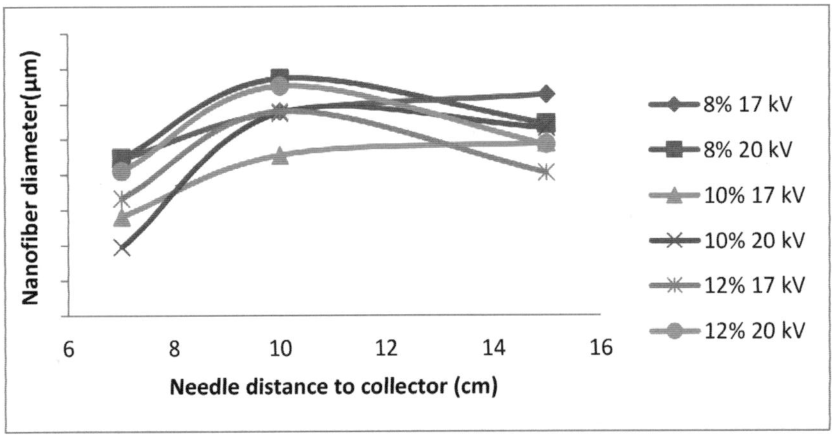

Figure 9. Effect needle distance to collector variation on PVA nanofiber diameter

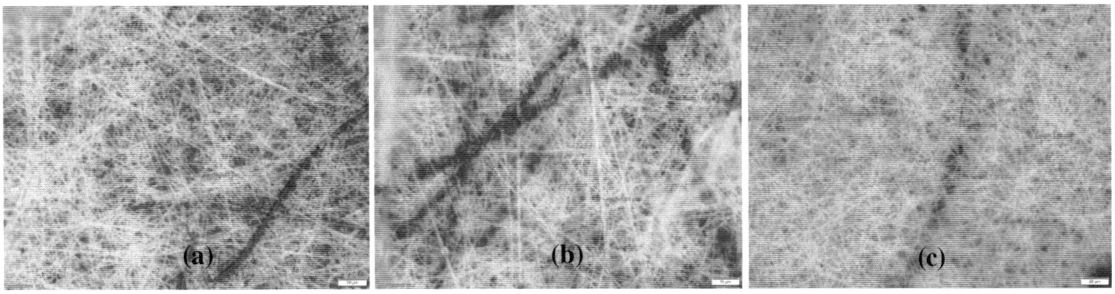

Figure 10. PVA nanofiber with 10% concentration, 20 kV (a) 7 cm distance, (b) 10 cm distance, and (c) 15 cm distance

Variations in voltage, concentration of PVA solution, and the distance between the needle and the collector all have an impact on the shape of the resultant fiber, as seen above. When the voltage variation is 20 kV, the concentration of PVA solution is 10%, and the distance between the needle and the collector is 15 cm, the results of the research on manufacturing PVA nanofibers in terms of morphology are the best.

4. Conclusion

Variations in the voltage used, the solution concentration, and the distance between the needle and the collector could all affect the condition of the nanofiber formed. The generated PVA nanofibers had the optimum morphological circumstances at a voltage of 20 kV, a concentration of 10% PVA solution, and a distance of 15 cm from the needle to the collector, according to the study's findings. The nanofibers in this version feature were strong, neat, homogenous fibers with a diameter of around 1.06 μm.

References

[1] Almetwally A A, El-Sakhawy M, Elshakankery M H and Kasem M H 2017 *J. Text. Assoc.* **78** 5
[2] Wahyudi T, Sugiyana D and Helmy Q 2011 *Arena Tekst.* **26** 55
[3] Kenry and Lim C T 2017 *Prog. Polym. Sci.* **70** 1
[4] Jain R, Shetty S and Yadav K S 2020 *J. Drug Deliv. Sci. Technol.* **57** 101604
[5] Shahi M, Moghimi A, Naderizadeh B and Maddah B 2011 *Sci. Iran.* **18** 1327
[6] Ahmed F E, Lalia B S and Hashaikeh R 2015 *Desalination* **356** 15
[7] Shirazi M M A, Bazgir S and Meshkani F 2020 Electrospun nanofibrous membranes for water treatment *Advances in Membrane Technologies* ed Amira Abdelrasoul (London, UK: Intech Open) chapter 6
[8] Herdiawan H, Juliandri and Nasir M 2013 *Pros. Sem. Nas. Sains Teknol. Nukl.* (Bandung: PTNBR BATAN) p 110
[9] Kamble P, Sadarani B, Majumdar A and Bhullar S 2017 *J. Drug Deliv. Sci. Technol.* **41** 124
[10] Panichpakdeea J, Larpkiattaworna S, Nuchchaponga S, Naruepaia B, Leekrajanga M and Somwongsa P 2019 *Mater. Today Proc.* **17** 2020
[11] Dzierzkowska E and Stodolak-Zych E 2019 *Eng. Biomater.* **149** 10
[12] Hulupi M dan Haryadi 2018 *Chimica et Natura Acta* **6** 101
[13] Widiyandari H dan Purwanto A 2015 *Reakt.* **15** 248

Journal of Physics: Conference Series **2110** (2021) 012011 doi:10.1088/1742-6596/2110/1/012011

Effects of Sintering Temperature and Holding Times on the Microstructure and Chemical Bond of Strontium Titanate ($SrTiO_3$)

D N Hikmah, D K Sandi, F Nurosyid, Y Iriani*

Physics Department, Faculty of Mathematics and Sciences, Universitas Sebelas Maret, Indonesia

*Email: yofent_iriani@staff.uns.ac.id

Abstract. Strontium Titanate ($SrTiO_3$) is one attractive material studied. In this study, $SrTiO_3$ has been fabricated via the co-precipitation technique. The samples were sintered at 800°C and 900°C with holding times of 2 h and 4 h for each temperature. The purposes of this study were to synthesize $SrTiO_3$ material using co-precipitation technique and to observe the microstructure and chemical bonds of the $SrTiO_3$ as the variations of the sintering temperatures and holding times. According to the X-Ray Diffraction (XRD) results, the sintering temperatures and holding times influenced the intensity values and peak broadening. The alteration in both parameters consequently changed the crystallite size and lattice strain of the $SrTiO_3$ material. Furthermore, the Fourier Transform Infrared (FTIR) results validated the $SrTiO_3$ material by existence of Sr-Ti-O chemical bonds. Also, the absorption peaks of O-H, C-H, and C=O chemical bonds in the $SrTiO_3$ declined due to the higher temperature and longer holding time demonstrating impurities declined. Therefore, according to this study, the sintering temperature of 900°C and the holding time of 4 h was the best parameter for fabricating $SrTiO_3$ powder.

1. Introduction

Strontium Titanate ($SrTiO_3$) is an ABO_3 cubic perovskite structure with Pm3m space group and lattice of $a = b = c = 3.9046$ Å [1]. It possesses an energy band of 3.2 eV with a density of 5.18 g/cm^3 at ambient temperature [2,3]. $SrTiO_3$is a perovskite material with semiconductor properties attributed to titanium oxide and enhanced by the strontium addition [4,5]. As a result, $SrTiO_3$ is used in the manufacture of electronic components. The use of strontium titanate for the manufacture of electronic components such as optical sensors, RAM, and transistors just received consideration in the early 2000s [6]. Besides, $SrTiO_3$ is non-toxic, economical, chemically and photochemically stable making it suitable for water splitting [1,2]. Recently, the demand for strontium titanate has been increasing rapidly. Therefore, the manufacture of $SrTiO_3$ has received much attention for researchers.

Commonly, the are several techniques used to synthesize $SrTiO_3$ material those are hydrothermal [1,5], solvothermal [7], sol-gel [4,8], and co-precipitation [9-11]. The co-precipitation procedure is low-price and relatively easy which results in homogenously smaller particles as compared to other techniques [12]. It mixes two or more different solids or liquids by stirring continuously to produce a precipitate [10]. $SrTiO_3$ fabrication with this technique has not been widely carried out, so studies on this still need to be conducted and developed. Hence, the goals of this study are to synthesize $SrTiO_3$

material using co-precipitation technique by varying the sintering temperatures and holding times and to investigate the effects of those variations on the crystallite structure and chemical bonds of $SrTiO_3$.

2. Method

$SrTiO_3$ was synthesized using the co-precipitation technique. The starting materials involved Strontium nitrate ($Sr(NO_3)_2$), Oxalic Acid Dihydrate ($C_2H_2O_4.2H_2O$) (Merck $\geq$ 99%), Titanium Tetrabutoxide ($Ti(OCH_2CH_2CH_2CH_3)_4$) (Sigma Aldrich, 97%), Isopropanol (IPA). First, Oxalic Acid Dihydrate was dissolved in IPA named Solution 1. Next, Titanium Tetrabutoxide was also dissolved into IPA called Solution 2. Solution 1 and 2 were then mixed producing Solution 3. Afterward, $Sr(NO_3)_2$ was dissolved in IPA and mixed into Solution 3. The mixed solutions were continuously stirred until producing a homogeneous solution. The solution was titrated using distilled water while stirring using a magnetic stirrer on the titrated solution. After that, the titrated solution was precipitated for 24 h at room temperature. The precipitated solution was washed with ethanol and distilled water twice. After washing, the solution was hydrolysed using an oven for 10 h yielding a dry precipitate. The resulting precipitate was milled to obtain homogeneous powder. The samples were sintered at temperatures of 800°C and 900°C with holding times of 2 h and 4 h for each temperature. The sample powder was tested via XRD to observe the crystal structure and FTIR to determine the chemical bonds. The crystal size of the sample was estimated using modified Scherrer equation as Equation 1 [13]. By plotting $\ln \beta$ vs $\ln {}^{1}/_{\cos\theta}$ as x-axis and y-axis, the value of $\ln {}^{k\lambda}/_{\beta}$ could be taken as the intercept of the plot. Thus, the crystal size could be attained.

$$D = {}^{k\lambda}/_{\beta \cos\theta}$$

$$\ln \beta = \ln {}^{k\lambda}/_{\beta} + \ln {}^{1}/_{\cos\theta} \quad (1)$$

3. Results and Discussion

Figure 1 depicts the XRD results of $SrTiO_3$ with sintering temperatures of 800°C and 900°C, respectively, in the terms of diffraction peaks and 2θ angles. After the peaks were compared with the ICDD database, it is known that the peaks are belong to $SrTiO_3$ phase matched with ICDD number #86-0178. However, other peaks marked (*) are apparent confirming impurities in the synthesized $SrTiO_3$ powder. The impurities are $SrCO_3$ based on the ICCD database #221437. Nonetheless, the impurity intensities decline from 800°C to 900°C and from 2 h to 4 h. Carbon and Hydrogen compounds can react with the main components of Strontium and Titanium to form unwanted compounds/ impurities such as $SrCO_3$, in which these compounds can be diminished by increasing sintering temperature and/or holding time. This denotes that with the higher sintering temperature and longer holding time, the reaction process for the formation of $SrTiO_3$ material is more perfect, leaving less impurities.

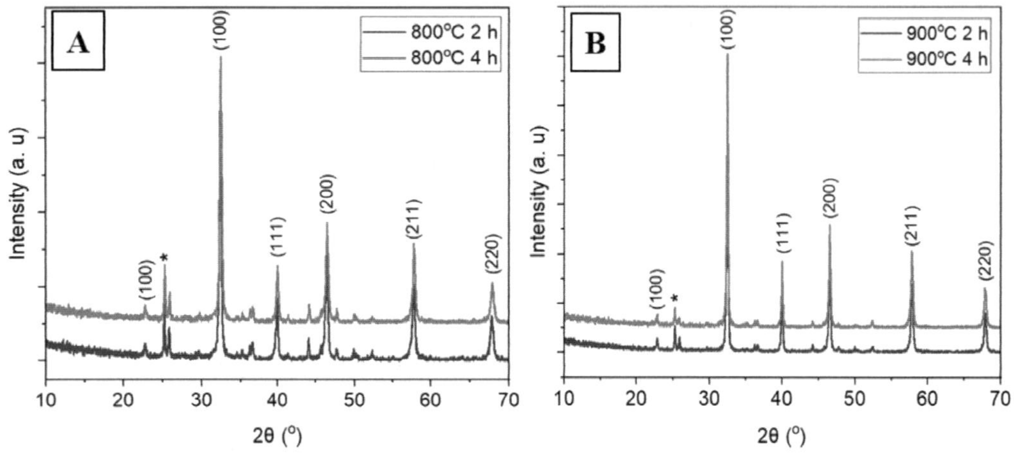

Figure 1. The XRD Peaks of $SrTiO_3$ Powder at (left) 800°C for Holding Times of 2 h and 4 h, and (right) 900°C for Holding Times of 2 h and 4 h

Table 1. Intensity of the 3 Highest Peaks of $SrTiO_3$ at Various Temperature and Holding Time

Samples		Intensity	
		2 h	4 h
Temperature	800°C		
Plane	(100)	718	718
	(200)	266	273
	(211)	187	202
Temperature	900°C		
Plane	(100)	1061	1112
	(200)	396	419
	(211)	257	267

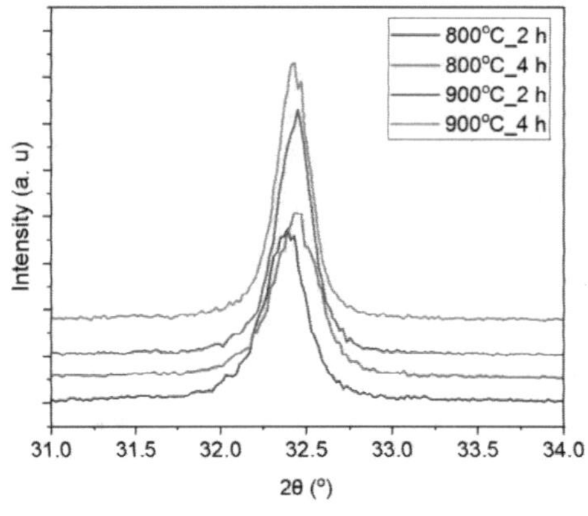

Figure 2. (100) peaks broadening of $SrTiO_3$ with variation of sintering temperatures and holding times

According to the XRD peaks, the variation of the sintering temperatures and holding times does not change the phase of $SrTiO_3$. It is showed by the absence of the angles shift on the diffraction peaks of each varied sample However, the change is apparent on the intensity of the diffraction peaks as shown in Table 1 and Figure 2. Table 1 exhibits the intensity values of the three highest peaks of the XRD patterns for $SrTiO_3$ with variations in temperature and holding time. It reveals that the higher sintering temperatures and longer holding time induce the higher intensity which indicates the better crystallinity of the samples. Besides, the temperature and holding times also affect the peak broadening as presented in Figure 2. The peak broadening indicates a change in the crystallite size and lattice strain of the $SrTiO_3$ materials exhibited by Figure 3. As the longer holding time, the crystallite size at 900°C expands while at 800°C it slightly shrinks. The lattice strain at 800°C decreases from 0.0835 to 0.0615 for the holding times of 2 h and 4 h, respectively. Meanwhile, at 900°C, it increases from 0.0872 to 0.0977 for the holding times of 2 h and 4 h, respectively.

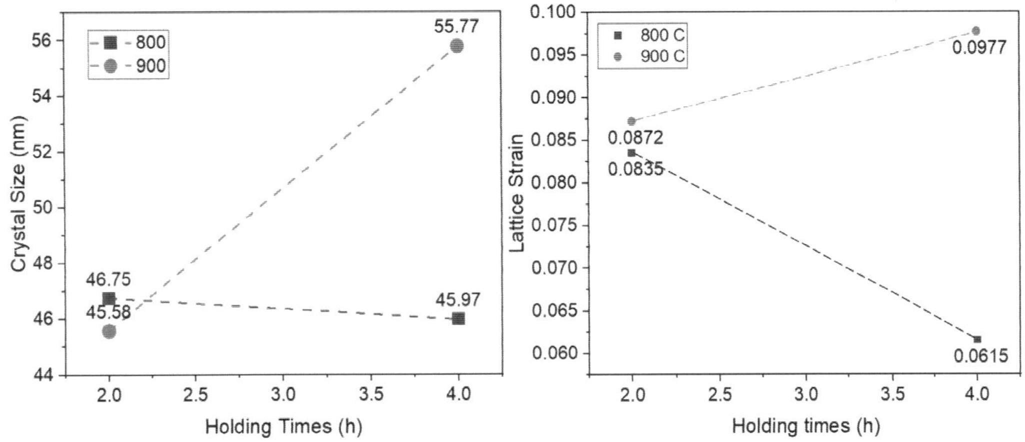

Figure 3. Crystallite size and Lattice strain of $SrTiO_3$ with variation of sintering temperatures and holding times

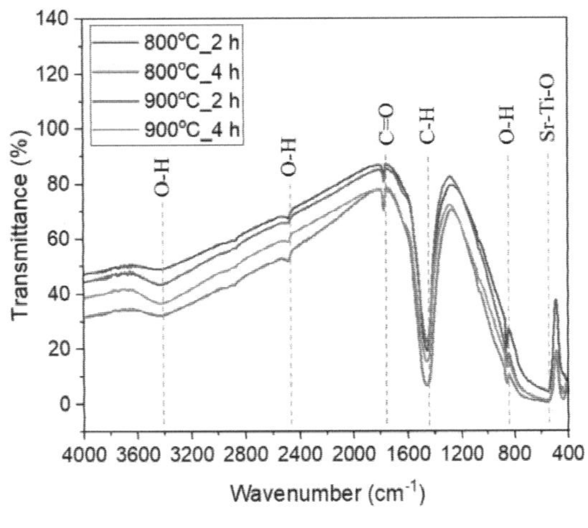

Figure 4. FTIR spectra of $SrTiO_3$ with variation of sintering temperatures and holding times

Figure 5 displays the FTIR spectra of $SrTiO_3$ with variation of sintering temperatures and holding times on the absorption band of 400 – 4000 cm^{-1}. The wavenumber of 557 cm^{-1} is included in the stretching vibration of the Sr-Ti-O bond [6]. In the wavenumbers 859.32 – 860.29 cm^{-1}; 2488.28 – 2490.21 cm^{-1}; and 3420.9 – 3422.83 cm^{-1} exhibit O-H molecules. The absorption peak of 1771.69 – 1772.66 cm^{-1} is due to the stretching vibration of C=O from the COOH group [5]. Based on

Figure 5, the sintering temperatures and holding times did not affect the wavenumber of each molecule significantly, but the transmittances or absorption values. It reveals that absorption peaks of O-H, C=O, and C-H reduce at higher temperature and longer holding time. Those molecules are the cause of the impurities production. Thus, it is in line with the XRD results that the higher temperature and longer holding time can reduce the impurities.

4. Conclusions

$SrTiO_3$ material has been synthesized using co-precipitation technique with variation of the sintering temperatures and holding times. According to the XRD results, the sintering temperatures and holding times did not affect the XRD peaks position but the intensity values and peak broadening. The change in both parameters could modify the crystallite size and lattice strain of the material. Moreover, based on the FTOR results, the all sample confirmed the $SrTiO_3$ material with presence of Sr-Ti-O chemical bonds. Besides, the absorption peaks of O-H, C-H, and C=O chemical bonds in the $SrTiO_3$ decreased as the higher temperature and longer holding time indicating impurities declined. Thus, in this study, the best parameter for fabricating $SrTiO_3$ powder was using the sintering temperature of 900°C for 4 h.

Acknowledgments

The authors would like to thank Kementerian Pendidikan, Kebudayaan, Riset dan Teknologi for financial support with a contract number of 221.1/UN27.22/HK.07.00/2021.

References

[1] Patial S, Hasija V, Raizada P, Singh P, Singh A A P K, and Asirin A M 2020 *J. Environ. Chem. Eng.* **8** 1
[2] Phoon B L, Lai C W, Juan J C, Show P L, and Pan G T 2019 *Int. J. Hydrog. Energy* **4** 14316
[3] Delecta-Vega A, Sepulveda-Ramos N, Crespo-Montoyo S, and Suarez O M 2019 *J. Compos. Sci.* **3** 7
[4] Roque-Ruiz J H, Meraz-Angel J, Farias R, Meléndez-Lira M, and Reyes-López S Y 2019 *J. Ceram. Sci. Technol.* **10** 29
[5] Xian T, Yang H, Di L, Ma J, Zhang H, and Dai J 2014 *Nanoscale Res. Lett.* **9** 327
[6] Xie T, Wang Y, Liu Chenglu, and Xu L 2018 *Mater.* **11**(4) 646
[7] Kuzmić N, Škapin S D, Nelo M, Jantunen H, and Matjaž Spreitzer 2021 *Front. Mater.* 2021. **8** 669421
[8] Morimoto Y, Nishiyama J, Takeda H, and Tsurumi T 2018 *Sci. Rep.* **8** 5031
[9] Zhang D, Wang Biao, Zhang L Y, and Wang M 2015 *J. Adv. Oxid. Technol.* **18** 221
[10] Setyadi A U L S, Nurosyid F, and Iriani Y 2019 *AIP Conf. Proc.* **2202** 020034
[11] Rofiko H, Iriani Y, and Suryana R 2017 *Indones. J. Appl. Phys.* **7** 25
[12] Fo Y, Wang M, Ma Y, Dong H, and Zhou X 2020 *J. Solid State Chem.* **292** 121683
[13] Subarwanti Y, Safitri R D, Supriyanto A, and Iriani Y 2017 *IOP Conf. Ser. Mater. Sci. Eng.* **176** 012044
[14] Ulfa U , Kusumandari K, and Iriani Y 2019 *AIP Conf. Proc.* **2202** 020036
[15] Nath D, Singh F, and Das R 2020 *Mater. Chem. Phys.* **239** 122021

Study of Electrical and Thermal Properties of PVA-Nanosilica as A Candidate for Supercapacitor Separators

N Hidayati[1] and Munasir[1,2]*
[1]Department of Physics, Faculty of Mathematics and Natural Sciences, Universitas Negeri Surabaya, Indonesia
[2]Material Physics Laboratory, Universitas Negeri Surabaya, Indonesia

*Email: munasir_physics@unesa.ac.id

Abstract. Separator in the supercapacitor that separates the cathode from the anode has an important role in a supercapacitor circuit. Polyvinyl Alcohol (PVA) is a substitute for polyolefin which is commonly used as a supercapacitor separator and PVA has more environmentally friendly properties. The addition of silica dioxide nanocomposites is useful for adding thermal stability and electrical insulators. The method used to make the separator is quite simple, namely by casting a PVA gel membrane with nanosilica on a glass plate and drying it at low temperature. PVA membrane with nanosilica variations different precursors are from sand by coprecipitation method and from TEOS. The properties of the two samples were characterized by scanning electron microscopy (SEM), thermo gravimetric analysis (TGA), and LCR meter. Nanosilica derived from sand and from TEOS have no much different characteristic.

1. Introduction

As the portable electronics industry develops, supercapacitors are becoming very useful components as efficient and sustainable energy storage. The supercapacitors can store electrical energy, can be easily carried and used whenever needed. Supercapacitors or double layer capacitors have a higher energy and power density than batteries [1]. The supercapacitor arrangement simply consists of an electrode (cathode and anode) with a layer of activated carbon, a separator, and an electrolyte [2].

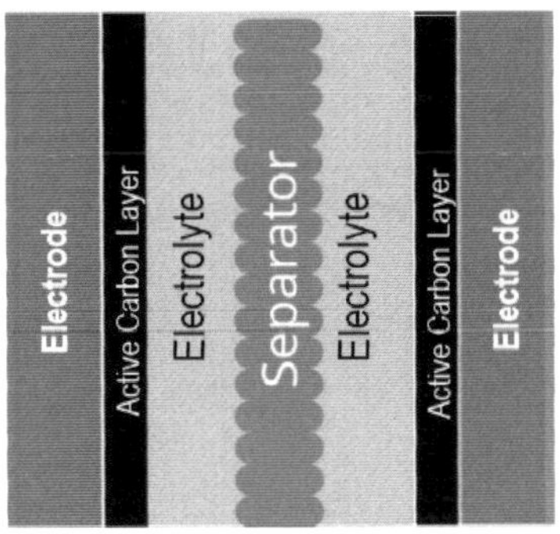

Figure 1. Supercapacitor structure

One of the components in the supercapacitor that plays an important role is the separator. The separator in the supercapacitor acts to separate the anode and cathode, so that when ion transport takes place during charging and discharging, it can avoid an internal short circuit [3]. Thickness, pore size, chemical and thermal stability, and good ion permeability are the requirements for the separator character in the supercapacitor. At high temperatures, separators with low thermal stability can cause various safety problems. The design of a separator with a porous structure and high thermal stability is very important for a supercapacitor performance. Polyolefin-based microporous membranes such as polypropylene and polyethylene are widely used as separator materials. However, there are things that need to be considered, such as thermal stability, porosity, and poor electrochemical performance [4].

For more than three decades the sol-gel reaction has been studied extensively as a method for preparing polymer composites. Solution of TEOS and PVA with sol-gel reaction get a composite because the loading TEOS acts as a cross-linking component making the composite denser [5]. Polyvinyl Alcohol (PVA) based membranes have important characteristics, such as uniform and porous surface morphology, symmetrical interconnected porous structure throughout the thickness of the separator, and excellent electrolyte wettability, resulting in superior electrolyte adsorption/retention capacity, excellent thermal degradation. lower, and higher ionic conductivity, compared to commercial polypropylene (PP) separators [6]. PVA with the incorporation of inorganic particles such as SiO2 can improve thermal stability and electrical insulator. In this study, nanosilica was added to the PVA-based separator in order to increase the electrical insulating properties and thermal stability of a separator. It is known that nanosilica acts as an electrical insulator with high chemical and thermal stability. Nanosilica can be synthesized from chemical precursors, namely Tetraethyl Orthosilicate (TEOS) to natural materials such as sand. Comparison of the separator characteristics of PVA with the addition of silica dioxide nanocomposite from chemical and natural precursors by polymerization method at low temperature is discussed in this article.

2. Method

2.1 Materials

The materials needed in this study were Polyvinyl Alcohol (PVA) from Merckmillipore, Tetraethyl Orthosilicate (TEOS, 98%) from Merck, and sand (76.8 wt% SiO_2). Other supporting materials used include aquades, ethanol (C_2H_5OH, 99%), hydrochloric acid (HCl, 37%), and sodium hydroxide (NaOH).

2.2 Synthesize method

2.2.1 Synthesize of SiO_2

76.8wt% SiO_2 sand was synthesized using coprecipitation or deposition methods. This method aims to make the solute to the bottom so as to form the required precipitate. In addition, coprecipitation requires a relatively short time (±12 hours) and a low working temperature [7]. In this study, previously sand was mixed with 7M NaOH to form a sodium silicate solution. Then it was allowed to stand for 24 hours while 2M HCl was titrated to form a silicite gel. The gel was washed using distilled water and then dried for 2 hours [8].

2.2.2 Nanocomposite Membrane

Nanocomposite membranes were fabricated with nanosilica from the synthesis of SiO_2 from sand and from TEOS compounds. The first membrane contains nanosilica powder that has been synthesized and then dissolved with a concentration of 1% using distilled water. Then the solution was stirred with 5wt% PVA solution for 12 hours. While the second membrane contains nanosilica from the TEOS compound, where 5wt% PVA solution is added with 5 ml of TEOS and 1 ml of HCl then stirred for 12 hours. PVA with nanosilica precursor from sand (PVA-SiO_2) and PVA with nanosilica precursor from TEOS (PVA-

TEOS) were poured on glass and dried at atmospheric pressure and room temperature for 2-3 days, respectively. After that the membrane can be peeled off slowly and dried at 60°C for 3 hours.

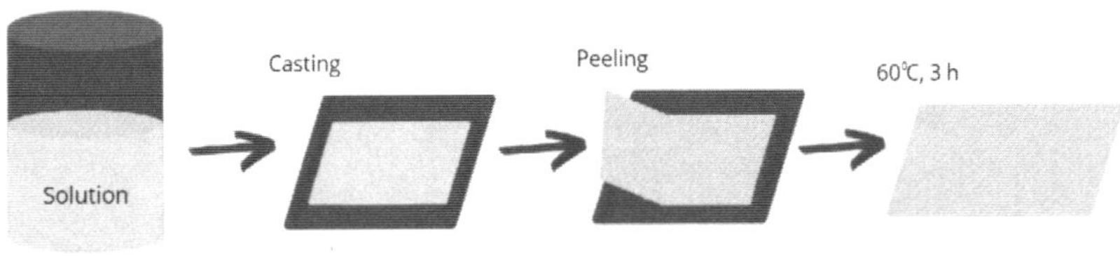

Figure 2. Membrane preparation process

2.3 Characterization.

The nanocomposite membranes were characterized using a Scanning Electron Microscope (SEM) (FEI Inspect-S50 type) to determine the transverse surface topography. Thermal stability testing was carried out using Thermo Gravimetric Analysis (TGA) (TA Instruments, DSC Q20, Waters LLC), in which the nanocomposite membrane was heated from ambient temperature to 600°C at a rate of 10^0C/min with nitrogen gas. The conductivity and resistance values are obtained from the LCR meter (Inductance, Capacitance & Resistance) with two probes.

3. Results and Discussion

3.1 Nanocomposite membrane characterization results

3.1.1. Scanning electron microscope (SEM)

Figure 3 shows the scan results of the two samples using SEM on the cross-surface. The technique used to analyze is a free software image at a magnification of 5000 times. The thickness of the two samples obtained is 18-60 μm. The microstructure of the PVA-TEOS sample looks like a sheet with more structured pores than the PVA-SiO_2 sample. The Si-O-Si bond polycondensation process is formed after the hydrogel turns into a membrane. This is because the hydrogen bonding ability between the -OH groups of PVA-TEOS is stronger than that of PVA-SiO [10]. PVA hydroxyl can react with the hydroxyl of the silanol group on the surface of the silica particles formed in the sol-gel process. This reaction lowers the surface energy of the silica particles and thus prevents aggregation. Meanwhile, the reaction between silica particles from sand and PVA could not be observed clearly.

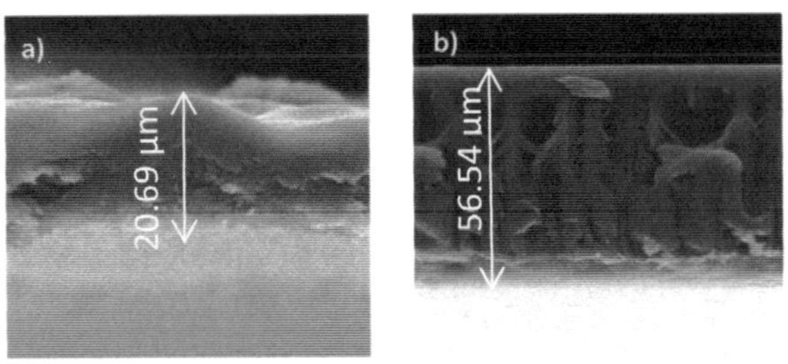

Figure 3. SEM on transverse surface a) PVA-SiO_2 b) PVA-TEOS

3.1.2. Thermo gravimetric analysis (TGA)

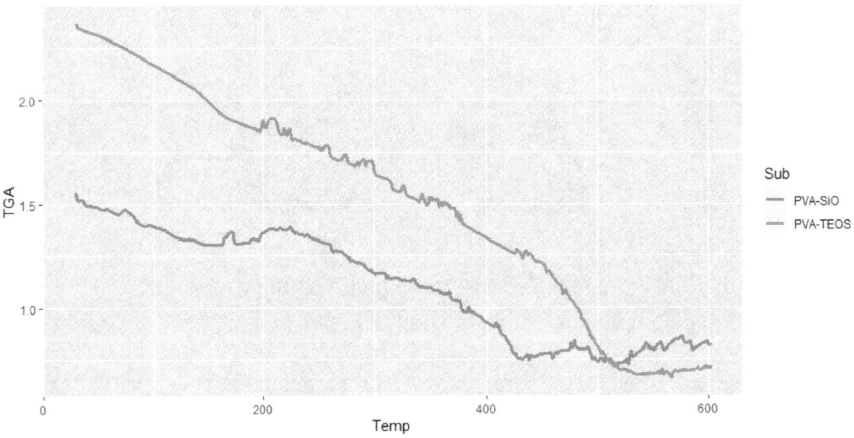

Figure 4. TGA results of the two samples

Based on the comparison of the two samples, the thermal and degradation behaviour of PVA-SiO_2 and PVA-TEOS can be known. The first decrease in the weight of the membrane between the ambient temperature to 200°C, which corresponds to the physical molecules of air and acid that occur. Most of the water molecules are in the direct state to the polymer chain via hydrogen but not in the free molecular state [9]. The first stage of weight loss in the PVA-SiO_2 sample experienced a loss of 10%, while the PVA-TEOS sample lost 19%. The second stage of PVA-TEOS membrane decomposition starts from 470-515°C, along the main PVA circuit. Meanwhile, the second stage of PVA-SiO_2 membrane weight reduction occurred at around 400-425°C. The third stage of weight loss occurs up to 600°C. PVA-SiO_2 membrane decomposes at high temperature which is more stable than PVA-TEOS. Nanosilica granules can increase thermal degradation which acts as a PVA thermal barrier [10]. The final weight residue remaining in PVA-SiO_2 is slightly more than that in PVA-TEOS.

3.1.3. Conductance and Resistance.

The membrane used as a separator in a supercapacitor has a low electrical conductivity and a high resistance value [11]. The resistance value is obtained using an LCR meter, while the conductivity value is calculated using Equation 1.

$$\text{Conductivity} = \frac{L}{R\,A} \tag{1}$$

Table 1. Conductivity and resistance

Separators	Conductivity Value (S/m)	Resistance Value (Ω)
PVA-TEOS	1.044×10^{-7}	5.65×10^{6}
PVA-SiO_2	1.012×10^{-7}	6.31×10^{6}

Low conductivity values can prevent short circuits in supercapacitors. The standard value of conductivity in a polymer membrane is 10^{-7} - 10^{-3} S/m [12]. In both membranes, the conductivity and resistance values are almost equal. The PVA-SiO^2 membrane obtained a slightly larger resistance value than the PVA-TEOS membrane, namely 6.31×10^{6}. Meanwhile, the conductivity value of the membrane after being calculated using Equation 1 obtained the same result of 10^{-7} S/m. The high resistance value is due to the small concentration of PVA used in the membrane. The less PVA in the membrane, the more water it contains. So the resulting resistance value is getting bigger.

3.2. Discussion of PVA-SiO_2 as a supercapacitor separator candidate

Based on the results of the characterization in previous studies, it was shown that the PVA-SiO_2 membrane was very potential as a membrane in supercapacitors. The electrical and thermal properties of PVA can be improved by dispersing small amounts of SiO_2 in the polymer matrix. The interfacial interaction between PVA chains and SiO_2 nanoparticles is the most decisive factor governing the properties of PVA-SiO_2 nanocomposites. Silanol and siloxane groups on the surface of the silica produce the hydrophilic properties of the nanoparticles [13]. In addition to the PVA-SiO_2 hydrogel composite molding method, the PVA-SiO_2 nanofiber composite made by electrospinning exhibits electrical and thermal properties that can replace commercial PP separators. This is indicated by the results of the thermal shrinkage at a certain temperature variation, where the membrane mass shrinks as shown in Figure 5.

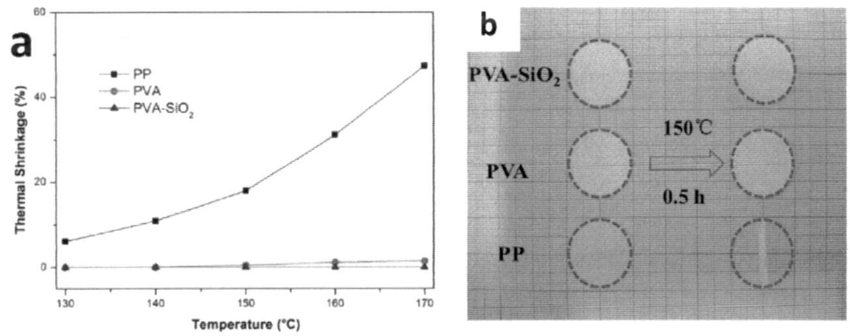

Figure 5. a) Thermal shrinkage results and b) Results after and before the membrane is heated [14]

At a temperature of 130-170°C PP membrane experienced a drastic thermal shrinkage, while PVA with PVA-SiO_2 showed almost no shrinkage. This can also be observed in Figure 5b, where PP shrinks into rolls after being heated at 150 °C for 0.5 hours (30 minutes).

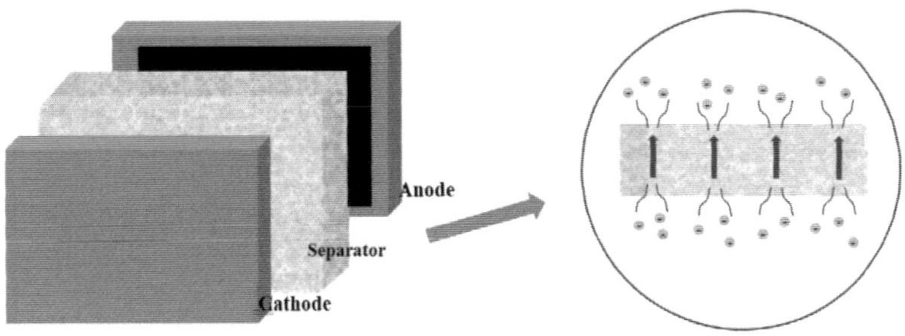

Figure 6. Schematic of the supercapacitor

The cartilage-like and porous structure of PVA makes it an ideal structure for a separator. The large concentration of silica content in the PVA hydrogel membrane can also affect the ionic mobility and the volume of the membrane cavity which affects the charge density in order to develop its electrical conductivity properties [15]. High ionic conductivity is essential for good cycling, depending on the pore structure, porosity and absorption ability of the electrolyte. Highly porous separators are beneficial for the absorption of large amounts of electrolyte and thus lead to higher ionic conductivity [16].

4. Conclusion

In this article, PVA membrane with 5% concentration has been made with the addition of nanosilica from different precursors as a separator. The resulting membrane thickness ranges from 18-60 μm. Nanosilica derived from sand and derived from TEOS have almost similar characteristics. It can be seen from the result of TGA which decrease the weight of the membrane between the ambient temperature to 200°C. The resistance and conductivity which are not much different. In the PVA-TEOS membrane, the resistance and conductance were $5.65 \times 10^6\ \Omega$ and 1.044×10^{-7} S/m, while in the PVA-SiO_2 were $6.31 \times 10^6\ \Omega$ and 1.012×10^{-7} S/m. Based on the test results of the two membranes, it can be concluded that the membrane is suitable to be used as a supercapacitor separator in terms of thermal and electrical properties.

References

[1] Simon P and Gogotsi Y 2012 *Acc. Chem. Res.* **46** 1094
[2] Deimede A and Voege G 2014 *Energy Technol.* **2** 275
[3] Huang X and Hitt J 2013 *J. Membr. Sci.* **163** 425
[4] Xiao W and Zhao L 2015 *J. Membr. Sci.* **487** 221
[5] Andreas A 2020 *Symmetry* **12** 960
[6] Lei B and Jin-wen W 2015 *Chinese J. Polym. Sci.* **31** 1546
[7] Munasir and Triwikantoro 2013 *J. Penelit. Fis. Apl.* **3** 12
[8] Hoepker N, Frey M and Cho D 2012 *Mater. Lett.* **68** 293
[9] Dirican M and Yanilmaz M 2014 *J. Electrochem. Soc.* **161** 14
[10] Moradian and Mazeyar P 2010 *Appl. Surf. Sci.* **256** 2792
[11] Wang J, Zenhua W, and Shao Z 2016 *RSC Adv.* **6** 97912
[12] Riska N 2016 *J. Kimia Akademika* **5** 44
[13] Shobhna C 2016 *AIP Conf. Proc.* **1728** 020053
[14] Shilin C and Zhixiong Z 2018 *Chem. Sel.* **3** 13365
[15] Yanilmaz M 2019 *J. Text. Inst.* **111** 1
[16] Martins P, Lanceroz S, and Costa C 2020 *Synthetic Polymer-Based Membrane for Lithium-Ion Batteries* (Amsterdam, Netherland: Elsevier Inc.)

Effect of Adding of Green Betel Leaf Extract and Honey on Mechanical Properties of Chitosan/PVA Hydrogel Wound Plaster

N L F Chamidah[1], A B Rahanti[1], M D Kadasih[1], F D Sharfina[2], and L Rohmawati[1]*

[1]Departement of Physics, Faculty of Mathematics and Science, Universitas Negeri Surabaya, Indonesia
[2]Departement of Biology, Faculty of Mathematics and Science, Universitas Negeri Surabaya, Indonesia

*Email: lydiarohmawati@unesa.ac.id

Abstract. Wound plasters on the market still have drawbacks, including lack of elasticity and stinging. Thus, there is a need for new materials from nature to be used as the base material for wound plasters, namely betel leaf extract and honey composited with PVA/chitosan. This study purpose to find out mechanical properties of wound plaster of chitosan/PVA with the composition of betel leaf extract and honey. The first stage is making hydrogel by mixing PVA and chitosan, and the second stage is extracting green betel leaves and honey by the maceration method. Then the two stages were mixed with a composition of 5wt%, 15wt%, and 25wt% green betel leaf extract. The results were tested for tensile strength to determine the elasticity properties of the wound plaster. In this study, the 25wt% betel leaf extract concentration had the best tensile strength value of 0.219 MPa, and the modulus of elasticity was 0.11 MPa.

1. Introduction

As a human organ in a protective layer, the skin is very susceptible to injury, such as wounds from stab wounds, burns, surgery or other causes. The effect can cause pain and scarring around the wound. Wound treatment is done by covered, one of which is using wound plaster. Through plaster, wound care will be more practical, comfortable and minimize contamination with the surrounding environment. However, wound plasters that are widely sold in the market still have drawbacks such as soreness and lack of elasticity so that when they are replaced with new plasters, they still cause wounds around the area covered by plaster, so there is a need for a better model of wound plaster such as hydrogel wound plaster. Hydrogel is a polymer wound dressing that is transparent, flexible, soft, and non-irritating [1] due to the release of compounds, namely as a transport medium. Currently, hydrogel wound plasters based on PVA, chitosan and starch have been developed [2]. Polyvinyl alcohol and chitosan are synthetic polymers as the main ingredients of hydrogels.

Chitosan has a high solubility in weak acids and high hydrophilicity based on amino and hydroxyl groups in the chitosan framework [3]. This chitosan can come in different forms and structures of materials such as hydrogel [4], film [5], nanofiber [6], and a 3D porous scaffold [7]. However, this hydrophilicity has an impact on low mechanical properties. So there needs to be modifications to reduce the shortcomings, such as using blending, composite and grafting techniques [8]. It is consistent with previous research that chitosan hydrogels have toughness and are shrinkable fast [9].

Consequently, the utilization of chitosan hydrogel in biomedical materials is still very limited [10]. In addition, other research has also indicated that the mechanical characteristic of the chitosan biopolymer after the chemical modification process can be significantly improved and combined with other polymers that are more effective for wound healing as wound dressings or plasters [11]. Polyvinyl alcohol (PVA) is one of the polymers used as a blending material for chitosan to improve thermal and mechanical stability. PVA properties, such as being easy to dissolve in water, mechanical stability and flexibility, easy to form into films and non-toxic, are the basis for using PVA for applications in the medical field [12].

Chitosan/PVA blending has been extensively studied and developed in various applications. The characteristics of the chitosan/PVA blend material that have been carried out include mechanical, thermal, morphological and pH sensitivity [13]. Crosslinkers are used in the manufacture of chitosan/PVA blends to improve their mechanical properties. The crosslinking material is glutaraldehyde (GD). Glutaraldehyde functions as a filler in the chitosan/PVA matrix. This material is usually applied to drug delivery systems. The results obtained represent that the amount of filler upgrades the mechanical stability of the matrix [14].

Besides GD, other materials used as fillers are essential oils and oxalic acid [15]. Many essential oils are contained in betel leaves, while oxalic acid can be found in honey. Previous studies have shown that essential oil has been shown to affect the physical properties of the wound plaster edible film [15]. Betel leaf and honey and improving the mechanical properties of chitosan/PVA blending can also minimize infection or other effects on wounds caused by synthetic materials because they are made more natural. This research aims to escalate the mechanical characteristic of the chitosan/PVA composite material with green betel leaf extract as filler and honey. The filler composition used is 5wt%, 15wt%, and 25wt%. It is hoped that the wound plaster can increase its elastic properties to minimize the emergence of new wounds when changing the plaster.

2. Method

2.1. Materials

The materials used in this research included green betel leaf, PVA, chitosan, honey, acetic acid (Merck), water demineralized, ethanol 96% (Merck), and aluminium foil.

2.2. Preparation of chitosan/PVA hydrogel wound plaster with green betel leaf extract and honey

Green betel leaves were washed with distilled water and then dried, and the texture of dry leaves was obtained. Dried green betel leaves were crushed using a blender, and the result was a fine powder sifted using a 200 mesh sieve. Then weighed and soaked with ethanol solution where the concentration of green betel leaf simplicia in ethanol solution was 5wt%, 15wt% and 25wt% respectively and soaked for three days. After the maceration process, the product was filtered and evaporated at 80°C for 4 hours. In addition to green betel leaf, honey is also extracted with a concentration of 100 % in the same way as betel leaf extraction.

Three grams of PVA were mixed in 100 ml of water demineralized and then stirring at 120°C. One percent acetic acid/demineralized water solution was mixed with 3 grams of chitosan and then homogenized at 70°C. The PVA and chitosan solutions were combined with the composition ratio of PVA: chitosan (3:1) stirred until the viscosity increased and green betel leaf extract and honey were added with the ratio of honey and betel leaf composition (1:1) and stirred again until homogeneous. The homogeneous membrane material was placed into a petri dish and oven at 85°C during 5 hours and cooled at room temperature.

2.3. Research Flowchart

The research stages used are outlined in the flowchart as Figure 1.

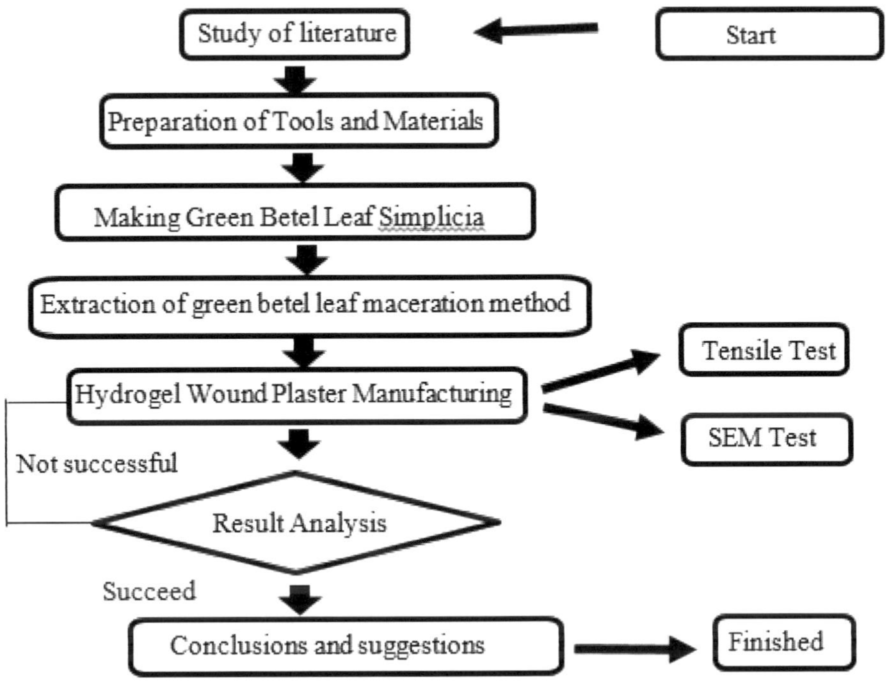

Figure 1. Flowchart of research stages

2.4. Instrumentation

Analysis of elasticity properties of chitosan/PVA hydrogel wound plaster with the addition of green betel leaf extract and honey using a tensile strength test at a speed configuration of 10 mm/minute using the SHIMADZU AG-10-TE10 TE autograph. The sample to be tested was cut to a size of 1x6 cm. Characterization of tensile strength gives results in the form of a graph of stress versus strain which is then analyzed and provides information in the tensile strength (MPa), modulus of elasticity (MPa). The sample tested was in the form of edible film hydrogel, where betel leaf extract and honey were homogenized with chitosan/PVA then evaporated at a specific temperature. The sample as tested by applying a special force (pull) until the sample stretches until the sample is disconnected. Analysis for tensile strength use equation (1) and elastic modulus value use in equation (2) [16].

$$F = \frac{P}{A} \quad (1)$$

$$E = \frac{\sigma}{\varepsilon} \quad (2)$$

F : Tensile strength (MPa)
P : Tensile force (N)
A : Cross-sectional area (mm^2)
E : Modulus of elasticity (MPa)
σ: Tensile strength (MPa)
ε: Strain

Surface structure analysis with green betel leaf extract and honey with the SEM test. The SEM test procedure, the best sample from the tensile strength test results, is placed on a round copper metal plate (sample holder). Furthermore, a coating process is carried out with a gold layer so that the sample has conductive properties. The SEM (Scanning Electron Microscope) test was carried out using the HITACHI FLEXSEM 1000 tool. The sample to be tested was cut to a size of 1x1 cm.

3. Results and Discussion

The physical appearance of the chitosan/PVA hydrogel with green betel leaf extract and honey looks transparent and greenish, where the higher the concentration of green betel leaf extract, the greener the colour is, as shown in Figure 2. The higher the concentration of betel leaf extract, the more betel leaf content, such as essential oil, where during the process of making green betel leaf extract, the amount used also increases according to the concentration. Membrane printing was done by evaporating the solvent at a temperature of 85°C, the transition temperature of PVA glass [16]. At this temperature, PVA will form bond hydrogen between the hydroxyl group of PVA and the carbonyl group of the amide resulting from a blend of chitosan and acetic acid. In addition, PVA, which functions as an interpenetrating network (IPN), will fill the hydrogel cavity so that the structure becomes tight and causes its mechanical strength to be more assertive when compared to pure chitosan. Based on the results of tensile strength measurements on chitosan/PVA wound plaster with betel leaf extract and honey with different concentrations of betel leaf extract, which can be observed in Table 1.

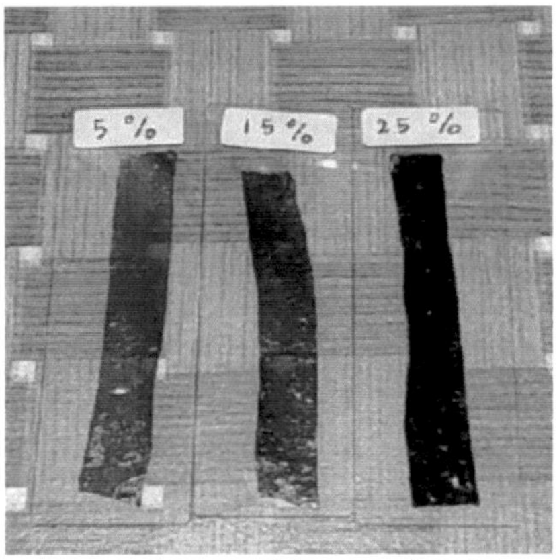

Figure 2. Chitosan/PVA hydrogel with green betel leaf extract and honey

The tensile strength test results in Table 1 show the best results in the sample with a concentration of green betel leaf extract and honey of 25wt% with a tensile strength value of 0.219 MPa and an elastic modulus of 0.111 MPa. It indicates that hydrogel wound plaster with betel leaf extract and honey on a concentration of 25wt% green betel leaf extract is quite elastic and not easily torn. In addition, compared with several other hydrogels, wound plasters such as carrageenan PVP hydrogel wound plaster with a tensile strength value of 0.193 MPa [18] and polydopamine/polyacrylamide hydrogel wound plaster with a tensile strength value of 0.051 MPa [19]. So that the chitosan/PVA hydrogel wound plaster with betel leaf extract and honey is a good composition for elastic wound plasters, safe and comfortable when applied and does not cause new wounds when changing new plasters.

Table 1. The results of the tensile strength test on chitosan/PVA hydrogel wound plaster with betel leaf extract and honey.

Sample concentration	Δ L (mm^2)	Force (N)	Tensile strength (MPa)	Strain	Modulus of elasticity (MPa)
5 wt%	39.066	0.256	0.102	0.928	0.110
15 wt%	58.672	0.460	0.138	1.314	0.105
25 wt %	80.949	0.526	0.219	1.973	0.111

The mechanical properties of a material can be analysed using a tensile test, wherewith this test, the elasticity value of edible plaster in hydrogel wound cab be determined. A high modulus of elasticity and tensile strength can protect the product from mechanical disturbances and is more comfortable to use. When the force applied to the hydrogel wound plaster of betel leaf extract and honey is large while the value of the increase in length is also more significant, it means that the wound plaster is relatively elastic, and it is proven that the hydrogel wound plaster film with a concentration of 25wt% green betel leaf extract can withstand a force of 0.526 N with an increase in length of 80.949 mm. The greater concentration of the sample in the hydrogel in Table 1, the greater the tensile strength value and the greater the elasticity. It indicated the increase in the value of delta L or the change in length in the sample being tested. This increase in length was due to the sample concentration of 25wt% having more betel leaf extract composition than the other samples, then adding betel leaf extract and honey to the chitosan/PVA hydrogel wound plaster could improve its mechanical properties. Surface morphology as well as pore formation and connectivity between hydrogel pores are very important for hydrogel wound plaster application in the formation of new tissues which can be observed through the SEM test [20].

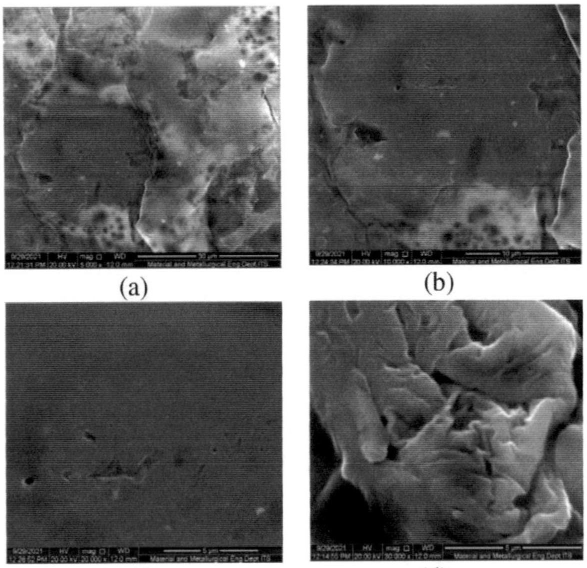
(a) (b) (c) (d)

Figure 3. SEM test results of hydrogel wound dressing, green betel leaf extract and honey with a concentration of 25 wt% enlargement (a) 500x (b) 1000x (c) 2000x (d) 3000x

Figure 3 shows the morphological structure of the hydrogel surface, which is porous, almost flat, smooth, and regular and there are few air bubbles. The presence of these pores indicates that the addition of green betel leaf extract concentration results in a larger surface area and is able to absorb excess exudate in the wound. The porous structure ensures good permeability, which is beneficial for wound healing. Meanwhile, the formation of air bubbles resulted in the mechanical properties of the hydrogel wound plaster getting smaller and more prominent.

Green betel leaf extract can permeate water and intensify the viscosity of the chitosan/PVA matrix. This viscosity affects its elastic properties, and the physical appearance of the wound plaster which is too runny cannot meet commercial requirements. The viscosity of the chitosan/PVA matrix will cause also the hydrogel wound plaster to be drier and break easily. The augment of betel leaf extract and honey increased the crystalline structure of the hydrogel [20]. Here, polymer chain arrangement occurs, filling the hydrogel's amorphous region with green betel leaf extract. When the hydrogel was given a tensile force, the first to receive the force is the amorphous region, and only then this crystalline structure will withstand the given tensile force to the maximum extent. Therefore, the more crystalline regions, the greater the hydrogel mechanical strength. The volatile oil compounds in green betel leaf extract that are soluble in water will strengthen the intermolecular bonds of the hydrogel wound plaster, causing the tensile strength value of the hydrogel wound plaster to strengthen. Following previous studies, adding essential oils into edible films with higher concentrations will affect its mechanical properties [20].

The increase in the elasticity value of a hydrogel film, according to Iqbal et al. [21], is due to the polymer matrix used, which can increase the formation of hydrogen bonds and the strengthening effect, causing the success of the polymer itself in forming hydrogen bonds so that the content contained in the green betel leaf extract can form high hydrogen bonds in the wound plaster film and has increased the elongation value when the polymer reaches its breaking point.

4. Conclusion

The results of this research that addition of green betel leaf extract and honey to chitosan/PVA hydrogel wound plaster affects its mechanical properties, including tensile strength test. The most excellent tensile strength value was obtained in the hydrogel with the most significant green betel leaf extract concentration, namely 25 % with a tensile strength value of 0.219 MPa. There is a need for further research on chitosan/PVA hydrogel wound plaster with several tests carried out such as SEM and anti-bacterial to determine the effectiveness of chitosan/PVA hydrogel with betel leaf extract and honey as wound plaster.

Acknowledgement

The researcher is grateful to the material laboratory at the Department of Physics, Universitas Negeri Surabaya, as a place for material synthesis and thanks to the Physics Laboratory of Universitas Airlangga as a place for mechanical testing of the results of this research.

References

[1] Saputra A A S, Dewi T, Ramadhani E K, Ibrahim N, and Wibisono G 2020 *Proc. Book Call Pap. Thalamus: Med. Res. Better Health* **2** 2721
[2] Baghaie S, Khorasani M T, Zarrabi Ali and Jamal M 2017 *J. Biomater. Sci. Polym. Ed.* **28** 2220
[3] Abraham A, P A Soloman, and V O Rejini 2016 *Procedia Technol.* **24** 741
[4] Ribeiro V P, Pina S, Oliveira J M and Reis R L 2018 *Adv. Exp. Med. Biol.* **1058** 305
[5] Nguyen T P, Nguyen Q V, Nguyen V H, Le T H, Vu Q N H, Vo D V N, Trint Q T, Kim S Y, and Le Q V 2019 *Polym. (Basel)* **11** 1933
[6] Selvaraj S, Fathima N N 2017 *ACS Appl. Mater. Interfaces* **9** 5916
[7] Cengiz I F, Helder P, Mendes J E, Kwon I K, Reis R L, and Oleivera J M 2019 *J. Mater. Sci. Mater. Med.* **30** 63
[8] Hu H, Xin J H, Hu H, Chan A, and He L 2013 *Carbohydr. Polym.* **91** 305
[9] Keihan R E, Radinekiyan F, Aliabadi H A M, Sukhtezari S, Tahmasebi B, Maleki A, and Madanchi H 2021 *Sci. Rep.* **11** 650
[10] Gu Y, Zhu J, Xue C, Li Z, Ding F, Yang Y, and Gu X 2014 *Biomater.* **35** 2253
[11] Jayakumar R, Prabaharan M, Kumar P T S, Nair S V, and Tamura H 2011 *Biotechnol. Adv.* **29** 322
[12] Marin E, Rojas J, and Ciro Y 2014 *Afr. J. Pharm. Pharmacol.* **8** 674
[13] Mulchandani N, Shah N, and Mehta T 2017 *Polym. Polym. Compos.* **25** 241
[14] Sionkowska A 2011 *J. Prog. Polym. Sci.* **36** 1254
[15] Sholehah M M 2016 *Karakteristik dan Aktivitas Antibakteri Edible Film dari Refined Karagenan dengan Penambahan Minyak Atsiri Lengkuas Merah* Undergraduate Thesis (Semarang: Universitas Diponegoro)
[16] Gu J and Catchmark J M 2014 *Cellul.* **21** 275
[17] Stevens and Malcolm P 2004 *Polymer Chemistry: An Introduction* (Jakarta: Pradnya Paramita)
[18] Utomo B S B, Fransiska D, and Dermawan M 2016 *J. Pascapanen Bioteknol. Kelaut. Perikan.* **11** 55
[19] Yang Z, Huang R, Zheng B, Guo W, Li C, He W, Wei Y, Du Y, Wang H, Wu D, and Hui W 2021 *Adv. Sci.* **8** 2003627

[20] Nuansa M F, Agustini T W, and Susanto E 2017 *J. Pengemb. Bioteknol.* **6** 54
[22] Mudassir I, Zafar H, Azhar M, Niazi M B K, and Aslam M W 2020 *Polym.* **12** 2112

Journal of Physics: Conference Series **2110** (2021) 012014 doi:10.1088/1742-6596/2110/1/012014

The Performance of Turmeric Paper as an Indicator of The Borax Content in Crackers

F U Ermawati*, B K Prahani, Dzulkiflih, M Yantidewi, and A Zainuddin

Physics Department, Faculty of Mathematics and Natural Sciences, Universitas Negeri Surabaya, Indonesia

*Email: frida.ermawati@unesa.ac.id

Abstract. Borax is one of food additives that has been prohibited from being used by the Indonesian Government because it is dangerous to health. However, the fact is that borax has been widely added to several food products, such as skewered meatballs, school children's snacks, bread, meatballs, tofu and crackers. Crackers containing borax is easily found in traditional markets. Short-term consumption of borax can cause various symptoms such as headache, fever, malaise, nausea or vomiting, pain in the upper abdomen and diarrhea. While long-term exposure to borax will have a more dangerous effect, which can cause stomach disorders, seizures to acute kidney failure which can lead to death. This paper reports a simple, easy and inexpensive test of borax content using a home-made turmeric paper on ten types of crackers purchased randomly in traditional market. The aim is that the borax content test can be carried out by anyone only by buying a few ingredients from the local market and pharmacy, and without the need to involve laboratory tests, but with accurate results when compared to laboratory tests. As a comparison, a flame test and a colour test on the borax content carried out in the laboratorium using the AOAC 18 ed. 2005 Official Method Boric Acid & Borates in Food 970.33 and the FI ed. V of 2014 General Identification Test <291> Borat method A on Page 1423 was also completed. Among the 10 crackers tested, six of them were detected to contain borax. These results are very consistent with the laboratory test results. The performance of the home made turmeric paper is therefore reliable.

1. Introduction

Borax is a mixture of high concentration white mineral salts containing Boron (B) and Oxygen (O) atoms which has the chemical designation Sodium tetraborate decahydrate with the molecular formula $Na_2B_4O_7 10H_2O$ or $Na_2[B_4O_5(OH)_4]8H_2O$, CAS number 1303-96-4, HS Code 28401990, has a molecular mass of 381.37 Daltons [1]. The types of borax commonly used include sodium biborate, sodium piborate, disodium tetraborate, sodium borate decahydrate, boric acid, disodium salt, antypionin, boracsu, boricin, jaikin, neobar, polybor. Borax is an odorless powder or crystal. This chemical is easily soluble in water, glycerol and alcohol, but insoluble in acid. If it dissolves in water it will form sodium hydroxide and boric acid (H_3BO_3).

Sources of borax in nature are from places containing evaporation deposits, such as salt mines and mud craters in Bledug Kuwu, Central Java. Borax is also known locally as *bleng* water if it is in the form of a clear solution, or *bleng* salt if it is in solid crystalline form. The method of making borax is similar to the traditional way of producing salt. *Bleng* mineral water is poured into bamboo, then dried. The result is a solid or liquid *bleng* which is an impure form of boric acid. Borax is widely used in various non-food industries, such as in the paper, wood, plastic, ceramic and glass industries [2]. Apart from being an inorganic preservative, borax is also a powerful microbial killer. Pyrex glasses or chemical laboratory glassware can have a very strong performance against heat because they are made with a mixture of borax. Borax's extraordinary preservative power is probably derived from active boric acid. Boric acid (H_3BO_3) is a weak monobasic Lewis boron acid that is often used as an antiseptic, insecticide, flame retardant, neutron absorber, or precursor to other chemical compounds such as for pharmaceutical products, soaps, mouthwashes, toothpastes and eye washes. However, globally, boric acid was reported to have been used as an additive in food products with the aim of extending the level of freshness of food and improving the quality of food texture. As in Malaysia, boric acid has been used in various noodle products and seafood-based processed foods [3,4]. In China, boric acid has been added on soybean products, rice products, aquatic products and other dried food products [5]; although most of the production of traditional staple foods in China (i.e. noodles and steamed bread) has also been reported to use natural additives including salt, wheat gluten and enzymes [6]. In the United States, boric acid has long been used in caviar and dried apricot products [7]. Consumption of foods containing boric acid has been reported to have adverse effects on human and animal health, ranging from poisoning with boric acid itself, to damage to the immune system and immune-related genes [8–11]. Detection of boric acid content in food has also been reported [12,13].

Like boric acid, borax is also very dangerous for human health if it is inhaled, drunk, ingested and then enters the human body in large quantities. In a short term, consumption of borax can cause various symptoms including dizziness, vomiting and diarrhea [14]. Long term consumption of foods containing borax will result in the accumulation of carcinogenic borax chemicals in human organs such as the liver, brain, kidneys and testes [1,2]. Consumption of high enough borax can cause symptoms of stomach cramps, kidney damage and even death if the dose of borax toxin has reached 10-20 grams. Pongsavee [15] reported that the accumulation of borax in the human body can cause damage to immune cells and chromosomes. The safe limit for using borax in food legally is 1 gram per 1 kilogram of food (1/1000) [2]. Due to its carcinogenic impact on human health, the Indonesian Government has officially banned the use of borax as an additive in food production processes [1]. Detection of borax content both qualitatively and quantitatively in food products has also been widely reported, including using curcumax reagent [16], natural indicator [17,18], optical fiber [19], UV-Vis spectrophotometry [20], by observing physical conditions and testing using $BaCl_2$ solution [21].

However, in many local communities, both in the city and in the countryside in Java, many food producers add borax to several products, such as meatballs [20,21], school children's snacks [22,23], bread [24], meatballs and *pempek* [25], tofu [26] and crackers. Tofu whose production process uses borax is sharp, very tasty and has a bitter taste on the tongue. Meatballs containing borax tend to be white, not brownish like the use of meat. When bitten, the meatballs return to their original texture, chewy and tend to be hard. The existence of crackers containing borax is easily found in traditional markets.

Based on that, in order to prevent the public from being exposed to the borax contained in traditional crackers, it is necessary to think about ways so that people can test the borax content independently, easily, simply and cheaply, but with results that remain accurate without the need to spend money on laboratory test fees, using certain chemicals, or advanced test facilities as reported above. In this present work, ten kinds of traditional crackers purchased at random from a stall in one of the traditional markets in Nganjuk were used as the test samples. What is meant by a turmeric paper is a filter paper that has been soaked in a liquid solution of turmeric starch for a while, then aerated. A sheet of filter paper can be purchased at a local pharmacy for a very cheap price. When turmeric paper is immersed in a liquid solution of crackers containing borax, it will change colour from yellow to brownish red due to the reaction that occurs between curcuma (contained in turmeric paper) and boric acid in borax. The redder the turmeric paper colour, the more borax content in the crackers. As is known, turmeric (*Curcuma domestica Val.*) is one of the native spices and medicinal plants from the Southeast Asian region which has many benefits such as a kitchen spice, natural colorant in food, cosmetics and as a family medicine [27]. The compound that is thought to play an important role in turmeric is curcuma [27,28]. According to Halim [29], curcuma can bind to boric acid which then forms a red rosocyanine component so that it can be used as a borax detection test.

2. Experiment

2.1 Turmeric paper preparation

Five hundred grams of fresh turmeric were cleaned, mashed, squeezed and filtered to produce a liquid solution of turmeric starch. 76 % alcohol as much as 10 % of the total turmeric extract solution to be made is added so that the ratio between the turmeric extract solution and alcohol was 9 : 1. A sheet of filter paper was cut into 3x3 cm^2 size and dipped in the turmeric extract solution for 5-6 minutes, drained and aerated. After drying, the turmeric paper was stored in a closed container. Turmeric paper was then ready to use.

2.2 Preparation of test crackers and borax content testing

Ten kinds of crackers were purchased from one of the traditional markets in Nganjuk East Java. The intended crackers were labelled "K (stands for krupuk) Berkah 1", "K. Emas 3", "K. Ikan Kakap 11", "K. Lempeng 14", "K. Berkah 13", "K. Batang 15", "K. Makaroni 12", "K. Ikan Nona 10", "K. Cap Ikan 9", and "K. Rahayu 8". Each was then crushed and placed in a separate small bowl, added with mineral water until all the flakes of the crackers was submerged and stirred homogeneously and left for a while before being tested for borax content. The turmeric paper was dipped in the cracker solution until it was submerged for about 5 minutes. Lifted and checked the colour. When the colour of the turmeric paper changes to reddish brown, it means there is an indication that the crackers contain borax. Confirmation of the borax content in all the test crackers were then carried out at the Testing Service Unit (TSU) of the Faculty of Pharmacy, Airlangga University, Surabaya using the following methods: (1) AOAC 18 ed. 2005 Official Method of Boric Acid & Borates in Food 970.33 and (2) FI ed. V Year 2014 General Identification Test <291> Borat way A Page 1423.

2.3 Qualitative test of borax at the TSU laboratory

For each test cracker, 100 grams of mass was needed to be ground and then ashed in the furnace. The ash was divided into two equal parts. The first part is for the flame test, while the second part is for the spot (colour) test. The flame test was carried out by acidifying the ash, i.e. by giving a few drops (1 ml) of concentrated sulfuric acid (H_2SO_2 36 N) to the ash in such a way that all the ash becomes wet. A few drops of methanol was then added and the sample was ignited. When the color is green, the sample contains borax because borax actually contains the element boron (B), which has a green flame colour when heated. The second ash is dissolved in HCI then the solution was smeared on a curcumin paper. When there is a brick red stain, the sample contains borax. The curcumin paper was prepared in the following way: 1 g of curcumin (pro analysis grade) was dissolved in 100 ml of Ethanol Absolute (pro analysis grade). Then the filter paper was immersed in the solution. Allow the curcumin paper to dry without using an oven or other heater. Figure 1 shows the flow chart of the borax test as described above.

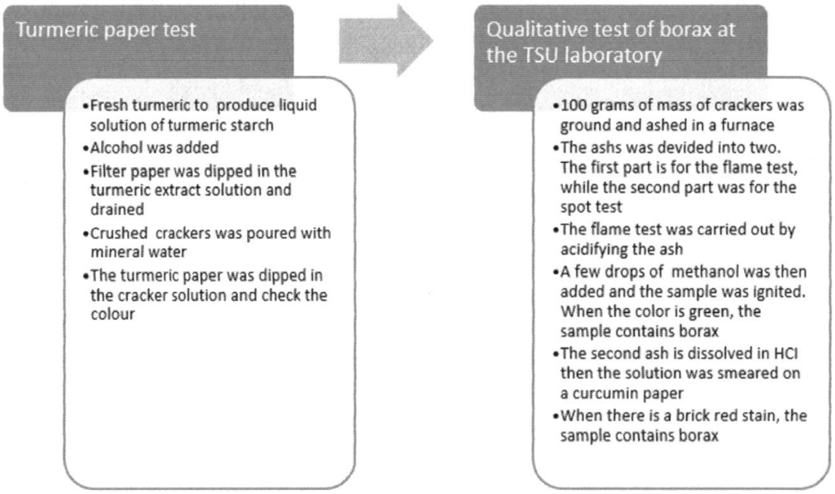

Figure 1. Flow chart showing the experimental procedure followed in this work

3. Results and Discussion

Figures 2-11 shows the results of the borax test in the ten kinds of crackers using the turmeric paper and their comparison using the results of the TSU laboratory test.

Figure 2a. The turmeric paper was soaked in the cracker “K. Berkah 1”. The paper colour turns reddish brown.

UNIT LAYANAN PENGUJIAN
FAKULTAS FARMASI
UNIVERSITAS AIRLANGGA

SERTIFIKAT PENGUJIAN
No. 1147/SA/V/2021*

1. No. Surat Permohonan : -
2. Tanggal Surat Permohonan : 5 Mei 2021
3. Tanggal Sampel dikerjakan : 5 Mei 2021
4. Nama Pemilik Sampel : FRIDA ULFAH ERMAWATI JURUSAN FISIKA FMIPA UNESA
5. Jenis/Nama Sampel/Kode : KRUPUK BERKAH NO 1
6. Keperluan Uji : -
7. Parameter yang diuji : borax
8. Hasil :

Jenis Pengujian	Metode	Hasil
Borax	AOAC 18 ed 2005, 970.33 FI IV, 2014	Positif

Surabaya, 17 Mei 2021
a.n Direktur
Kepala Bidang Teknis

Drs. Hariana, M.Sc.

Figure 2b. The results of the borax test on "K. Berkah 1" crackers from the TSU laboratory: Positive → contains borax.

Figure 3a. The turmeric paper was soaked in the cracker “K. Emas 3”. The paper colour remains yellow, does not change.

UNIT LAYANAN PENGUJIAN
FAKULTAS FARMASI
UNIVERSITAS AIRLANGGA

SERTIFIKAT PENGUJIAN
No. 1149/SA/V/2021*

1. No. Surat Permohonan : -
2. Tanggal Surat Permohonan : 5 Mei 2021
3. Tanggal Sampel dikerjakan : 5 Mei 2021
4. Nama Pemilik Sampel : FRIDA ULFAH ERMAWATI JURUSAN FISIKA FMIPA UNESA
5. Jenis/Nama Sampel/Kode : KRUPUK EMAS NO 3
6. Keperluan Uji : -
7. Parameter yang diuji : borax
8. Hasil :

Jenis Pengujian	Metode	Hasil
Borax	AOAC 18 ed 2005, 970.33 FI IV, 2014	Negatif

Surabaya, 17 Mei 2021
a.n Direktur
Kepala Bidang Teknis

Drs. Hariana, M.Sc.

Figure 3b. The results of the borax test on "K. Emas 3" crackers from the TSU laboratory: Negative. → doesn’t contain borax.

Figure 4a. The turmeric paper was soaked in the cracker "K. Ikan Kakap 11". The paper colour turns reddish brown.

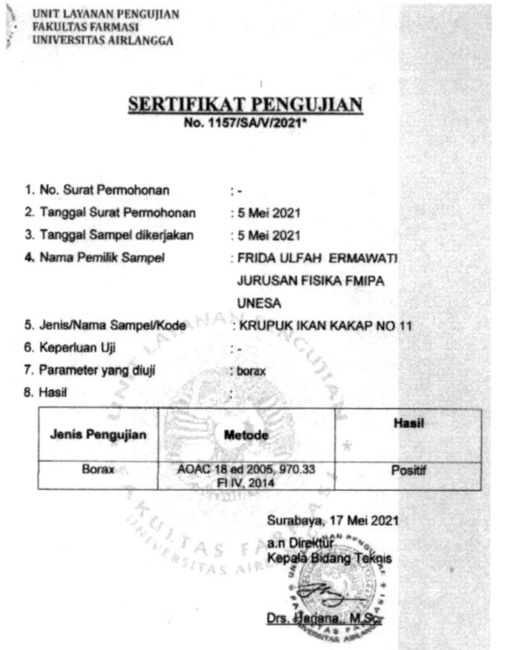

UNIT LAYANAN PENGUJIAN
FAKULTAS FARMASI
UNIVERSITAS AIRLANGGA

SERTIFIKAT PENGUJIAN
No. 1157/SA/V/2021*

1. No. Surat Permohonan : -
2. Tanggal Surat Permohonan : 5 Mei 2021
3. Tanggal Sampel dikerjakan : 5 Mei 2021
4. Nama Pemilik Sampel : FRIDA ULFAH ERMAWATI JURUSAN FISIKA FMIPA UNESA
5. Jenis/Nama Sampel/Kode : KRUPUK IKAN KAKAP NO 11
6. Keperluan Uji : -
7. Parameter yang diuji : borax
8. Hasil :

Jenis Pengujian	Metode	Hasil
Borax	AOAC 18 ed 2005, 970.33 FI IV, 2014	Positif

Surabaya, 17 Mei 2021
a.n Direktur
Kepala Bidang Teknis

Drs. Hariana, M.Sc.

Figure 4b. The results of the borax test on "K. Ikan Kakap 11" crackers from the TSU laboratory: Positive → contains borax.

Figure 5a. The turmeric paper was soaked in the cracker "K. Lempeng 14". The paper colour turns reddish brown.

UNIT LAYANAN PENGUJIAN
FAKULTAS FARMASI
UNIVERSITAS AIRLANGGA

SERTIFIKAT PENGUJIAN
No. 1160/SA/V/2021*

1. No. Surat Permohonan : -
2. Tanggal Surat Permohonan : 5 Mei 2021
3. Tanggal Sampel dikerjakan : 5 Mei 2021
4. Nama Pemilik Sampel : FRIDA ULFAH ERMAWATI JURUSAN FISIKA FMIPA UNESA
5. Jenis/Nama Sampel/Kode : KRUPUK LEMPENG 2 NO 14
6. Keperluan Uji : -
7. Parameter yang diuji : borax
8. Hasil :

Jenis Pengujian	Metode	Hasil
Borax	AOAC 18 ed 2005, 970.33 FI IV, 2014	Positif

Surabaya, 17 Mei 2021
a.n Direktur
Kepala Bidang Teknis

Drs. Hariana, M.Sc.

Figure 5b. The results of the borax test on "K. Lempeng 14" crackers from the TSU laboratory: Positive → contains borax.

Figure 6a. The turmeric paper was soaked in the cracker “K. Berkah 13”. The paper colour remains yellow, doesn’t change.

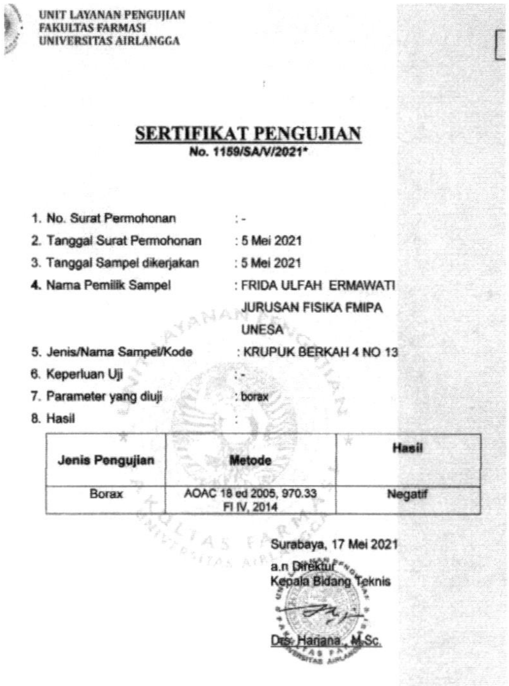

UNIT LAYANAN PENGUJIAN
FAKULTAS FARMASI
UNIVERSITAS AIRLANGGA

SERTIFIKAT PENGUJIAN
No. 1159/SA/V/2021*

1. No. Surat Permohonan : -
2. Tanggal Surat Permohonan : 5 Mei 2021
3. Tanggal Sampel dikerjakan : 5 Mei 2021
4. Nama Pemilik Sampel : FRIDA ULFAH ERMAWATI JURUSAN FISIKA FMIPA UNESA
5. Jenis/Nama Sampel/Kode : KRUPUK BERKAH 4 NO 13
6. Keperluan Uji : -
7. Parameter yang diuji : borax
8. Hasil :

Jenis Pengujian	Metode	Hasil
Borax	AOAC 18 ed 2005, 970.33 FI IV, 2014	Negatif

Surabaya, 17 Mei 2021
a.n Direktur
Kepala Bidang Teknis

Drs. Hariana., M.Sc.

Figure 6b. The results of the borax test on "K. Berkah 13" crackers from the TSU laboratory: Negative → doesn’t contain borax.

Figure 7a. The turmeric paper was soaked in the cracker “K. Batang 15”. The paper dan the liquid colour both turns very strong brownish yellow.

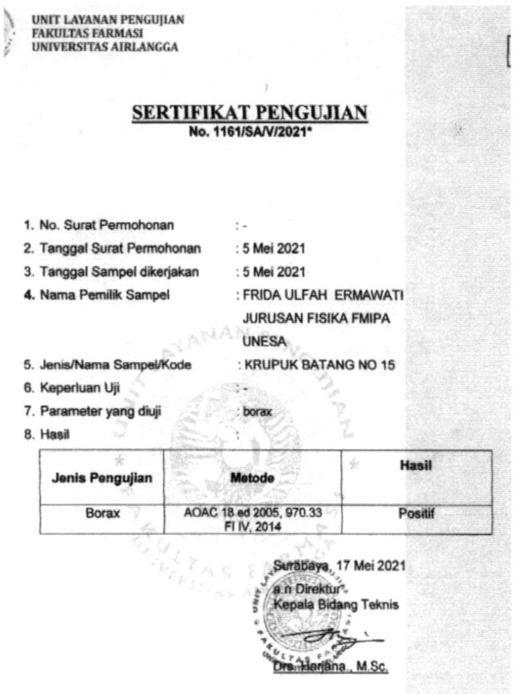

UNIT LAYANAN PENGUJIAN
FAKULTAS FARMASI
UNIVERSITAS AIRLANGGA

SERTIFIKAT PENGUJIAN
No. 1161/SA/V/2021*

1. No. Surat Permohonan : -
2. Tanggal Surat Permohonan : 5 Mei 2021
3. Tanggal Sampel dikerjakan : 5 Mei 2021
4. Nama Pemilik Sampel : FRIDA ULFAH ERMAWATI JURUSAN FISIKA FMIPA UNESA
5. Jenis/Nama Sampel/Kode : KRUPUK BATANG NO 15
6. Keperluan Uji : -
7. Parameter yang diuji : borax
8. Hasil :

Jenis Pengujian	Metode	Hasil
Borax	AOAC 18 ed 2005, 970.33 FI IV, 2014	Positif

Surabaya, 17 Mei 2021
a.n Direktur
Kepala Bidang Teknis

Drs. Hariana., M.Sc.

Figure 7b. The results of the borax test on "K. Batang 15" crackers from the TSU laboratory: Positive → contains borax.

Figure 8a. The turmeric paper was soaked in the cracker "K. Makaroni 12". The paper colour turns yellowish brown.

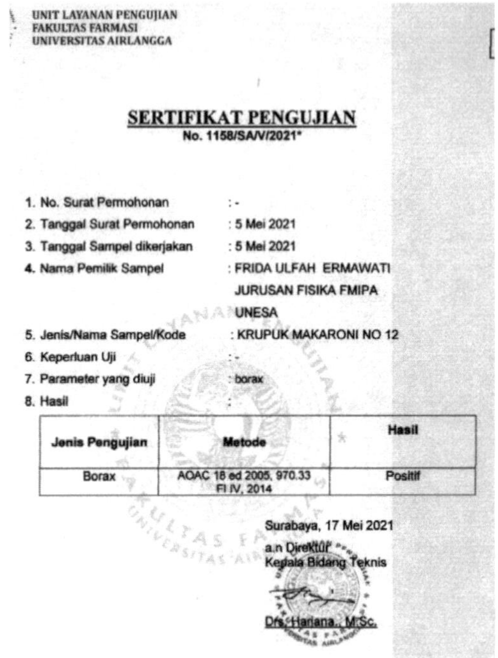

UNIT LAYANAN PENGUJIAN
FAKULTAS FARMASI
UNIVERSITAS AIRLANGGA

SERTIFIKAT PENGUJIAN
No. 1158/SA/V/2021*

1. No. Surat Permohonan : -
2. Tanggal Surat Permohonan : 5 Mei 2021
3. Tanggal Sampel dikerjakan : 5 Mei 2021
4. Nama Pemilik Sampel : FRIDA ULFAH ERMAWATI JURUSAN FISIKA FMIPA UNESA
5. Jenis/Nama Sampel/Kode : KRUPUK MAKARONI NO 12
6. Keperluan Uji : -
7. Parameter yang diuji : borax
8. Hasil :

Jenis Pengujian	Metode	Hasil
Borax	AOAC 18 ed 2005, 970.33 FI IV, 2014	Positif

Surabaya, 17 Mei 2021
a.n Direktur
Kepala Bidang Teknis

Drs. Hariana., M.Sc.

Figure 8b. The results of the borax test on "K. Makaroni 12" crackers from the TSU laboratory: Positive.

Figure 9a. The turmeric paper was soaked in the cracker "K. Ikan Nona 10". The paper colour turns reddish brown.

UNIT LAYANAN PENGUJIAN
FAKULTAS FARMASI
UNIVERSITAS AIRLANGGA

SERTIFIKAT PENGUJIAN
No. 1156/SA/V/2021*

1. No. Surat Permohonan : -
2. Tanggal Surat Permohonan : 5 Mei 2021
3. Tanggal Sampel dikerjakan : 5 Mei 2021
4. Nama Pemilik Sampel : FRIDA ULFAH ERMAWATI JURUSAN FISIKA FMIPA UNESA
5. Jenis/Nama Sampel/Kode : KRUPUK IKAN NONA NO 10
6. Keperluan Uji : -
7. Parameter yang diuji : borax
8. Hasil :

Jenis Pengujian	Metode	Hasil
Borax	AOAC 18 ed 2005, 970.33 FI IV, 2014	Positif

Surabaya, 17 Mei 2021
a.n Direktur
Kepala Bidang Teknis

Drs. Hariana., M.Sc.

Figure 9b. The results of the borax test on "K. Ikan Nona 10" crackers from the TSU laboratory: Positive → contains borax.

Figure 10a. The turmeric paper was soaked in the cracker "K. Cap Ikan 9". The paper colour turns remains yellow, doesn't change.

UNIT LAYANAN PENGUJIAN
FAKULTAS FARMASI
UNIVERSITAS AIRLANGGA

SERTIFIKAT PENGUJIAN
No. 1155/SA/V/2021*

1. No. Surat Permohonan : -
2. Tanggal Surat Permohonan : 5 Mei 2021
3. Tanggal Sampel dikerjakan : 5 Mei 2021
4. Nama Pemilik Sampel : FRIDA ULFAH ERMAWATI JURUSAN FISIKA FMIPA UNESA
5. Jenis/Nama Sampel/Kode : KRUPUK CAP IKAN NO 9
6. Keperluan Uji : -
7. Parameter yang diuji : borax
8. Hasil

Jenis Pengujian	Metode	Hasil
Borax	AOAC 18 ed 2005, 970.33 FI IV, 2014	Negatif

Surabaya, 17 Mei 2021
a.n Direktur
Kepala Bidang Teknis

Drs. Hariana., M.Sc.

Figure 10b. The results of the borax test on "K. Cap Ikan 9" crackers from the TSU laboratory: Negative → doesn't contain borax.

Figure 11a. The turmeric paper was soaked in the cracker "K. Rahayu 8". The paper colour turns remains yellow, doesn't change.

UNIT LAYANAN PENGUJIAN
FAKULTAS FARMASI
UNIVERSITAS AIRLANGGA

SERTIFIKAT PENGUJIAN
No. 1154/SA/V/2021*

1. No. Surat Permohonan : -
2. Tanggal Surat Permohonan : 5 Mei 2021
3. Tanggal Sampel dikerjakan : 5 Mei 2021
4. Nama Pemilik Sampel : FRIDA ULFAH ERMAWATI JURUSAN FISIKA FMIPA UNESA
5. Jenis/Nama Sampel/Kode : KRUPUK RAHAYU NO 8
6. Keperluan Uji : -
7. Parameter yang diuji : borax
8. Hasil

Jenis Pengujian	Metode	Hasil
Borax	AOAC 18 ed 2005, 970.33 FI IV, 2014	Negatif

Surabaya, 17 Mei 2021
a.n Direktur
Kepala Bidang Teknis

Drs. Hariana., M.Sc.

Figure 11b. The results of the borax test on "K. Rahayu 8" crackers from the TSU laboratory: Negative → doesn't contain borax.

As shown in Figures 2, 4, 5, 7, 8 and 9, the turmeric papers have changed colour to reddish brown or brownish yellow. Even in Figure 7, the cracker solution also changes the colour to very strong brownish yellow! Meanwhile, the results of the TSU laboratory test on the associated crackers confirmed that the crackers were positive, i.e. those crackers indeed contain borax. While the colour of the turmeric papers in the other crackers in Figures 3, 6, 10 and 11 all remained yellow, unchanged, indicating that the crackers in the Figures 3, 6, 10 and 11 did not contain borax. The results of the TSU laboratory test confirmed the correctness of these indications because the TSU's results were negative. Previously, the borax test by means of a flame test and a colour test had been reported by Hardiana on commercially produced bread and traditionally made bread in Blang Pidie District [24]. Among all the test samples, none were indicated to contain borax. No borax test was performed with turmeric paper as reported by the authors. As far as the author is aware, there are no publications that report the results of testing the content of borax in food using turmeric paper independently, as reported by the authors, without the involvement of laboratory tests. Based on a series of tests for the borax content in the ten crackers using the turmeric paper reported in this work, where the test results are very consistent with the TSU laboratory test results, it can be concluded that turmeric paper has acted as a good indicator of the borax content in the crackers.

4. Conclusion

A series of experiments on the borax content in the crackers using turmeric papers have been completed. The results are in close agreement with the TSU laboratory test results. Turmeric paper is therefore very suitable to be used as the first indicator of the borax content in food, not limited in crackers.

Acknowledgement

This work is funded by the 2021 FMIPA UNESA PNBP funds based on the Contract No. 35446/UN38.3/PM.06.03/2021 dated July 1, 2021.

References

[1] Balai-Besar-POM 2019 *Apa itu Boraks?* Retrieved from https://www.pom.go.id/new/view/more/artikel/14/Apa-itu-Boraks-.html
[2] Damopolii R R 2018 *Mengenal Boraks dan Dampak Penggunaannya* Retrieved from http://sib3pop.menlhk.go.id/index.php/articles/view?slug=mengenal-boraks-dan-dampak-penggunaannya
[3] Amit Z 2020 Borneo *J. Resour. Sci. Technol.* **10** 70
[4] Yiu P H, Jian S, Amartalingam R and Choon-Fah J B 2008 *Am. J. Agric. Biol. Sci.* **3** 476
[5] WenZhu Z, Yi Z and Yu G 2020 *J. Food Saf. Qual.* **11** 4242
[6] Wang X, Ma Z, Li X, Liu L, Yin X, Zhang K, Liu Y and Hu X 2018 *Chem. Biol. Technol. Agric.* **5** 1
[7] Holak W 1972 *J. Assoc. Off. Anal. Chem.* **55** 890
[8] See A, Salleh A, Bakar F, Yusof N, Abdulamir A and Yook Heng L 2010 *Am. J. Appl. Sci.* **7** 620
[9] Kumar G and Srivastava N 2011 *Iran. J. Environ. Health. Sci. Eng.* **8** 361
[10] Abd-Elhakim Y M, Hashem M M M, Abo-EL-Sooud K, Ali H A, Anwar A, El-Metwally A E, Mahmoud E A and Moustafa G G 2020 *Gene* **742** 144590
[11] Hadrup N, Frederiksen M and Sharma A K 2021 *Regul. Toxicol. Pharmacol.* **121** 104873
[12] Siti-Mizura S, Tee E and Ooi H 1991 *J. Sci. Food Agric.* **55** 261
[13] Ha C T, Lien N T H, Anh N D and Lam N L 2017 *J. Electron. Mater.* **46** 6843
[14] BPOM Indonesia 2015 *Pusat Informasi Obat dan Makanan: Asam Borat* Retrieved from http://ik.pom.go.id/v2016/
[15] Pongsavee M 2009 *J. Occup. Med. Toxicol.* **4** 27
[16] Astuti E and Nugroho W 2017 *J. Sain Vet.* **35** 42
[17] Rahman N 2019 *J. Phys. Conf. Ser.* **1363** 12092
[18] Do T A S, Thach-Nguyen R, Lam H H and Dang-Bao T 2020 *J. Phys. Conf. Ser.* **1711** 12005
[19] Yunianto M, Andaresta D, Anwar F and Saraswati T 2021 *J. Phys. Conf. Ser.* **1825** 12078
[20] Rahmah S and Mauludina K R 2019 *Int. J. Pharm. Pharm. Res.* **16** 337

[21] Sari I P, Yanti F A, Saefullah D I and Yunianto B T 2021 *J. Sustain. Sci. Technol.* **1** 44
[22] Amir S 2014 *Analisis Kandungan Boraks Pada Pangan Jajanan Anak di SDN Kompleks Lariangbangi Kota Makassar* Undergraduate Thesis (Makassar: Universitas Hasanuddin)
[23] Santi A U P 2017 *J. Ilm. PGSD* **1** 57
[24] Hardiana, Safrida Y D, Andriani A, Raihanaton and Maulidda S 2020 *Lantanida J.* **8** 29
[25] Siregar V A 2018 *Analisa Boraks Pada Bakso dan Pempek Serta Analisa Formalin Pada Ikan Teri dan Ikan Tongkol Secara Kualitatif dan Kuantitatif* Undergraduate Thesis (Medan: Universitas Sumatera Utara)
[26] Fuad N R 2014 *Identifikasi Kandungan Boraks Pada Tahu Pasar Tradisional di Daerah Ciputat* Undergraduate Thesis (Jakarta: Universitas Islam Negeri Syarif Hidayatullah)
[27] BPS 2019 *Statistik Tanaman Biofarmaka Indonesia* 2018 (Jakarta: Badan Pusat Statistik)
[28] Jayaprakasha G K, Jaganmohan Rao L and Sakariah K K 2006 *Food Chem.* **98** 720
[29] Halim A A, Roslan N A, Yaacub N S and Latif M T 2013 *Sains Malays.* **42** 1293

Journal of Physics: Conference Series **2110** (2021) 012015 doi:10.1088/1742-6596/2110/1/012015

Prezi Mind Mapping Media in Physics Learning: A Bibliometric Analysis

I Zakhiyah, N Suprapto*, and W Setyarsih

Department of Physics, Universitas Negeri Surabaya, Indonesia

*Email: nadisuprapto@unesa.ac.id

Abstract. This study aims to determine the trend of the author's topics and keywords that are widely used related to Media Prezi Mind Mapping, to determine the contribution of Indonesian researchers in related research topics at an international scope, and to find out future research opportunities with associated topics mainly in physics learning. This research method is library research with descriptive analysis method through the bibliometric approach. The article's metadata was obtained from the search results of the Google Scholar database, which found 16.600 articles for 2016-2020 with the keyword 'Mind Mapping in Learning'. Meanwhile, based on the Scopus database, 7 Scopus indexed articles were obtained for 2014-2020 with the keyword 'Prezi Mind Mapping'. All metadata obtained is saved in *.csv* and *.ris* format for further visualization by *VOSviewer.* The results of data analysis show that trend fans and contributions Indonesian researchers still very little. It can pose a challenge for future research to prove how far the existence level of Prezi Mind Mapping is to continue to be developed. In addition, there are recommendations for collaboration on mind mapping research topics in learning with one of the related keywords, namely STAD (Student Team Achievement Division), hoping to produce a more efficient learning model, especially in physics, to be applied in future research.

1. Introduction

Rapid technological developments require educators to continue to innovate in learning, one of which is through learning media [1]. Learning media is anything that can be used to transmit messages from sender to receiver to stimulate students' thoughts, feelings, interests and attention in such a way that the learning process can occur [2]. The use of media in the teaching and learning process plays an important role, namely as a tool to create a fun and exciting process so that students can easily understand the material presented by the teacher [3]. Learning physics itself is very demanding of students to master the concept or basic knowledge. Therefore, the need for appropriate learning media to improve students' understanding of basic concepts. One of the efforts to achieve this is by applying a concept map as a learning medium.

A concept map is a schematic chart to describe a conceptual understanding in a series of statements. According to Biktimirov and Nilson [4], a concept map is a network that connects interrelated concepts and ideas that are interconnected with each other. Concept maps use the brain's ability for visual recognition to get the most out of a combination of colours, images, and symbols. Concept maps are more visually stimulating, so it will be easier to remember information [5]. An application that can be used as a learning medium in the form of a concept map is the Prezi application.

Prezi is a tool to develop various visuals, audio, and animations with engaging narrative formats. Presentations via Prezi are created on a canvas by visualizing ideas, highlighting details, and viewing the entire concept map, allowing the user to create a cinematic journey or concept map flow. The materials presented on the canvas can be pulled, shifted, rotated left and right to look attractive without the need to change slides. Prezi uses the Zooming User Interface (ZUI), which can show the menu in detail and give a deep impression on the message's recipient [6]. In addition, Prezi is an application based on Adobe Air, so videos and flash animations can be run more efficiently than when using PowerPoint.

This research on the Prezi Mind Mapping media has been carried out at the secondary to tertiary level of education. From the previous study results, the application of Prezi media in physics subjects has a significant effect on improving student learning outcomes in the cognitive, affective and psychomotor domains, marked by increased student learning outcomes [7]. However, this study only reveals the effect of using Prezi media on improving students' learning abilities and developing the media in learning. The average value of student learning outcomes after using Prezi media showed higher results than the intermediate student learning outcomes before using Prezi media [8].

The opportunity to use Prezi Mind Mapping learning media provides space for educators to optimize teaching and learning activities to improve students' understanding and learning abilities. One way that can be used is to integrate Prezi media into the learning. Prezi media was chosen based on the advantages and features of the zooming presentations as a medium to support learning that is interesting, interactive, and easily understood by students. Based on various previous studies, research on Prezi Mind Mapping media is in the form of knowing its effectiveness and development in learning. However, the survey of bibliometric literature review has not been discussed. The purpose of this study is to provide a bibliometric literature review to find recommendations for spare future trends related to Prezi Mind Mapping media in physics learning.

Finally, the research questions include:

a. How is the trend of research topics related to Media Prezi Mind Mapping during 2014-2020?
b. What are the most used author keywords in research topics related to Media Prezi Mind Mapping during 2014-2020?
c. How is the contribution of Indonesian researchers in research topics related to Media Prezi Mind Mapping in an international scope?
d. What are the future research opportunities with the topic of Media Prezi Mind Mapping in physics learning?

2. Method

The research method used is library research with descriptive analysis method through a bibliometric approach. The bibliometric analysis aims to measure the development of scientific article publications and scientific contributions [9-13]. The bibliometric study can look at the distribution of publications to evaluate the contribution of articles to the advancement of knowledge of various literature using a statistical approach and can provide a broader understanding of the entire discipline [14]. In this study, researchers focused on international journal publications, namely Scopus and Google Scholar. Google Scholar is considered an alternative that provides extensive scientific literature in the discipline and its sources [15].

The article metadata was obtained from the Scopus and Google Scholar database searches since March 2021, with keyword criteria adjusted to the research focus. Furthermore, the Google Scholar database obtained 16,600 articles for 2016-2020 with the keyword 'Mind Mapping in Learning'. Meanwhile, the Scopus database obtained 7 Scopus indexed articles for 2014-2020 with the keyword 'Prezi Mind Mapping'. Specifically for metadata from Google Scholar, only Top 40 papers were selected randomly but still based on their suitability for titles and keywords analyzed manually. All acquired metadata is saved in *csv* format (comma separated values) and *ris* (research information systems). The preparation of initial data statistics can be done especially on data from the Scopus database, namely by grouping data through Microsoft Excel based on source, top author, country, institution, and top citation for *.csv* data format.

Furthermore, metadata mapping for the *.ris* format was carried out with the *VOSviewer* application. This application can produce output that provides three map visualizations: network, overlay, and density, both based on keywords and author. *VOSviewer* can work efficiently with large data sets and offer various exciting and visually apparent analyses through a comprehensive mapping visualization feature [16]. The Figure 1 is the flow of data collection and processing for bibliometric analysis.

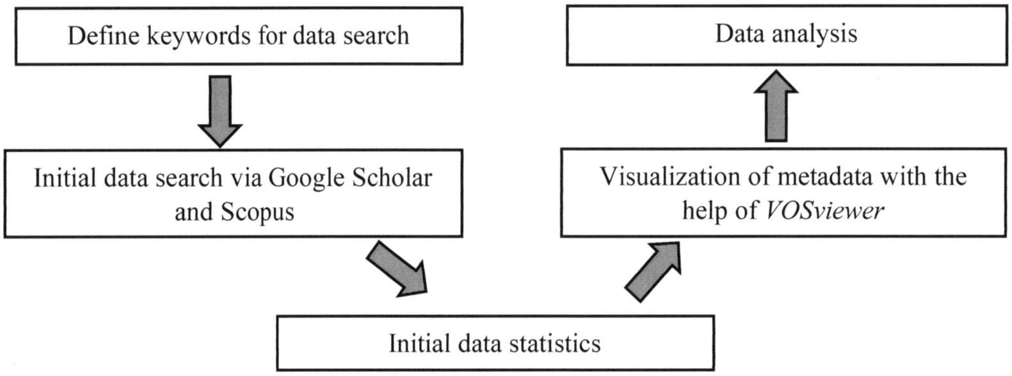

Figure 1. The flow of data retrieval and processing

The last stage is data analysis which refers to the *VOSviewer* mapping. Data analysis is done by looking at the visualization results both in-network, overlay, and density. In addition, it can also be seen based on the size of the nodes formed in the visualization results. Nodes with a large size indicate the dominant cluster on the research topic. On the other hand, tiny groups are shown by visualizing small nodes. This cluster shows the level of linkage between the results of the *VOSviewer* mapping and the research topic.

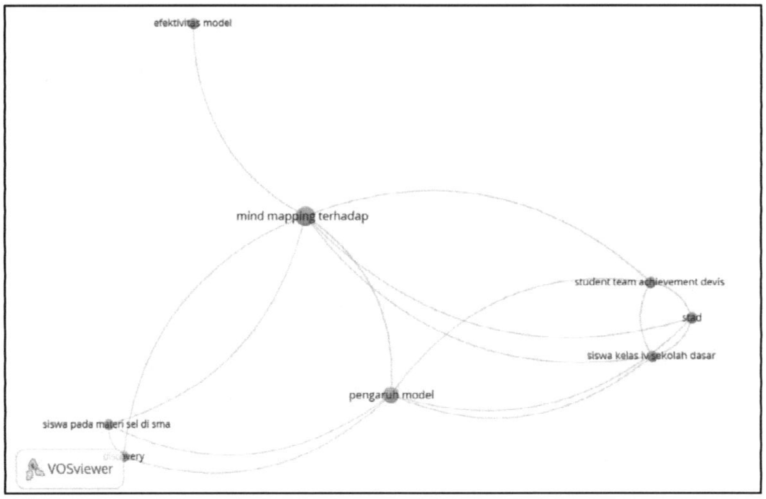

Figure 2. Network visualization the most used keyword for the Google Scholar database

3. Results and Discussion

Based on metadata search results on Google Scholar, 16,600 articles were obtained for 2016-2020 with the keyword 'Mind Mapping in Learning'. Meanwhile, based on the Scopus database, 7 Scopus indexed articles for 2014-2020 with the keyword 'Prezi Mind Mapping'. Metadata saved in *.csv* (Comma Separated Values) format and *.ris* (Research Information Systems) will then be prepared to be visualized using *VOSviewer* software. First, based on the Google Scholar database, after 40 articles sourced from journals were randomly selected but still adjusted to the titles and keywords analysed, the mapping was obtained.

The visualization displays 40 articles that match the criteria, namely keywords, titles, and abstracts. Each term in the visualization image represents a keyword, title, or abstract related to the research. Based on the entered metadata, the first step of the metadata will be set to 1 author has 1 document resulting in 101 terms that will be selected to be readjusted with the research topic. After the first result is obtained, the data is re-limited to produce eight related terms or keywords, as shown in Figure 2. There are three different network visualization clusters, as shown in Figure 2. The first cluster (red) has three such as student team achievement division, STAD, and grade six elementary school students. The second cluster (green) has three phrases: students on cell material in high school, discovery, and model influence. While in the last group (blue), there are two relevant emergence terms, namely model effectiveness and mind mapping. The keywords in each cluster represent deep research entitled 'Mind Mapping in Learning', which was carried out for the last five years (2016-2020). In addition, with the visualization overlay display in Figure 3.

The yellow colour shows the latest research conducted in 2018, while the light green to purple colours represent research that has been carried out several years before. It can be identified that each existing cluster is connected by title and abstract. From the groups displayed, it can be detected that there are still few topics to discuss over the last five years to become opportunities for developing future research topics. A cluster with a large node form is a cluster with dominant keywords, while for a small node, it describes a minor group. This means that the topic of mind mapping in learning is related to the keyword mind mapping (the most dominant cluster is dark green) and the influence of the learning model (the dominant group is light green).

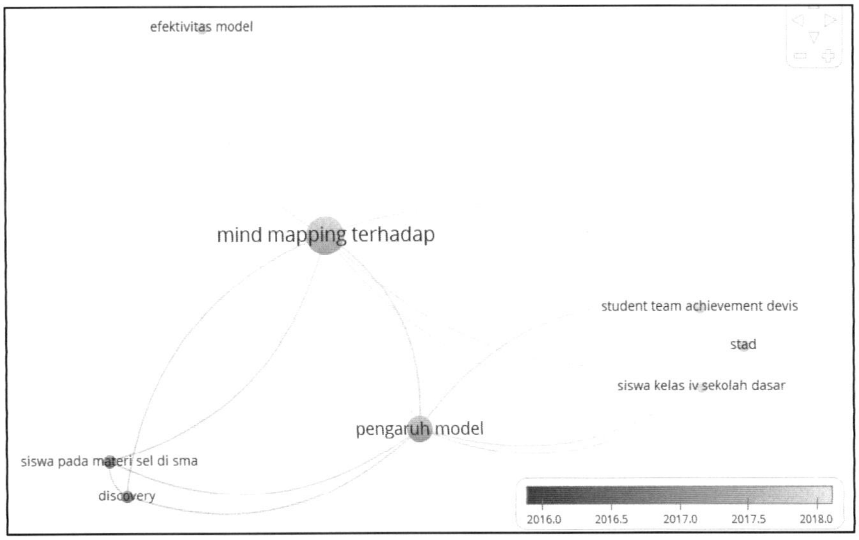

Figure 3. Overlay visualization of the most used keywords for the Google Scholar database

Based on the bibliometric analysis described by *VOSviewer*, the trend of research topics that have the opportunity to be carried out in the future is the keyword 'Student Team Achievement Division', which is in the little yellow cluster with small nodes. This keyword is connected to two other clusters: the light green cluster (the influence of the model) and the dark green cluster (mind mapping in learning). So based on these results, there are recommendations for mind mapping research topics in education that can collaborate with the 'Student Team Achievement Division (STAD) is a cooperative learning model that emphasizes student interactions and activities so that they can support each other in mastering lessons that have an impact on learning outcomes.

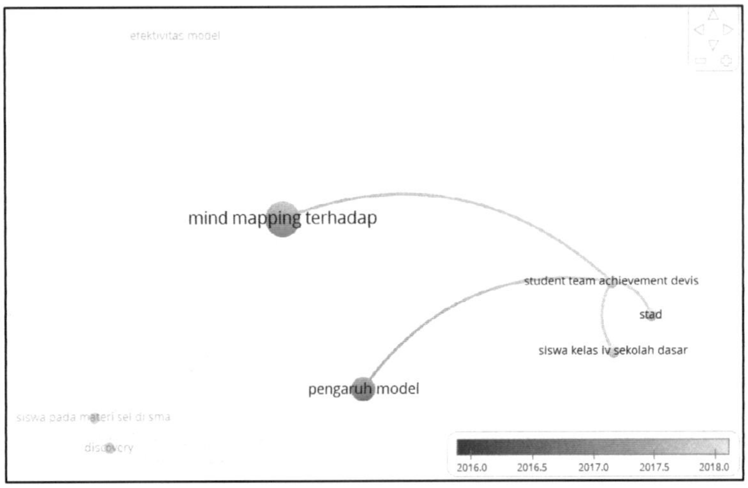

Figure 4. Opportunity for student team achievement division in further research

According to Slavin, STAD learning teaches students to work together on a common task and coordinate joint efforts to complete it [17]. If it has collaborated with Mind Mapping as a tool made by students to express their thoughts when interacting with each other, it is possible to produce an effective learning model to be applied.

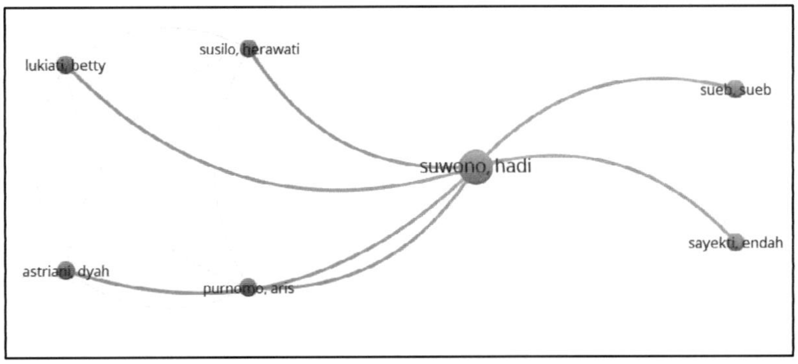

Figure 5. Top Author visualization for metadata based on google scholar database

Based on the 85 authors detected in the metadata for interconnected co-authorships, seven authors are connected, according to Figure 5, from the two resulting clusters, namely the red cluster consisting of four authors and the green group composed of three authors. It can be seen that Hadi Suwono is the top author with the dominant visualization display, namely the large green circle (centre) for the topic of mind mapping in learning based on the Google Scholar database.

As for the visualization of 85 authors without being limited by connections to each other, as shown in Figure 6. it can be seen that more clusters are generated. The visualization of Figures 5 and 6 indicates that the Hadi Suwono cluster from the State University of Malang is the top author for this research topic. He was then followed by another group of light blue, purple, dark green, etc.

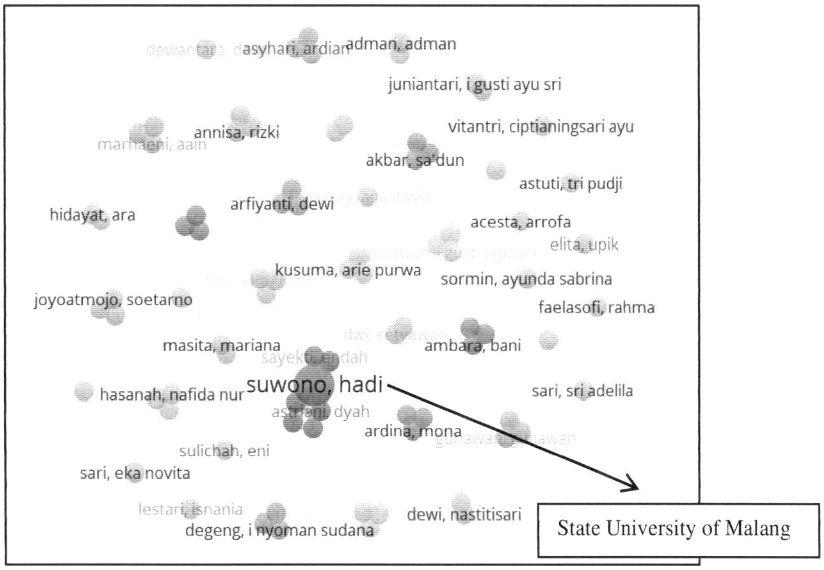

Figure 6. Overall author visualization for metadata based on Google Scholar database

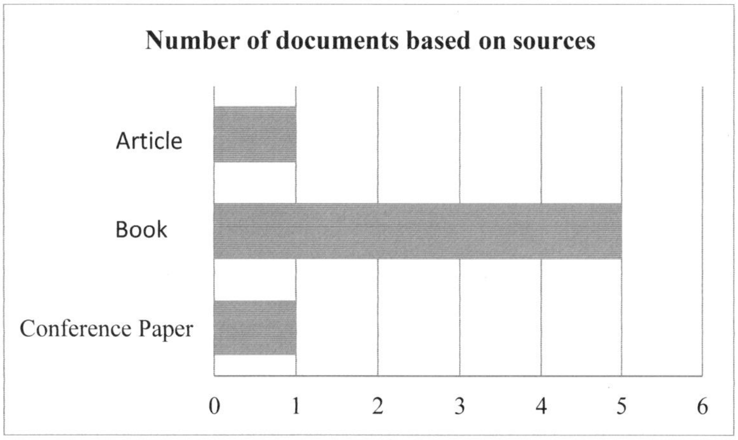

Figure 7. Number of documents by sources

Furthermore, Figure 7 is the result of grouping metadata with the help of Microsoft Excel for Scopus indexed documents for the all years category that has been saved in *.csv* form (Comma Separated Values). There are five documents sourced from books and one copy sourced from articles and conference papers each. This number is categorized as very low for searches with the Prezi Mind Mapping keyword. In other words, this topic has not become a trend of international interest. This can pose a challenge for future research to prove how far the existence level of Prezi Mind Mapping is to continue to be developed. On the other hand, the opportunity to create this research topic is still quite good, as evidenced by several other articles taken from several international journals showing exciting discussions. Some of these points can be seen in Table 1.

Table 1. Articles related to the topic of prezi mind mapping from several international journals

Research Background	Literature Review	Research Methods
	Lam [18]	
Common methods used to deliver didactic education were through PowerPoint; Prezi was used as an alternative. It has been described as innovative and dynamic, allowing for zooming in and out to focus on general and specific content.	Prezi is a cloud-based presentation editor that allows the user to create presentations that zoom. The goal is to make presentations more dynamic in nature and not as linear in structure.	A quantitative methodology was chosen comprising of a survey which examined themes of Prezi's effectiveness, and results were analysed via descriptive statistics.
Strasser [19]		Gabor [20]
Presentation software has centred around PowerPoint style editors for at least the last 20 years. While using PowerPoint in the classroom was considered to be cutting edge, it is now seen as the epitome of boring	Prezi is an internet-based presentation software that allows you to store your presentations in the cloud. There are many advantages to using this software, including the fact that it is cloud-based and that it is more exciting than the standard presentation software. Prezi is very visually-oriented and always enhances a presentation that is well-designed.	In this case the means could be compared with Independent Samples T-test, which showed up a difference between the means.
Gabor[20]	Rachman [21]	
Teachers need to use different presentation tools in the College to make presentations fun for students.	Prezi is an online presentation program that offers storage of presentation in the cloud. It is very different from PowerPoint in that its approach is non-linear. You are faced with a canvas instead of slides for creating your presentation. This means that you see the entire presentation at once instead of in discrete segments.	Particularly, qualitative research was the primary method in collecting the data. Meanwhile, the interview was held to gain more information about students' experiences in the using of Prezi in their learning activities.
Rachman [21]		Chou, Chang, and Lu [22]
The important role Information and Communication Technologies (ICT) play in education has been increasing recognized by educators of all levels. To help prepare future teachers who can make good use of ICT, teacher education institutions and programs are expected to not only "model the new pedagogies and tools for learning," but also "develop strategies and plans to enhance the teaching-learning process"	Prezi, in particular, is a significant tool to present visual and verbal textual materials to the students to make the processes of teaching and learning more effective and retainable (Anderson, 2013).	A quasi experimental pre- and post-test control group design was adopted to fulfil the research purpose. Results were analysed via descriptive statistics.

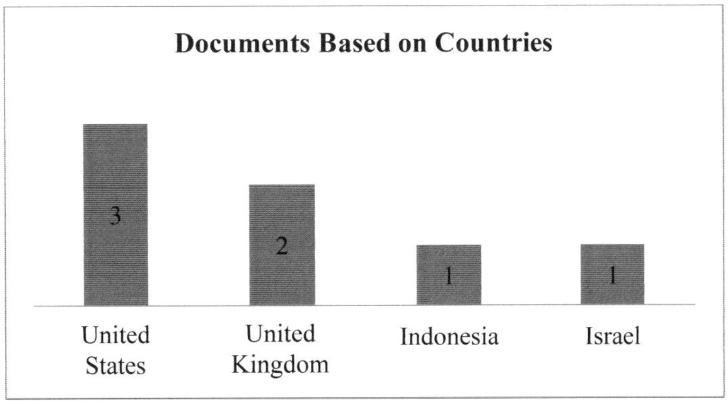

Figure 8. Number of documents by country

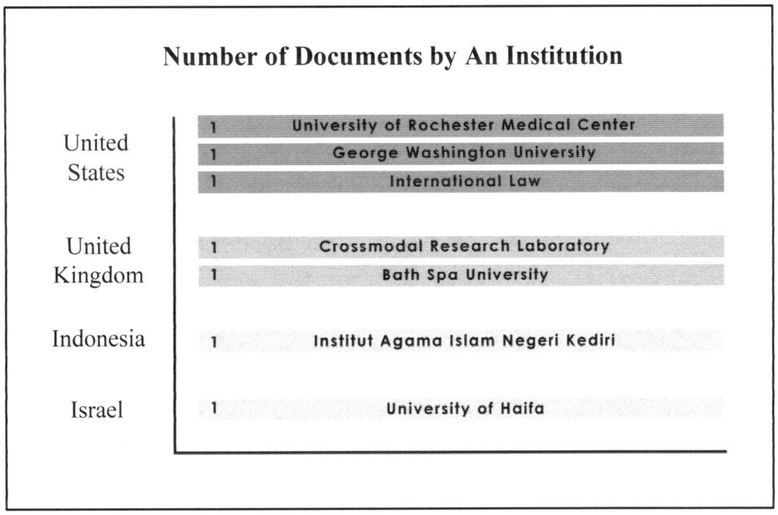

Figure 9. Number of documents by an institution

In addition, the countries with the highest number of contributors to articles in this research topic are the United States with three documents, followed by the UK with a total of two papers, and Indonesia and Israel with one piece each. Based on the institution for the United States, it consists of the University of Rochester Medical Centre, George Washington University, and International Law. For the UK, it consists of the Cross-modal Research Laboratory and Bath Spa University. For Indonesia, it is the Kediri State Islamic Institute, and for Israel, it is the University of Halfa. In this case, it can be seen that the contribution of Indonesian researchers is still very lacking in the research topic of Prezi Mind Mapping. That way, the opportunity to develop this topic is still vast for Indonesian researchers at the international level.

Based on Figure 10 which shows the top citations from the seven Scopus indexed documents, it can be seen that the most sources are Spence, Piqueras-Fiszman, with 166 authorities originating from George Washington University, United States of America. After that, in second place was Kober with a not so much difference, namely 143 from the Cross-modal Research Laboratory, England. Suppose it is reviewed in Figure 10 the names of Indonesian researchers still have a small number of citations for this research topic, namely only two for Yasin, Huda, Komarudin, Suherman, Septiana, Palupi. This again shows that there is still very little of the contribution of Indonesian researchers in this research topic.

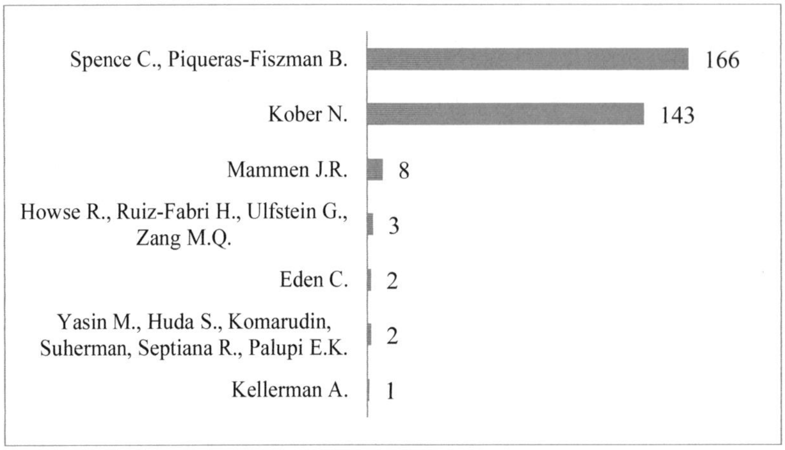

Figure 10. Top authors' citations

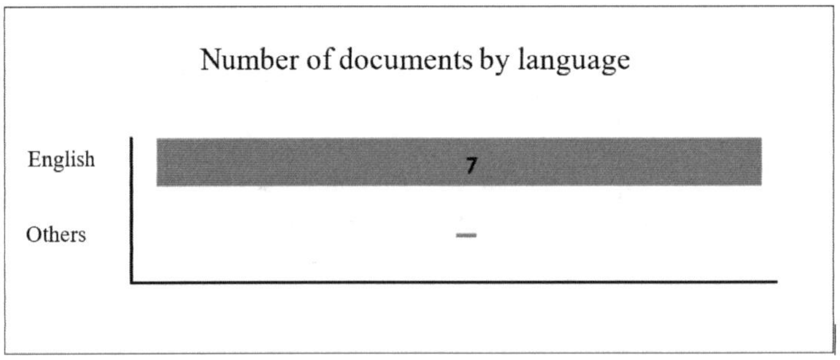

Figure 11. Number of documents by language

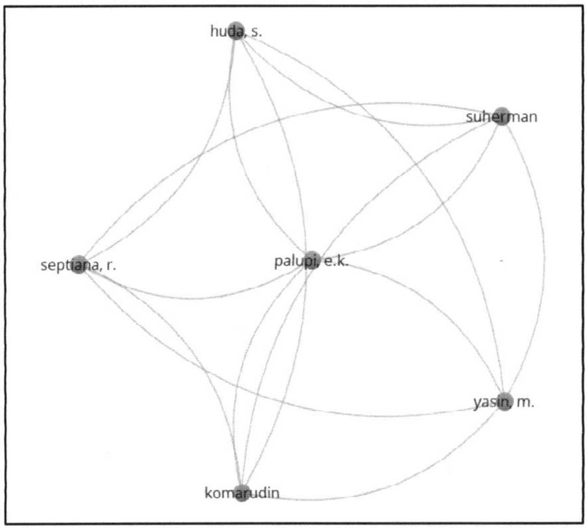

Figure 12. Network visualization for interconnected authorship

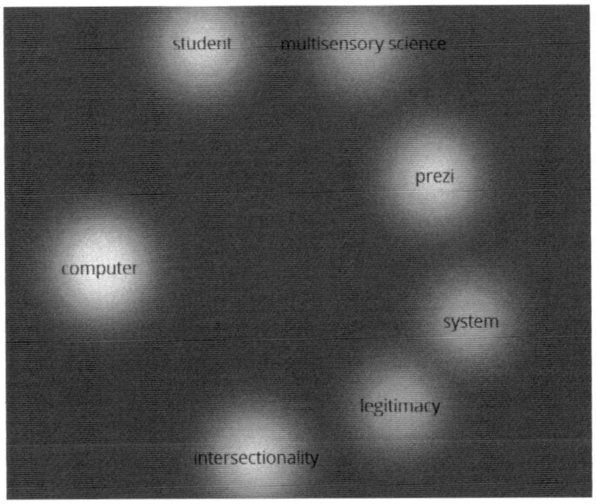

Figure 13. Density visualization of the most used keywords overall for the scopus database

Figure 12 on network visualization for connected authorships shows several names of authors from Indonesia. As for other authorships, it can be seen in Figure 15. Furthermore, based on the results density visualization above, it can be seen that the most used keywords in the last five years for seven clusters are computer, Prezi, student, and intersectionality as the dominant cluster. Therefore, this data shows that the topic of this research still has a slight research trend.

Network visualization shows the authorship group related topics as many as seven clusters. The red cluster is the dominant cluster, then the green, blue, and so on sets. Other than that, based on the results density visualization above, the development of authorship research can be seen every year. Data related to research on Prezi Mind Mapping in the last five years, both authorship in Figure 15 and related keywords in Figure 14, shows that again there are still very few results. The concept of mind mapping in education has been known for a long time. This is a research opportunity regarding the application of mind mapping, especially in the use of Prezi media, given that the application of Prezi Mind Mapping media in previous studies has proven to improve critical thinking skills, improve understanding of concepts, as well as to improve the achievement of competence in learning outcomes quite well. In particular, in physics, according to research from Rodhi [23], the use of Prezi-based learning media can improve critical thinking skills on heat material. Other than that, the Prezi Mind Mapping media development can improve understanding of the physics concept of optical instrument material and can improve student learning outcomes in the cognitive, affective, psychomotor domains [7, 24]. Thus, the use of Prezi Mind Mapping media in Physics learning can become a research topic in the future to further test its effectiveness.

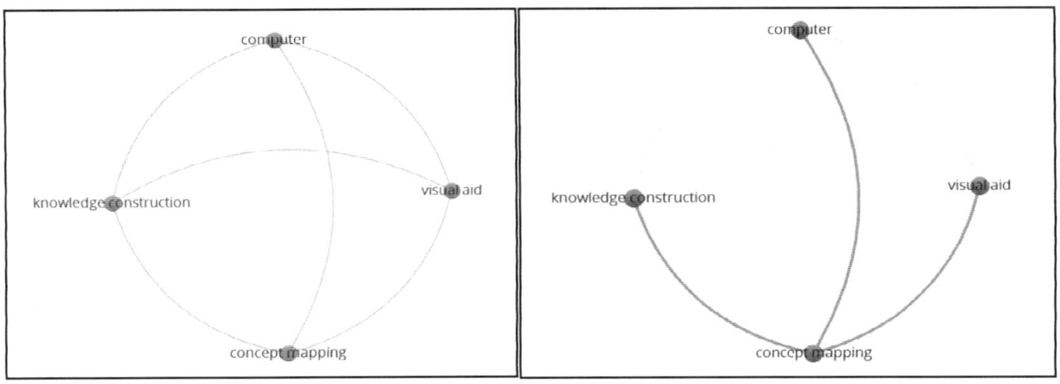

Figure 14. Network visualization of related keywords

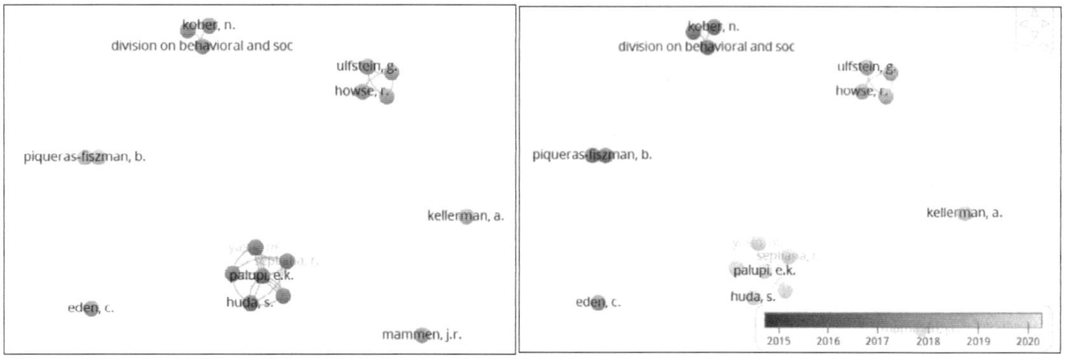

(a) (b)

Figure 15. (a) Network visualization for authorship (b) Overlay visualization for authorship

4. Conclusion

Research topics related to Prezi Mind Mapping media in learning have a relatively small trend of enthusiasts based on the Google Scholar database, even very few based on the Scopus database. It can pose challenges for future research to prove how far the existence level of Prezi Mind Mapping is to continue to be developed. This research topic uses several dominant keywords, including computer, knowledge instruction, visual aid, concept mapping, influence model, and STAD. Based on this, the researcher recommends the collaboration of mind mapping research topics in learning with one of the keywords: STAD, in future research. STAD is a cooperative learning model that emphasizes student interactions and activities to support each other in mastery. If collaborated, then Mind Mapping media can act as a tool made by students to express their thoughts when interacting with each other more effectively. It is possible to produce a more efficient learning model to be applied. Other than that, the contribution of Indonesian researchers in this research topic in the international scope is still very lacking where the dominance is still in the United States and the UK, so this is a challenge and opportunity to develop the research topic of Prezi Mind Mapping Media in the future, especially in physics learning.

References

[1] Mahyuddin R S, Wati M, and Misbah 2017 *Berk. Ilm. Pendidik. Fis.* **5** 229
[2] Sadiman A S 2008 *Media Pendidikan* (Jakarta: PT. King Grafindo Persada)
[3] Sujarwo A and Kholis N 2016 *J. Pendidik. Teknik Elektro* **5** 897
[4] Davies M 2011 *High. Educ.* **62** 279
[5] Ginting M K B 2013 *Basastra: J. Kaji. Bhs. Sastra Indones.* **2** 823
[6] Diamond S 2010 *Prezi for Dummies* (Canada: Wiley Publishing)
[7] Melida D, Masril, and Hufri 2014 *Pillar Phys. Educ.* **4** 113
[8] Zannah P Z, Mulhayatiah D, and Alatas F 2014 *Edusains* **6** 211
[9] Hakim L 2020 *J. Ilmu Manaj.* **8** 176
[10] Suprapto N, Prahani B K, and Deta U A 2021 *Libr. Philos. Practic.* **5599**
[11] Suprapto N, Prahani B K, and Deta U A 2021 *Libr. Philos. Practic.* **5928**
[12] Suprapto N, Sukarmin, Puspitawati R P, Erman, Savitri D, Ku C-H, and Mubarok H 2021 *Int. J. Eval. Res. Educ.* **10** 4
[13] Yanuarti E A and Suprapto N 2021 *Stud. Phil. Sci. Educ.* **2** 7
[14] Yulianingsih K A, Sukarsa K G, Suciptawati L P 2012 *e-Journal Mat.* **1** 59
[15] Baneyx A 2008 *Arch. Immunol. Ther. Exp.* **56** 363
[16] Liu Z, Yin Y, Liu W and Dunford M 2015 *Scientometr.* **103** 135
[17] Slavin R E 2015 *Cooperative Learning Theory, Research, and Practice* (Bandung: Nusa Media)
[18] Lam A 2014 *J. Med. Imaging Radiat. Sci.* **45** 174
[19] Strasser N 2014 *J. Coll. Teach. Learn.* **11** 95

[20] Kiss G 2016 *Turk. Online J. Educ. Technol.* **15** 126
[21] Rachman D 2016 *Scr. J. Linguist. Engl. Teach.* **1** 83
[22] Chou P N, Chang C C, and Lu P F 2015 *Comp. Educ.* **91** 73
[23] Rodhi M Y and Wasis 2014 *J. Inov. Pendidik. Fis.* **3** 137
[24] Istiawan R, Mosik M, and Sopyan A 2016 *Unnes Phys. Educ. J.* **5** 87

STEM in Disaster Learning Media: A Literature Review

A I Agusty*, F A Alifteria, and M Anggaryani
Physics Education, Universitas Negeri Surabaya, Indonesia

*Email: alifia0708@gmail.com

Abstract. What is a natural disaster? How much do we know about it? Modern technology undoubtedly helped people learn about natural phenomena that sometimes can be misdirected into disaster. Natural disasters can cause quite an enormous impact, such as property damage and the number of injured or dead victims. Disaster education that applied science, technology, engineering, and mathematics could be a potential strategy to prepare the local community in managing the risk of natural disasters. This study is part of literature research on the integration of STEM in disaster education focused on disaster learning media to increase students' awareness of disaster prevention, impacts, and mitigation. This study aims to identify national and international journals related to STEM, disasters, and learning media; and evaluate the effectiveness of implementing STEM integrated learning media. In 2011-2021 there are 305 publications and 15 articles reviewed in this study. The results indicate that STEM-based disaster learning media can develop students' ability to solve problems related to disaster mitigation and disaster knowledge. The development of learning media to teach disaster education need to be reviewed and adapted to the geographical conditions of Indonesia, several topics about disaster as recommendation for learning is disaster that often afflict Indonesia.

1. Introduction

In the first-quarter of 2021, Indonesia has been threatened by 1,205 natural disasters dominated by hydro meteorological disaster with 501 events of floods, 339 of hurricane, and 233 of landslides [1]. The numbers of disasters in Indonesia can be caused by the geographical location which on the Pacific Ring of Fire and between three large tectonic plates: Indo-Australia plates, Eurasia plates, and Pacific plates [2,3]. These conditions can threaten to occur earthquakes, tsunamis, and landslide. Due to these geographical conditions, it has a big potential for disaster to occur with various characteristic, so that the management has their own handling for every disaster [4]. Natural events that causes disaster is difficult to predict even the movement of plates and through geological calculations can be ascertained, but it's difficult to avoid the impact of disasters. Improper disaster mitigation and management can cause the high number of deaths, losing property, damaging the environment, and affect the psychological condition of local community. The disasters that occur by July 2021 in Indonesia damaging 767 units of property, and 4 death of victims [5]. Those impact of disaster can be reducing by applying the technologies to announce the early-warning to their local community in the disaster prone-area, but it costs a lot. The effective and simpler way to reduce the impact of disaster and also increasing risk awareness of local community can be done through school by developing disaster knowledge integrated with several subjects to students [3,6].

Science, Technology, Engineering, and Mathematics (STEM) in Indonesia has supporting students facing the Fourth Industrial Revolution (Industry 4.0) in these years. The rapid development on the issue of STEM education provides students to increase their abilities, skills, knowledge of fact, procedures, concept, metacognitive skill [2], and science process skills [7] to be ready as future human resources in the STEM fields. The integration of STEM education in teaching and learning should be applied to all levels of education [8]. STEM education also impacted into student's motivation and directly involving students into the manufacturing process [9,10]. The implementation of STEM is related to the teachers as an important role to integrate STEM into their class. Teacher can integrate STEM as a learning strategy, learning approach, the method of learning, assessment, textbooks, teaching materials, and learning media [11,12]. In these pandemic COVID-19 era, textbooks are not an effective way to deliver the learning material for students because of School from Home (SFH) regulation, students need an interactive way to learn some subject. Teacher that integrate STEM usually gives an explanation through listening material, e-book, interactive learning media, or a video [7].

Natural disaster in Indonesia need to be manage seriously and intensively especially in disaster-prone areas, the alternative way is through education sector. By introducing students to disaster education including disaster mitigation, disaster prevention, and disaster preparedness, it can enhance students' awareness of disaster in early stage [13] and they can act to saving the nature. It also has the same important role as giving the early warning to local communities before the disaster. Disaster knowledge or education can be taught in schools, it also affected students understanding in dealing with disaster [14] using an interactive learning media. The communities' disaster preparedness knowledge is still lack even they lives in those area for a long time [13]. The high risk they will face when a disaster occurs, both physical and environmental losses need to be increased through a disaster education. Disaster can be taught by integrating science learning process that have strong relation to disaster such as physics with an appropriate learning model [14]. Using learning media can simulate disaster in greater way such as given information process with a physics-analogy warning [14], explain the disaster management policy, visualize the disaster prone-area in Indonesia. Developing disaster learning media itself gives teacher a challenge to explore their creativity to be able to build a meaningful understanding of natural disaster [15].

The interpretation of learning media is to deliver learning meaning [16,17]. Learning media is designed by a teacher to adjust the material with their students' imagination. The advantage using learning media is to carry out learning process effectively and more efficient by considering the learning goals [18]. By implementing audio-visual feature to learning media can boost students' attention to the maximum level in learning process. Integrating STEM into learning media in disaster education related topic should be the innovative way to teach students disaster awareness so they will perceive that learning in attractive, fun, and more profound [17]. By simulating the process of disaster using visualization, audio, and video can engage students' imagination towards disaster in various aspects. Therefore, this research aims to identify national and international journal in STEM, disaster, learning media topic using literature review method and evaluate the effectiveness of implementing STEM integrated learning media.

2. Method

This research method applied in this study is literature review for reducing a potential bias that possibly occurred when it comes to scientific procedures in diverse research approach. This literature review uses secondary data from several journals national and international, which relate to Disaster, learning media and STEM. The searching process focused on the literature written in English and the period of 2011-2021. Researchers using "Disaster", and "Learning Media" and "STEM" keywords in database. There are 305 articles that match with the keywords found on *Google Scholar*, the researcher chooses 15 articles that are very relevant to STEM-based disaster learning media and physics topics. The selected articles are articles that focus on discussing STEM-based disaster learning media at the elementary school to high school level especially in physics topic so the discussion only focus in the implementation of STEM in Indonesia disaster learning system.

After collecting some journals, the data were analysed using descriptive qualitative analysis through literature review. The articles in this literature review were analysed, extracted and synthesized and then summarized the results related to "Disaster", "Learning media", and "STEM". The process of analysing the articles resulted in the essence of the research, namely: the name of the researcher, the year of publication, the title of the study, the results and conclusions of the study. The extracted digest is then entered into a table so that the extraction results are easy to read [19].

Figure 1. Step of literature review research

The author discusses three steps to conduct literature review research which exemplified by several activities in qualitative data analysis as follows:

1. Identification. Identify data obtained from journal reviews, to obtain the required information.
2. Examination. From the data that has been obtained then summarized, write down the main points and focus on the discussion that will be discussed so that the objectives and results of data analysis have a clear picture.
3. Interpretation. The data that has been analyzed is presented an explanation in the form of a description and drawn conclusions from the data obtained so as to produce answers to the formulation of the problem.

3. Results and Discussion

3.1. Studies related to STEM, Disaster, Learning Media

The integration of STEM in disaster learning media has been a new strategy for Indonesia education sector to enhance students' skill in STEM and also increasing students' awareness of disaster. This disaster studies topic and the use of learning media as a tool to deliver disaster education from 2011-2021 there are 305 articles from national and international journal that match with the keyword and 15 article meets the criteria of these study including the author(s), year, and learning media type as shown in Table 1. Publication in the 2019-2021 are mostly concern in the case-study of the studied articles in Indonesia disaster, and the technology used in learning media to deliver disaster learning is still accordance with current research trends.

Table 1. Studies related to STEM, disaster, and learning media

No	Article Title, Author, & Year	Learning Media Type	Substance
1	Development of Digital Teaching Materials on Earthquake Themes to Improve STEM Literacy. A Hikmawati, D Puspitasari, D Ardianto, and S Kurniasih. 2020 [7]	Digital Learning	The development of digital teaching materials on earthquake using ADDIE model was collected using validation sheet, valid question about STEM literacy, and questionnaires. The result that digital teaching materials designed can improve STEM literacy in the medium category
2	STEM project-based learning on student's STEM literacy: the case of teaching earth layer and disaster. Kartini, F.S., Widodo, A., & Winarno, N. 2021 [2]	-	In 2019, Indonesia not only threatened by COVID-19 but also natural disaster. STEM-Project based learning are designed to teach students about

			earth layer and disaster topic. It is recommended to enhance students STEM Literacy
3	STEM Education Implementation in Indonesia: A Scoping Review. Farwati, R., Metafisika, K., Sri I., Sitinjak, D.S., Salikha, D. F., & Salfarina, S. 2021 [11]	-	There are 597 articles based on research conducted in 2015-2020 about STEM education in Indonesia. The implementation of STEM education has support teacher in many ways and it can be a trigger as a new strategy in education sector
4	MEMS and IoT applications in ISLE-based STEM physics learning media for mechanic's topic with LabVIEW integration. Irwandi, I., Sari, I. M., Oktavia, R., & Syukri, M. 2020 [20]	Application	In the 21st century, technology such as sensor and IoT (Internet of Things) have not been optimally used as a media for physics learning. From the research, students' knowledge about mechanics was develop while conducted to the experiments. The learning process was very fun using STEM approach.
5	Development of Module Disaster Mitigation Based on STEM for Secondary School. Septaria, K., Dewanti, B. A., Afidah, M. I. E. 2020 [21]	Module	The validation result of disaster mitigation module is 95.75% in average point which categorized as very valid. The module itself is interesting and can improve student learning outcomes towards COVID-19 disaster.
6	Preparedness of Preservice Elementary Teachers' Nature of Science (NoS) Understanding for Better STEM Learning. Subuh, A. 2019 [22]	-	Main purpose of learning science is to enhance students' literacy. The better understanding of students literacy can be an easier way to solving a problem mainly in NOS Understanding using STEM learning
7	The movement of STEM education in Indonesia: Science teachers' perspectives. Nugroho, O.F., Permanasari, A. & Firman, H. 2019) [10]	-	Teachers are an important role to integrate STEM into their learning process. This research was examining teacher perception on STEM education. The result is teacher have a good understanding to STEM education
8	KOASE: Disaster Mitigation Learning Media in Elementary School. Noviana, E. Kurniaman, O., Affendi N. 2020 [23]	Comic	Developing KOASE comic to teach elementary students was appropriate to teach disaster mitigation as a medium for learning. It has a very good

			category according to teacher and students responds.
9	Multimedia in Disaster Risk Reduction. Sejati, P. M., Budiningsih, C. A. 2019 [16]	Attractive learning media	Multimedia is a new medium to create interactive learning process to the students. According to the research, it is necessary to use attractive learning media to improve students' comprehension. By applying attractive learning media, students also can improve their disaster risk knowledge.
10	STEM-based learning analysis to improve students' problem solving abilities in science subject: A literature review. Astuti, N. H., Rusilowati, A., & Subali, B. 2021 [24]	-	Physics is a subject that requires students' problem solving abilities. From this literature review study, combining STEM into physics education can improve students' problem solving abilities.
11	Google Earth Pro as a Learning Media for Mitigation Adaptation of Landslide Disaster. Rany, T. D., Kuswanto, H., Abdillah, A. J. 2020 [25]	Google Earth-Pro	Determining the learning outcomes using Google Earth Pro as media-assisted inquiry model in landslide disaster mitigation. The Google Earth Pro-based media has a higher point to improving students learning outcomes than using Power point media.
12	The Encouragement of 21st Century Skills Through the Integration of STEM Activities in Basic Education. CA. T. 2019 [26]	-	Digital technology is some issue that challenge teacher having the 4C skills. In the 21st century the young generation must sharpen their minds to support the education for facing the digital technology to support learning process in STEM activity.
13	STEM at home: provide scientific activities for students during the Covid-19 pandemic. Zulirfan, Z., Yennita, Y., & Rahmad, M. 2020 [27]	Science Kit Project	Because of COVID-19 pandemic, students' scientific activities were decrease because of school from home regulation. This research was developing module and kit of STEM Project so the students can have scientific literacy in their own home.
14	Developing of a learning media for smartphones for Disaster Mitigation Education Wahyuningtyas, N., Ruja, I. N., Yahya, M. H. 2020 [28]	smartphone-based	Community need to learn about disaster mitigation as an important prevention towards disaster. Using mobile encyclopedia in a smartphone

			application for volcanic eruption disaster education. This media able to give others learning media option for teacher.
15	STEM Education in Indonesia: Science Teachers' and Students' Perspectives. Permanasari, A., Rubini, B., & Nugroho, O. F. 2021 [29]	-	Using survey approach to analyze teacher and students' perceptions towards STEM. Teacher shows positive feedback towards STEM education, but they have not yet applying STEM Education to their learning process. It causes student understanding toward STEM in weak level.

Based on the literature using STEM, learning media, and disaster theme. the most commonly keywords are represented by the pie diagram shown in the Figure 2.

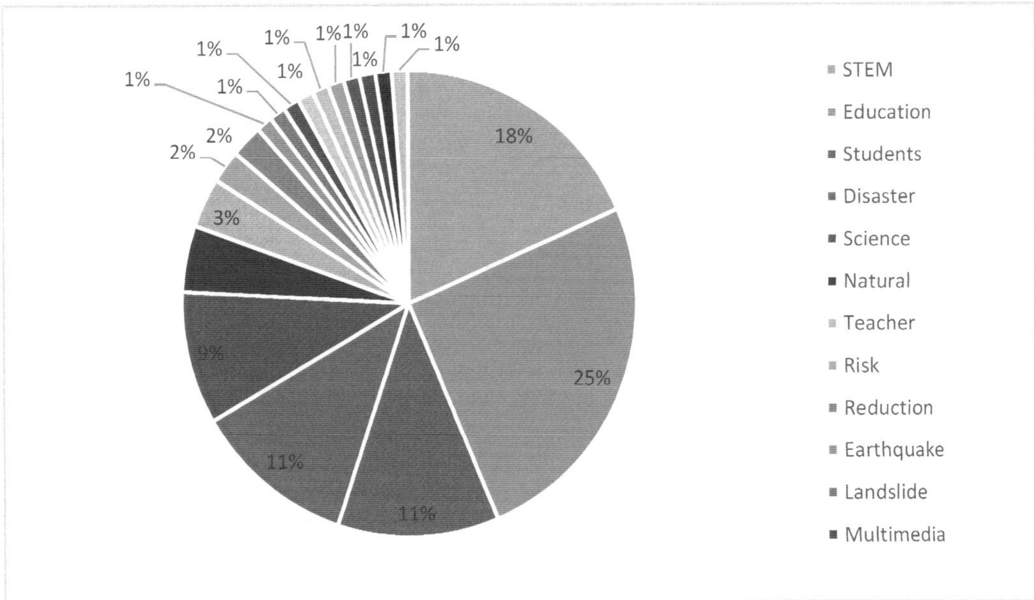

Figure 2. The frequently keywords shown in literature

The words "Education" and "STEM" are the most frequently word arise in those 15 articles with 28% and 20% percentage. Based on those point the involvement of students and teacher in STEM research was rarely conducted, especially for teacher that have lower points which 3%. It is also relevant with [29] that the understanding level of teacher towards STEM still in weak category. Those points indicate that the implementation and learning approach using STEM was rarely done by teacher in their learning to introduce STEM to students. Related to the keywords about disaster which "Disaster" have 11% percentage and some keywords related to it such as, "Earthquake", "Natural" and "Landslide" have a lower point. The specification of disaster in STEM Education in Indonesia represent the geographical condition of Indonesia are vulnerable to disaster so by applying STEM Education about disaster to students can teach disaster education meaningfully, focused, right on the target and have bigger impact for teacher, students, and local community that have poor understanding toward disaster. The implementation of STEM in learning process especially the use of technology is commonly used as learning media in these keywords, such as "Multimedia", "Digital", "Comic", and "Smartphone-based" are still rarely implemented to students so that the use of technology in learning that uses as an approach and modeling STEM in their learning needs improvement. STEM and learning media can improve students' critical thinking skills [30] and STEM literacy skills [27].

3.2. The Implementation of STEM in Disaster Education

Disaster learning is a main topic relate to disaster prevention and simplify as disaster risk reduction [31]. Disaster education also focusing on the strategy to construct the school-community disaster resilience [32,33]. Disaster resilience is often to integrate with school curriculum to teach students about disaster education. Enhancing school community members' knowledge, risk awareness level, and readiness in responding the future disaster are the main expectation after school integrate disaster education into their school curriculum [32]. Disaster education is one of the internal solutions in community to prevent and minimize the impact of hazard. Disaster education has been made in various format such as disaster mitigation community-base, disaster education leads to communities-awareness of disaster, and disaster management based on their local wisdom [34].

The implementation of disaster education into school curriculum has been made by the educators to prevent and teach students about disaster. Through the observation indicate that teacher is not fully-well understand about disaster risk reduction concept [16]. Disaster risk reduction has not been into the main curriculum of Indonesia education. The disaster education usually implemented into science subject such as science [7], physics [3, 13, 35], and biology [36]. Physics is a subject that studies about the universe and its causes. Earthquake can be taught by integrate wave topic with the process of it.

This study confirmed that STEM as an approach in learning process was an effective way to increasing students' awareness, knowledge, and preparedness towards disaster [32], changing the atmosphere of learning in fun way [20], and able to carry out in online classes [34]. Based on Table 1, it is known that the STEM-integration in disaster education learning media which can be in the form of interactive media including text, video, images, and audio can increase students' disaster awareness. Moreover, students critical thinking abilities can also improve by learning using attractive media [30] and the weak level of understanding towards STEM can increase, so that students ready to solve daily life problem related to STEM. The integration of STEM in preventing disasters has also been carried out by [8] using STEM-D (Science, Technology, Engineering, Mathematics, and Disaster). Local community literacy about STEM and disaster can improving by various studies from non-STEM fields, such as the studies from psychology sector to manage post-disaster traumatic for the community after the disaster. However, the integration of STEM in disaster learning media should be implemented by developing learning media which can include audio-visual such as Virtual Reality (VR) technology. Indeed, VR technology can improve students learning interest towards physics topic [29].

4. Conclusion

The results of this literature review study indicate that the implementation of STEM as an approach in the development of learning media is an effective way to improve and increasing students and teacher understanding, and also give students interesting vibes of learning. Teacher and knowledge about STEM still need to be improved. Through this research, it is known that the use of STEM as an approach to develop learning media is limited. The development of learning media to teach disaster education need to be reviewed and adapted to the geographical conditions of Indonesia, several topics about disaster as recommendation for learning is disaster that often afflict Indonesia such as earthquakes, natural disasters, and landslides.

References

[1] Jati R 2021 *Bencana Hidrometeorologi Dominan Sepanjang Awal Januari Hingga Akhir April 2021* Retrieved from: https://bnpb.go.id/berita/bencana-hidrometeorologi-dominan-sepanjang-awal-januari-hingga-akhir-april-2021-
[2] Kartini F S, Widodo A, and Winarno N 2020 *J. Phys. Conf. Ser.* **1806** 012221

[3] Rany T D, Kuswanto H, and Abdillah A J 2020 *J. Phys. Conf. Ser.* **1440** 012029
[4] Desfandi M 2014 *Sosio Didaktika: Soc. Sci. Educ. J.* **1** 191
[5] Muhari A 2021 *Review Bencana Juli 2021: Bertemunya Bencana Hidrometeorologi Basah dan Kering* Retrieved from: https://www.bnpb.go.id/berita/review-bencana-juli-2021-bertemunya-bencana-hidrometeorologi-basah-dan-kering
[6] Latupeirisa V P S and Pujianto 2020 *J. Phys. Conf. Ser.* **1440** 012093
[7] Hikmawati A, Pursitasari I D, Ardianto D, and Kurniasih S 2020 *J. Phys. Conf. Ser.* **1521** 042053
[8] Sampurno P J, Sari Y A, and Wijaya A D 2015 *Int. J. Learn. Teach.* **1** 74
[9] Wiguna B J P K, Suwarma I R, and Liliawati W 2018 *J. Phys. Conf. Ser.* **1013** 012082
[10] Nugroho O F, Permanasari A, and Firman H 2019 *J. Pendidik. IPA Indones.* **8** 417
[11] Farwati R, Metafisika K, Sari I, Sitinjak D S, Solikha D F, and Solfarina S 2021 *Int. J. STEM Educ. Sustain.* **1** 11
[12] Farwati R, Metafisika K, Sari I, Sitinjak D S, Solikha D F, and Putra E E 2021 *STEM Education Dukung Merdeka Belajar (Dilengkapi dengan Perangkat Pembelajaran Berbasis STEM)* (Riau: CV. DOTPLUS Publisher)
[13] Labibah U N, Wilujeng I, Sulaiman S, and Rahmawati L 2019 *J. Ilm. Pendidik. Fis. Al-BiRuNi* **8** 229
[14] Septikasari Z and Ayriza Y 2018 *J. Ketahanan Nas.* **24** 47
[15] Anggaryani M 2020 *J. Phys. Conf. Ser.* **1805** 012008
[16] Sejati P M, Budiningsih C A, and Pujianto 2019 *J. Phys. Conf. Ser.* **1233** 012090
[17] Rusman K D and Riyana C 2012 *Pembelajaran Berbasis Teknologi Informasi dan Komunikasi* (Jakarta: Raja Grafindo Perkasa)
[18] Maleeva H P 2015 *Pengembangan Media Pembelajaran Menggunakan Media Video Materi Gunung dan Kebencanaan kelas VII SMP Muhammadiyah 4 Sambi, Boyolali 2014/2015* (Surakarta: Universitas Muhammadiyah Surakarta)
[19] Nugroho W D, Indah W, Alanish, Istiqomah N, Cahyasari I, Indrastuti M, Sugondo P, and Isworo A 2020 *J. Bionursing* **2** 101
[20] Irwandi, Sari I M, Oktavia R, and Syukri M 2020 *J. Phys. Conf. Ser.* **1462** 012066
[21] Septaria K, Dewanti B A, and Afidah M I E 2020 *J. Penelit. Pendidik. IPA* **5** 61
[22] Subuh A 2019 *Proc. 1st Int. Semin. Sci. Technol. Eng. Math. Learn. Int. Forum* p 26
[23] Noviana E, Kurniaman O, and Affendi N 2020 *Tadris: J. Kegur. Ilmu Tarb.* **5** 11
[24] Astuti N H, Rusilowati A, and Subali B 2020 *J. Innov. Sci. Educ.* **10** 79
[25] Suharini E, Ariyadi M H, and Kurniawan E *Int. J. Inf. Educ. Technol.* **10** 820
[26] Talib C A 2019 *Proc. 1st Int. Sem. Sci. Technol. Eng. Math. Learn. Int. Forum* p 1
[27] Zulirfan Z, Yennita Y, and Rahmad M 2020 *J. Phys. Conf. Ser.* **1655** 012068
[28] Winangun M M 2019 *Proc. 1st Int. Sem. Sci. Technol. Eng. Math. Learn. Int. Forum* p 426
[29] Agusty I A and Anggaryani M 2020 *J. Pendidik. Fis.* **9** 134
[30] Suliyanah, Deta U A, Kurniawan F K, Lestari N A, Yantidewi M, Jauhariyah M N R, and Prahani B K 2020 *J. Phys. Conf. Ser.* **1805** 012038
[31] Nuochi R, Sato S, and Imamura F 2015 *J. Disaster Res.* **10** 1117
[32] Adiyoso W and Kanegae H 2012 *J. Disaster Mitig. Cult. Herit. Historic Cities* **6** 165
[33] Shiwaku K 2009 *Disaster Manag. Glob. Chall. Local Solut.* **16** 576
[34] Setyowati D L 2019 *Pendidikan Kebencanaan* (Semarang: Universitas Negeri Semarang)
[35] Rahmawati L, Labibah U N, and Kuswanto H 2020 *J. Phys. Conf. Ser.* **1440** 012042
[36] Susanto L H, Istiana R, Retnowati R, Ekamilasari E, Ichsan I Z, Sigit D V, Rahman M M, Babu R U M, Darussyamsu R, and Rosyid A 2021 *J. Pendidik. Biol. Terap.* **6** 40

Research on Physics of Photography: A Bibliometric Study (2000-2020)

N Suprapto*, A Kholiq, B K Prahani, and U A Deta

Physics Education Programme, Universitas Negeri Surabaya, Indonesia

*Email: nadisuprapto@unesa.ac.id

Abstract. The purpose of this study is to analyse physics on photography research and see its trends to find the opportunity for further research. A bibliometric analysis was used in this study. The search results from the Scopus database were extracted using the *VOSviewer* software. A total of 432 documents related to physics of photography were analysed and mapped. The results mapped the trend of documents across years, language, countries, affiliation, type, sources, authors, and key words. The results have also indicated two major clusters and one minor cluster in researching on physics of photography. The first cluster (green colour) was photography as an application. The second cluster (red) was photography specified into high-speed photography and modern photography. The third cluster (blue) was technology of photography in connecting with competition. A proposed model for future research is also illustrated. Thus, it can be an opportunity for further researchers to conduct research related to this area.

1. Introduction

Photography consists of two Greek words, namely (*phōs*), which means light and (*graphê*), which means line or image [1,2]. Literally, Photography can mean "drawing with light" [1-3]. Photography itself can be defined by using various perspectives, both from art, science, and activities or activities. In essence, Photography is a process of recording certain light or electromagnetic waves on a specific medium (in the form of chemical or electronic media) into an image that is permanent (durable). Epistemologically the word photography was introduced by "Sir John Herschel in a lecture at the Royal Society London, on March 14, 1839, causing the word photography to be known in the world today" [4]. The term Photography was also first written by Johann von Maedler in the German newspaper *Vossische Zeitung* in 1839 [4].

Photography as a science refers to the fields of pure science, such as mathematics, chemistry, physics and computers, which are the basis for the existence of photography from its early days to the present day. The working mechanism of the camera, the dynamics of exposure and the development of the chemical medium of film or digital must be properly balanced and used in such a way by photographers to produce images according to their needs [5].

Physics in photography has a fundamental role that provides a fundamental influence in the development of photography. The existence of lenses has been known since antiquity, but in writing, the lens was introduced in Ancient Greece, in Aristophanes' play The Clouds (424 BC) which mentions a burning-glass (a convex lens used to focus sunlight to create fire) [6]. Next, Seneca the Younger (3 BC) describes the magnifying effect of a round glass filled with water. The development of scientific lenses in their application in photography was published by Arab Muslim scientist Abu Ali al-Hasan Ibn Al-Haitham (965-1038), who wrote a book entitled *Kitab al-Manazir* (Book of Optic), which discusses the first optical theory and main. Haitham explained that the lens in the human eye forms an image on the retina. The spread of lenses did not occur until the invention of glasses in Italy in the 1280s [6].

On the inside of the camera, the working mechanisms are several mechanical mechanisms that are designed in such a way as to be able to adjust the exposure time and open the diaphragm gap to allow the entry of a certain amount of light according to the needs of image formation on peer media (both film and digital media) in a dark room. The prominent role of physics in making a working system for regulating light entering the camera is measuring light using a light meter.

Photographic film is one of the fundamental things that put chemistry in photography. Film in photography is a plastic sheet coated with an emulsion (light-sensitive silver halide salt bonded by gelatine) with variable crystal size, determining the degree of reaction sensitivity to light, contrast, and film resolution. When the emulsion is exposed to sufficient light, it forms an invisible image. The subsequent chemical processing of this reaction allows the formation of images that appear both negatively and positively through a process called film development [7].

Photography could also be a way to explain some physical concepts through images. During last year, we started an elective course: "physics of photography" with the main output as the project to practice our capacity to take good pictures, create something new, and explain concepts at once. We discovered that it is straightforward to interpret a physical concept through shapes, body and colours; it is beneficial to translate physics into something more achievable by "science-outsiders". It was very challenging at the start but fascinating and satisfying when people understand we are talking, for example, about gravity with a given picture [8].

On the other hand, through the course "literacy of physics", we emphasise the use of bibliometric analysis in exploring the idea of physics research through library research. By combining the essence of these two courses, this bibliometric study was conducted. Therefore, through this bibliometric analysis, the purposes of the research were analysing the publication output, document sources, and language sources, publication distribution of countries and institutes, top authors, publication patterns, and the visualisation of research trends on physics of photography.

2. Method

This study adopted a bibliometric study [9-12]. The data were retrieved from the Scopus Database. The data collection was conducted on August 2021. The keywords used in data searching were: "TITLE-ABS (*physics AND photography*)" and resulted in 699 documents. After the filtering process for the year 2000-2020, finally, the authors analysed 432 documents. These documents were downloaded in *.csv* and *.ris* format to be processed using *VOSviewer* to visualise and analyse the trends in the bibliometric form [13].

3. Results and Discussion

3.1 Publication output, document sources, and language sources

Based on a search using the Scopus Elsevier database obtained 432 documents related to the study of physics of photography. The publications devoted to the physics of photography research from 2000 to 2020 are demonstrated in Figure 1. The number of documents across the year has fluctuated. The record of documents was 44 in 2012. Starting from 2017, the trend raised slowly. It can be predicted that the number of articles in 2021-2025 will increase dramatically.

Meanwhile, the number of articles based on sources indicated the dominance of articles in the journal (256 documents). It was followed by a conference proceeding (146) and a book (19). The book series and trade journal accounted for the fewest sources of documents. Meanwhile, all documents in this domain used a hundred percent of English as a paper language.

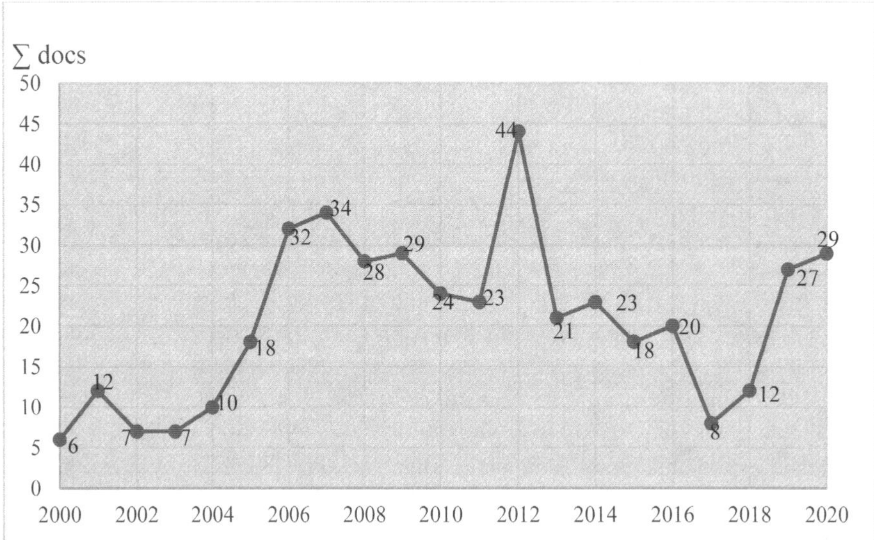

Figure 1. The documents trend of photography physics (2000-2020)

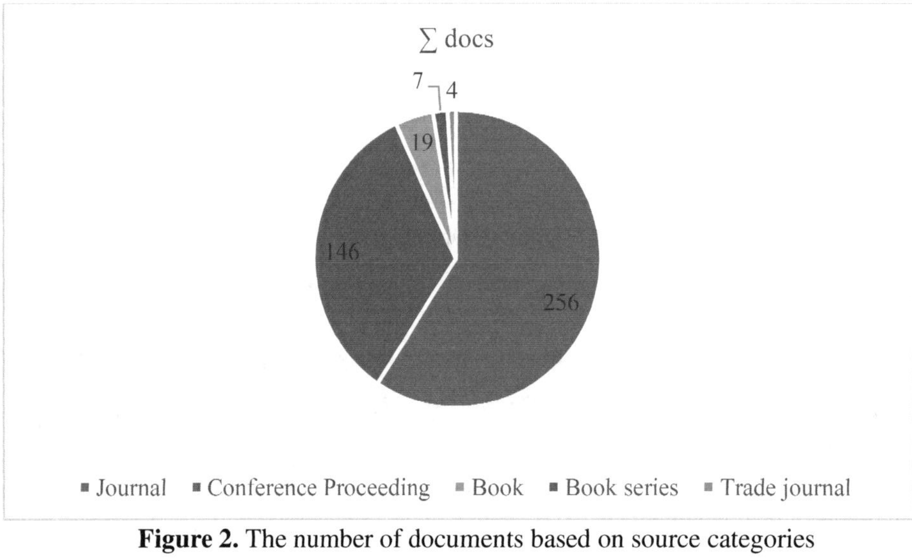

Figure 2. The number of documents based on source categories

3.2 Publication distribution of countries and institutes

Based on the number of documents across countries, the USA's dominance was clear, with 142 documents from 2000 to 2020. In the second and third position were UK and China with 59 and 57 documents, respectively. The countries such as Japan, Germany, France, and Canada contributed with 11 to 23 documents in the last decade. The remained countries were Russia, Australia, India, and Iran resulted in less than ten documents.

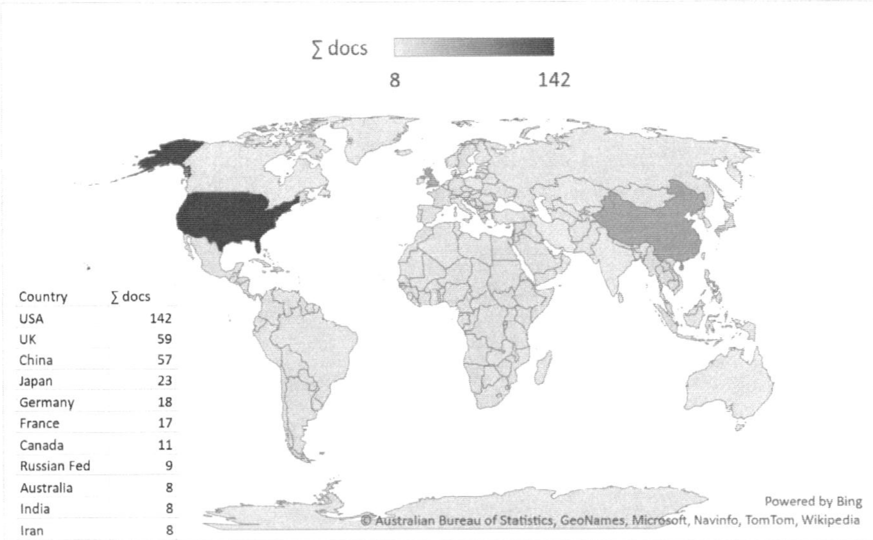

Figure 3. Top countries that researching on physics of photography

Department of Physics, University of Cambridge is the most affiliation that researching on physics of photography. Institutions from the USA, the UK, China, and Japan were also dominated among those respective countries. China Academy of Engineering Physics and Southwest Institute of Fluid Physics Mianyang from China contributed to the second position, followed by Georgia Institute of Technology from the USA with nine documents in the third rank (see Table 1).

Table 1. Number of documents based on affiliation

Affiliation	Number of documents
Department of Physics, University of Cambridge	26
China Academy of Engineering Physics	12
Southwest Institute of Fluid Physics *Mianyang*	12
Georgia Institute of Technology	9
The University of Manchester	8
University of Colorado Boulder	6
Los Alamos National Laboratory	6
Amirkabir University of Technology	6
Imperial College London	6
The University of Tokyo	6
The Royal Institute of Technology KTH	5
A-Star, Institute of High-Performance Computing	5
CNRS Centre National *de la Recherche Scientifique*	5
Chinese Academy of Sciences	5

3.3 Top authors in researching on physics of photography

In terms of the most productive authors, Table 2 indicates the top authors researching the physics of photography. Kontis and Proud were the most productive authors on this topic. In the second rank, there were Jaberi, Tadjfar, Thadhani, and Zare-Behtash with six documents in the last decade. Moreover, Figure 4 indicates all those top authors, their co-authorships, and the most influenced authors on the physics of photography. It was clear that Proud, Wang W, Kontis, Jaberi were the leader of their clusters.

Table 2. Top authors on physics of photography

Author	Number of documents
Kontis, K.	8
Proud, W.G.	8
Jaberi, A.	6
Tadjfar, M.	
Thadhani, N.N.	
Zare-Behtash, H.	
Hertzberg, J.	5
Klaseboer, E.	
Khoo, B.C.	4
Li, J.	
Xiao, Z.F.	

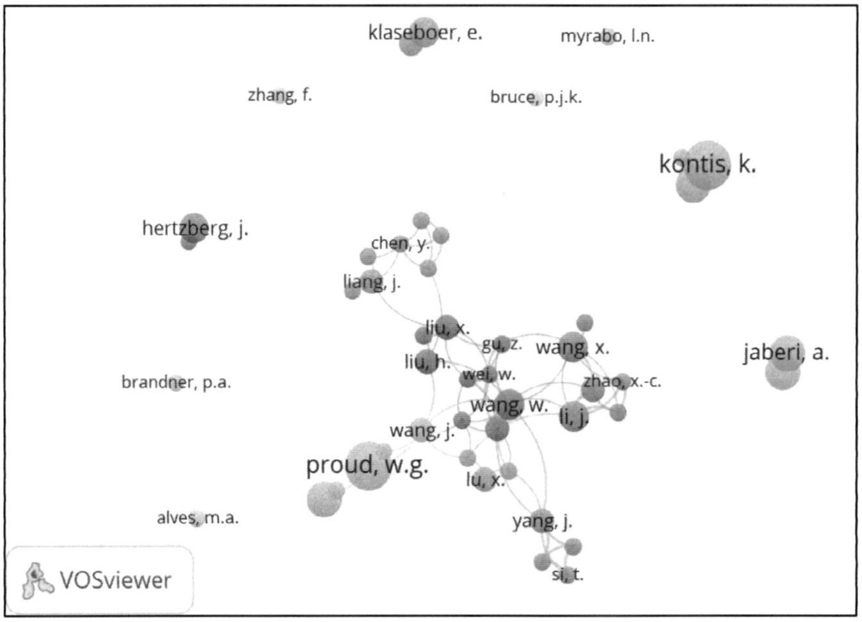

Figure 4. Top authors, co-authorships, and the most influenced authors on physics of photography

3.4 Publication patterns: source titles (journal or proceeding)

Table 3 illustrates the most contribution journal or proceeding on the research on physics of photography. AIP Conference Proceedings was a leading conference series that contains articles in this domain. Meanwhile, the Journal of Applied Physics and Physics of Fluids was also the leading journal covering this topic in their content as the second and the third rank. The remaining journals were Applied Physics Letters, Physics of Plasmas, Physics Teacher, Australasian Physical Engineering Sciences in Medicine, Review of Scientific Instruments, Japanese Journal of Applied Physics, and Physics Education (see Table 3).

Table 3. Number of physics of photography documents (2000-2020) across source titles

Sources	Number of documents
AIP Conference Proceedings	71
Journal of Applied Physics	28
Physics of Fluids	24
Proceedings of SPIE The International Society for Optical Engineering	15
Applied Physics Letters	10
Physics of Plasmas	9
Physics Teacher	9
Australasian Physical Engineering Sciences in Medicine	7
Review of Scientific Instruments	7
Japanese Journal of Applied Physics	6
Physics Education	6

Furthermore, the following Figure 5 lists the most keywords used by the authors in researching the physics of photography. It was clear that the term 'high-speed photography' and 'photography' frequently appeared in the Scopus database. The term 'shock waves, 'Schlieren photography', and 'students have also coloured the research on photography in physics. Schlieren photography is "a visual process used to photograph the flow of fluids of varying density" [14,15]. This type of photography was invented by the German physicist August Toepler in 1864 to study supersonic motion, and it is widely used in aeronautical engineering to photograph the flow of air around objects [14].

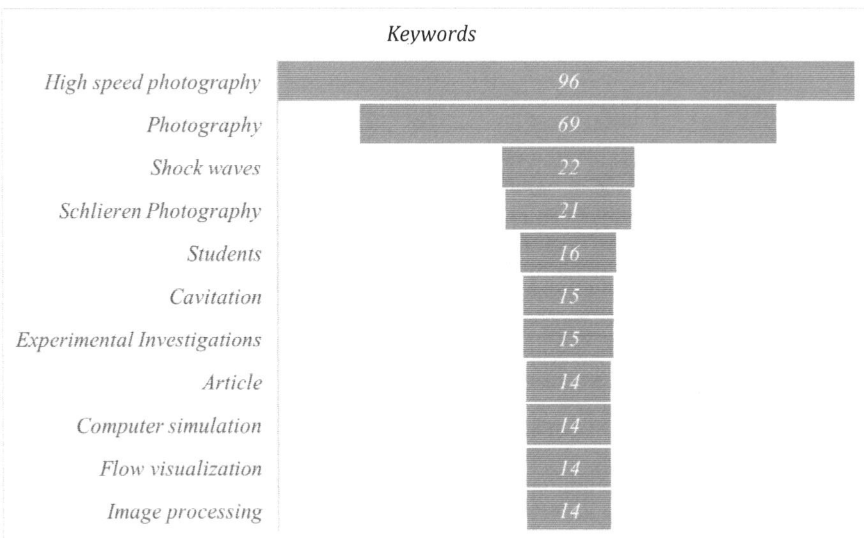

Figure 5. Top keywords research on physics of photography during 2000-2020

3.5 Visualisation of research trends on physics of photography based on vosviewer software

From 432 documents related to photography in physics in the Scopus database, the researchers visualised the research trends on this topic supported with *VoSViewer* software. This effort helps find the novelty of the research on this domain. Figure 6 specifies the whole picture of research on the physics of photography. Researchers worldwide produced two primary clusters indicated with green and red and one minor cluster (blue). The first cluster (green colour) was photography as an application. The second cluster (red) was photography specified into high-speed photography and modern photography. The third cluster (blue) was the technology of photography in connecting with the competition.

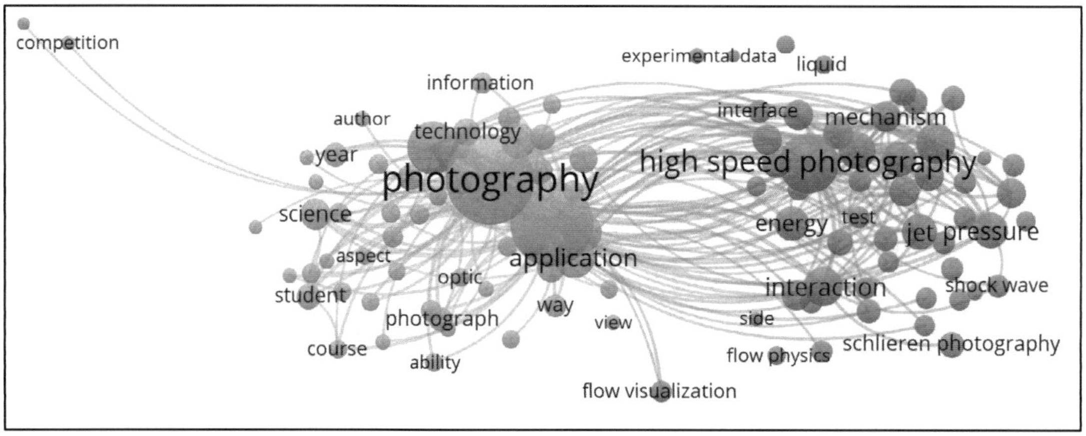

Figure 6. The whole picture of research on photography physics during 2000-2020

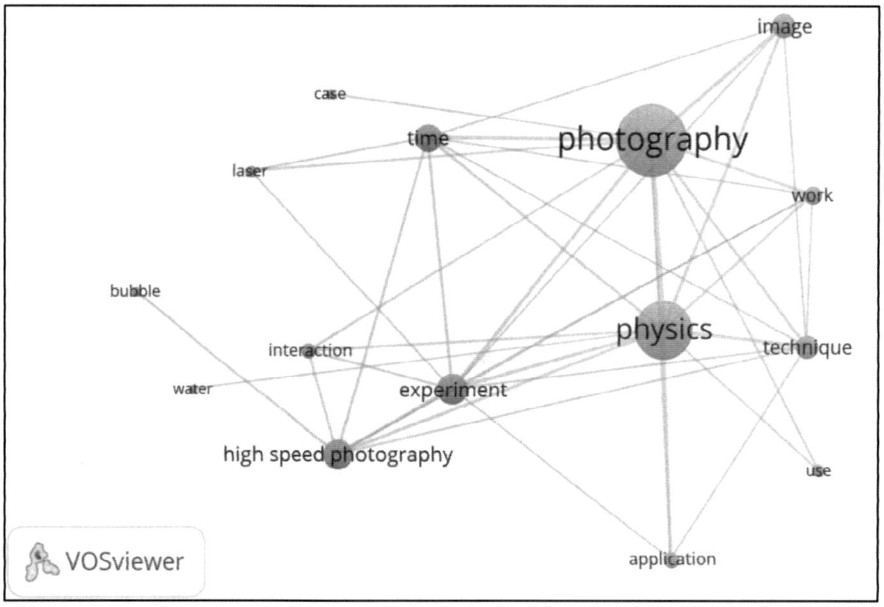

Figure 7. Simple proposed model of network visualisation on physics of photography

Finally, the authors offered the simple proposed model of research on the physics of photography. Figure 7 indicates the relationships among the critical variables in researching photography in physics through network visualisation. The figure has also illustrated main variables and supporting variables and the link strength among those variables.

4. Conclusion

Research trend on physics of photography research was performed through this bibliometric research. The authors indicated some significant points regarding the research on the physics of photography in the last two decades. The number of documents across the year was fluctuated, which were dominated by articles in the journal. The USA contributed the most documents on physics of photography. Meanwhile, the Department of Physics, University of Cambridge, was the most institution that

contributed to this topic. Then, the visualisation of research trend on physics of photography resulted in two major clusters and one minor cluster:
1. Photography as an application.
2. Photography is specified into high-speed photography and modern photography.
3. Technology of photography in connecting with competition.

The authors have also proposed a simple proposed model of research on physics of photography. This model will give a new lens and new direction on the future research of photography physics.

Acknowledgements
The Directorate of Resources supported this research, Directorate General of Higher Education Indonesia, through Fundamental Research Scheme with the contract number 133/SP2H/LT/DRPM/2021 and SPK number B/12084/UN38.9/LK.04.00/2021.

References
[1] Kemevor A K and Essuman J 2017 *J. Soc. Sci. Res.* **3** 13
[2] Galer M 1999 *Essential skills: Location photography* (Butterworth-Heinemann: Oxford)
[3] Graham R and Bolton V 1990 *Focus on photography* (Hodder & Stoughton: Great Britain)
[4] Delisio M 2014 *Looking at Photography* Retrieved from http://michaeldelisio.com/looking-at-photography/
[5] Hirsch R 2014 *Why People Make Photographs* Retrieved from http://www.photovisionmagazine.com/articles/whyphoto1.html
[6] Grundberg A 2020 *History of photography* Retrieved from https://www.britannica.com/technology/photography/Photographys-early-evolution-c-1840-c-1900
[7] Präkel D 2010 *The fundamental of creative photography* (AVA Publishing: United Kingdom)
[8] Biondi S 2017 *The balance between Physics and Photography* Retrieved from https://www.interactions.org/blog/balance-between-physics-photography
[9] Suprapto N, Prahani B K, and Deta U A 2021 *Libr. Philos. Practice* **5599**
[10] Suprapto N, Prahani B K, and Deta U A 2021 *Libr. Philos. Practice* **5928**
[11] Suprapto N, Sukarmin, Puspitawati R P, Erman, Savitri D, Ku C-H, and Mubarok H 2021 *Int. J. Eval. Res. Educ.* **10** 4
[12] Yanuarti E A and Suprapto N 2021 *Stud. Philos. Sci. Educ.* **2** 7
[13] Eck N J V and Waltman L 2011 *VOSviewer Manual* Retrieved from https://www.vosviewer.com/documentation/Manual_VOSviewer_1.6.16.pdf
[14] Bradley J W, Oh J, Olabanji O T, Hale C, Mariani R, and Kontis K 2011 *IEEE Trans. Plasma Sci.* **39** 2312
[15] Bunjong D, Pussadee N, and Wattanakasiwich P 2018 *J. Phys. Conf. Ser.* **1144** 012097

Bibliometric Analysis on Online Physics Learning during COVID-19 Pandemic: Contribution to Physics Education Undergraduate Program

B Jatmiko[1], B K Prahani[1*], N Suprapto[1], S Admoko[1], U A Deta[1], N A Lestari[1], M N R Jauhariyah[1], M Yantidewi[1], and D Muliyati[2]

[1] Physics Education, Universitas Negeri Surabaya, Indonesia
[2] Physics Education, Universitas Negeri Jakarta, Indonesia

*E-mail: binarprahani@unesa.ac.id

Abstract. Research will focus on generating a profile of research trends in online physics learning during the COVID-19 pandemic. The main research is bibliometric. Scopus document data collection in August 2021. The 1007 document results with keywords online physics learning. Then focused on 2020 and 2021* as 277 document results of which these 2 years are the COVID-19 pandemic period. In general, the research results show that the COVID-19 pandemic has had a positive impact on accelerating online learning in developing countries. Another finding is online learning which is proven to be effective for improving student learning outcomes in physics learning. Educational transformations today and after the COVID-19 pandemic can occur in online education, changes in learning from home, changes in roles, and approaches. The research implications are supported by recent research: (1) These findings will provide an empirical basis for the development of physics education research, especially online physics learning during the COVID-19 pandemic; (2) Deep learning system, machine learning system, and education computing can still become a research trend in current research, especially in the integration and adaptation in physics learning; (3) Retracing the trend and novelty of physics education research, especially in the physics education undergraduate program.

1. Introduction

The coronavirus pandemic has had a wide impact on a number of sectors, including education. Recent research has found COVID-19 to be dangerous and can continue to mutate [1-5]. The latest developments on the COVID-19 virus pandemic in the world have been uploaded again, obtained from worldometers.info on Friday, August 13, 2021 get the number of cases of the COVID-19 virus in the world reached 206,181,372 cases and the addition of new cases was + 712,590 cases, only in the last 24 hours. Meanwhile, the number of patient deaths in the last 24 hours of the COVID-19 virus in the world increased by +10,379 people, so that the accumulated cases of patients dying from the COVID-19 virus reached 4,347,175 people. The number of patients who have been declared cured is 185,041,837 people and around 16,792,360 active cases who are still undergoing treatment in hospitals and self-isolation at their respective homes.

The COVID-19 pandemic has also had an impact on physics learning in schools. Schools have implemented online physics learning during COVID-19 pandemic. Online physics learning has become an alternative way in the COVID-19 pandemic. This is supported by the results of recent studies that prove the success of online physics learning variations during the COVID-19 pandemic [6-14].

Online physics learning is very difficult to implement massively in developing countries. Limited resources and financial support can hinder online physics learning in developing countries. Apparently, the COVID-19 pandemic have a positive impact on accelerating the implementation of online physics learning. This opinion is in line with the results of recent research which states that online-based physics learning can improve student learning outcomes in countries affected by the COVID-19 pandemic [15-23].

However, in previous studies, there had been no researches that specifically profiled research trends in online physics learning during the COVID-19 pandemic. Therefore, research will focus on generating a profile of research trends in online physics learning during the COVID-19 pandemic. The main research implication is for contributing to the physics education undergraduate program, in particular the profile of online physics learning during the COVID-19 pandemic.

2. Method

The main research method is bibliometric analysis [24-33]. In general, there are six steps of the research procedure as shown in Figure 1.

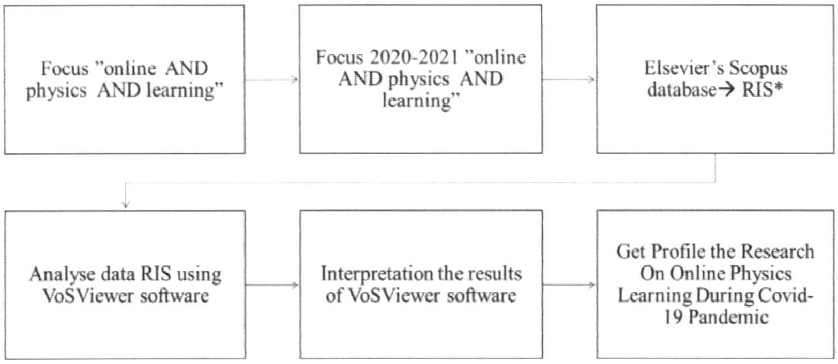

Figure 1. Research procedure

The Scopus document data was collected in August 2021. The 1007 document results with keywords online physics learning, then focused on 2020 and 2021* as many as 277 document results of which these 2 years are the COVID-19 pandemic period. The VOSViewer software [34-41] was used to support and figure out the research trend on online physics learning during the COVID-19 pandemic.

3. Results and Discussion

Among those 277 papers related online physics learning during COVID-19 pandemic research in the Scopus database, then the researchers visualised the research trends on this topic assisted with the VOSViewer software (see Figure 2). This effort is useful for finding the novelty of the research on this domain.

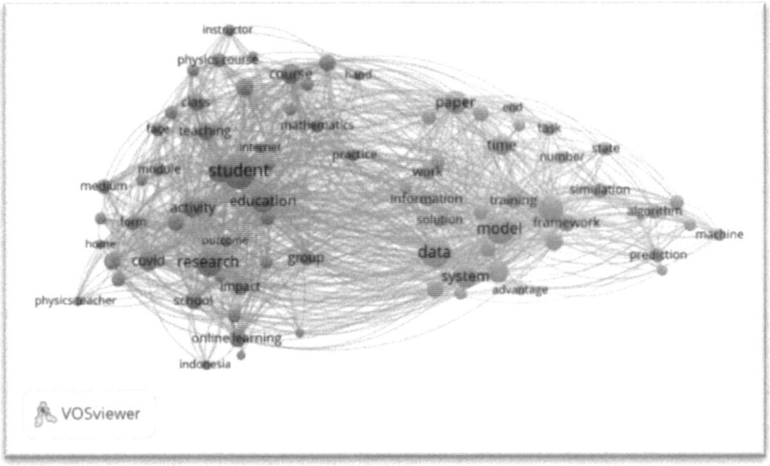

Figure 2. Network visual of research on online physics learning during COVID-19 pandemic

Figure 2 indicates the whole picture research on online physics learning during COVID-19 pandemic. Researchers on the world produced two clusters. The first cluster (red colour) was online physics learning during COVID-19 pandemic as a transformation related to student, education, online learning, physics teacher, physics course, research, activity, and school. The second cluster (green) was online physics learning during COVID-19 pandemic focus on simulation, information, solution, model, data, task, system, and machine (IoTs).

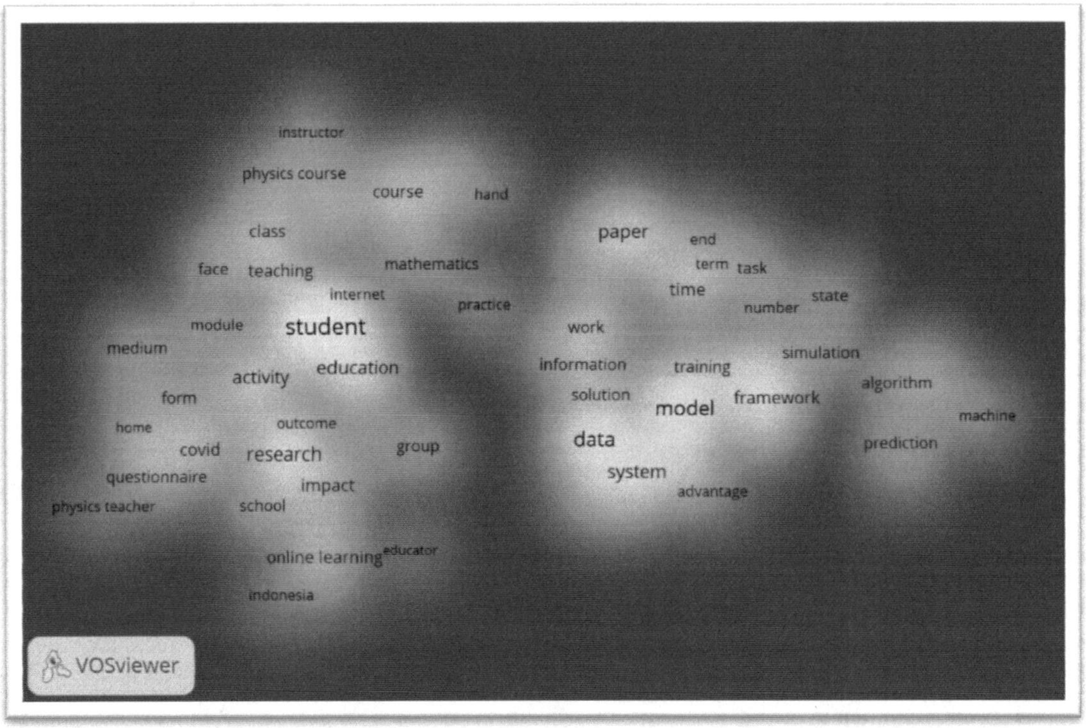

Figure 3. Density visual of research on online physics learning during COVID-19 pandemic

Figure 3 shows the density of research on online physics learning during the COVID-19 pandemic. The dominance of yellow is based on student, research, data, system, model, framework, education, and online learning. The findings indicated it was some parameters or interrelationships among variables to capture the trend and novelty of researching on online physic learning research during COVID-19 pandemic in Figure 4.

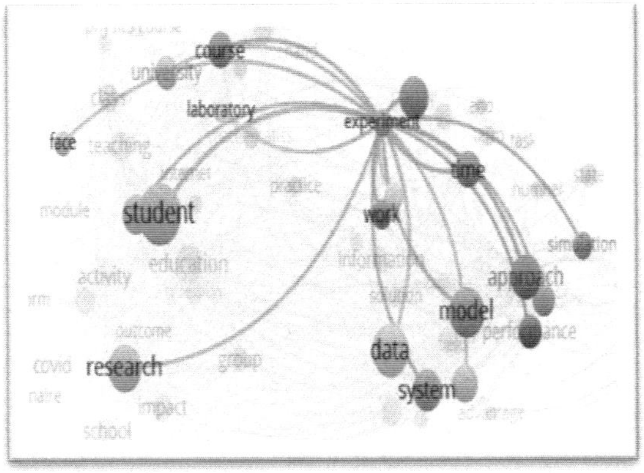

(4a) Experiment

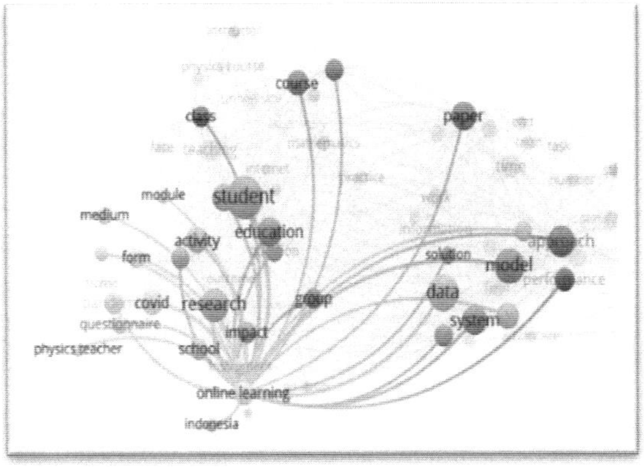

(4b) Online learning

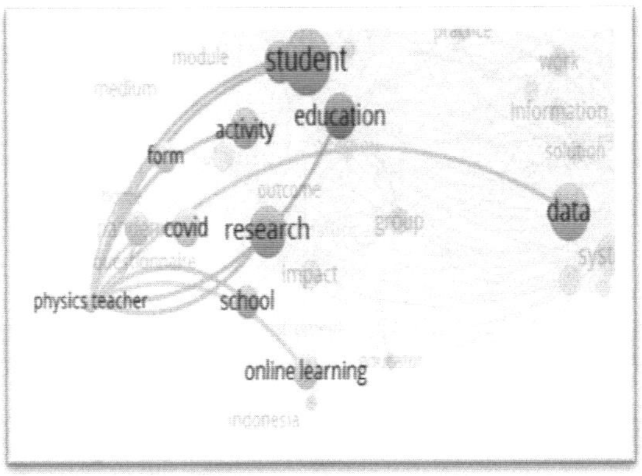

(4c) Physics teacher

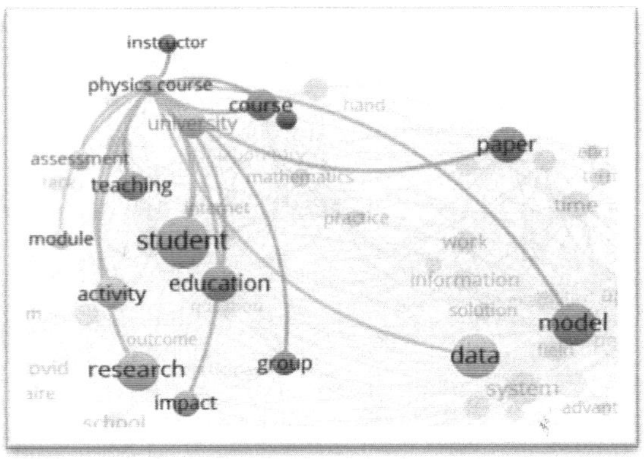

(4d) Physics course

Figure 4. Research trend on online physics learning during COVID-19 pandemic

Figure 4 shows the findings indicated it was some parameters or interrelationships among variables to capture the trend and novelty of researching on online physic learning research during COVID-19 pandemic, such as researching experiment (Figure 4a), online learning (Figure 4b), physics teacher (Figure 4c); and physics course (Figure 4d) on online physics learning research during COVID-19 pandemic. Research trends as of August 14, 2021 (Scopus Database Documents) related to experiments (6,845,363), online learning (123,697), physics teacher (16,565); and physics course (13,486). This finding has a positive contribution to the development of physics education, especially in future research trends and physics learning innovation in general. Other findings point to the top keywords trends as of August 14, 2021 (Scopus Database) connecting to online physics learning during COVID-19 pandemic such as: Students (80), E-learning (77), Online Learning (48), Physics (45), Learning Systems (38), Deep Learning (22), Education Computing (22), Machine Learning (18), Curricula (17), and Engineering Education (16). The research implications are supported by recent research:

- These findings will provide an empirical basis for the development of physics education research [42,43], especially online physics learning during the COVID-19 pandemic;

- Deep learning system [44], machine learning system [45,46], and education computing [47-49] can still become a research trend in current research, especially in the integration and adaptation in physics learning;
- Retracing the trend and novelty of physics education research [50-53] especially in the physics education undergraduate program.

4. Conclusion

The findings indicated there were some parameters or interrelationships among variables to capture the trend and novelty of researching on online physics learning research during COVID-19 pandemic, such as researching experiment, online learning, physics teacher, and physics course on online physics learning research during COVID-19 pandemic. Fundamental implications supported by recent research is retracing the trend and novelty of online physics learning during the COVID-19 pandemic in the physics education undergraduate program. The research limitation is that it only involves Scopus documents. Further research can be done by involving document data from the Web of Science (WoS).

5. Acknowledgment

Thanks for the financial supporting and conducting research in 2021 from the LPPM, Universitas Negeri Surabaya; Number: 855/UN38/HK/PM/2021 dated June 24, 2021.

6. References

[1] Romero L, Pao L Z, Clark H, Hinckley A, and Siza C 2020 *Morb. Mortal. Wkly. Rep.* **69** 1895
[2] Wadman, M *Science* 2021 **370** 1384
[3] Fedi G, Guidi S, and Ragazzo F 2021 *Italian J. of Medicine* **14** 219
[4] Hou Y J, Chiba S, Halfmann P, Kawaoka Y, and Baric R S 2021 *Science* **370** 1464
[5] Middleton P, Perez-Guzman P N, Cheng A, Thursz M R, and Nayagam S 2021 *Sci. Rep.* **11** 2455
[6] Surahman, E., and Sujarwanto, E 2021 *J. Phys. Conf. Ser.* **1869** 012159
[7] Dark M L 2021 *Phys. Educ.* **56** 055015
[8] Marisda D H, and Ma'Ruf, M 2021 *J. Phys. Conf. Ser.* **1806** 012034
[9] Nurliani R, Sinaga P, and Rusdiana D 2021 *J. Phys. Conf. Ser.* **1806** 012043
[10] Surmaini, Syafe'I I, and Diani R 2021 *IOP Conf. Ser. Earth Environ. Sci.* **1796** 012099
[11] Sani D M, Sukarmin, and Suharno 2021 *IOP Conf. Ser. Earth Environ. Sci.* **1796** 012020
[12] Klein P, Ivanjek L, Dahlkemper M N, Küchemann S, and Susac A 2021 *Phys. Rev. Phys. Educ. Res.* **17** 010117
[13] Bergeler E, and Read M F2021 *J. Sci. Educ. Technol.* **30** 97
[14] Abdullah H, Malago J D, and Arafah K 2021 *J. Pendidik. IPA Indones.* **10** 220
[15] Ahamad A N, Samsudin M A, Ismail M E, and Ahmad N J 2021 *Eurasia J. Math. Sci. Technol.* **17** 1
[16] Xu H, Ma J, Tan P, Xuan J, and Ni M 2020 *Energy and AI* **1** 100003
[17] Kustijono R, and Hakim S R 2020 *J. Phys. Conf. Ser.* **1569** 042065
[18] Purba L S L 2020*J. Phys. Conf. Ser.* **1567** 022039
[19] Kustijono R, Hakim S R, and Wiwin E 2020 *J. Phys. Conf. Ser.* **1491** 012063
[20] Wang Y, Wang L, Liang H, Zhao L, and Huang Y 2020 *Eur. J. Phys.* **41** 035701
[21] Rizal R, Rusdiana D, Setiawan W, and Siahaan P 2020 *J. Pendidik. IPA Indones.* **9** 600
[22] Ummah R, Sulisworo D, Raharjo W, Maruto G. and Nurul Huda Abd R. 2020 *Univers. J. Educ. Res.* **8** 69
[23] Liliana R A, Raharjo W, Jauhari I, and Sulisworo D 2020 *Univers. J. Educ. Res.* **8** 59
[24] Ferreira J J, Fernandes C I, Schiavone F, and Mahto R V 2021 *Technol. Forecast. Soc. Change* **173** 121077
[25] Aalirezaei A, and Kabir G 2021 *Innov. Infrastruct. Solut.* **6** 230

[26] Ojeda-Pereira I, and Campos-Medina F 2021 *Resour. Policy* **74** 102272
[27] Yurtcu M, and Güzeller C O 2021 *Participatory Educ. Res.* **8** 426
[28] Marcal J, Bishop T, Hofman J, and Shen J 2021 *Chemosphere* **284** 131267
[29] Zhang J Z, Srivastava P R, Sharma D, and Eachempati P 2021 *Expert Syst. Appl.* **184** 115561
[30] Yin J, Wan J, Zhu J, Pan Y, and Zhou H 2021 *Chinese Med.* **16** 53
[31] Morin C, and Gaboury I 2021 *BMC Complement. Med. Ther.* **21** 196
[32] Milán-García J, Caparrós-Martínez J L, and Rueda-López N, de Pablo Valenciano J 2021vGlob. Health **17** 74
[33] Marín-Marín J -A, Moreno-Guerrero A -J, Dúo-Terrón P, and López-Belmonte J 2021 *Int. J. STEM Educ.* **8** 41
[34] Yao K -K, Zhang H, and Zhang N -Y 2021 *IOP Conf. Ser. Earth Environ. Sci.* **783** 012101
[35] Peng Y, Wu Z, Zhuang Q, and Zhang J 2021 *Orthop. J. China* **41** 694
[36] Mardiastuti A, Masy'Ud B, Ginoga L N, Sastranegara H, and Sutopo 2021 *IOP Conf. Ser. Earth Environ. Sci.* **771** 012004
[37] Cavalcante W Q F, Coelho A, and Bairrada C M 2021 *Sustainability* **13** 4987
[38] Yahan S, Yunxia W, and Daoyang F 2021 *Chinese J. Tissue Eng. Res.* **25** 2385
[39] Putri C R, Soleh S M, Saregar A, Anugrah A, and Susilowati N E 2021 *IOP Conf. Ser. Earth Environ. Sci.* **1796** 012056
[40] Han L, and Gong Z 2021 *E3S Web Conf.* **237** 04016
[41] Chen Y, Cheng L, Lian R, Song Z, and Tian J 2021 *Biosci. Trends* **15** 64
[42] Aiken J M, De Bin R, Lewandowski H J, and Caballero M D 2021 *Phys. Rev. Phys. Educ. Res.* **17** 020104
[43] Walsh C, Stein M M, Tapping R, Smith E M, and Holmes N G 2021 *Phys. Rev. Phys. Educ. Res.* **17** 010119
[44] Rahman A, Hossain, M S, Alrajeh N A, and Alsolami F 2021 *IEEE Internet Things J.* **8** 9603
[45] Han Y, Modaresnezhad M, and Nemati H 2021 *Knowl. Based Syst.* **227** 107164
[46] Wang K, and Guo P 2021 *Cognit. Comput.* **13** 724
[47] Colomo-Palacios R, Yahia N B, Larrucea X, and Casado-Lumbreras C 2020 *Revista Iberoamericana de Tecnologias del Aprendizaje* **15** 336
[48] Lachney M, and Yadav A 2020 *Commun. ACM* **63** 18
[49] Graves A L, and Light A D 2020 *Comput. Sci. Eng.* **22** 50
[50] Yun E 2020 *J. Baltic Sci. Educ.* **19** 388
[51] Exarhos S 2020 *Phys. Teach.* **58** 461
[52] Jo K 2020 *New Phys. Sae Mulli* **70** 1101
[53] Kanim S, and Cid X C 2020 *Phys. Rev. Phys. Educ. Res.* **16** 020106

Elementary School Teacher's Training and Coaching as Trustees of the National Science Olympiad (OSN) in Surabaya Using Science Practicum at SDN I Airlangga Surabaya

Asnawi[1]*, D Iriyani[2], T Prakoso[3], and L E Setijorini[4]

[1]Department of Physics, Universitas Negeri Surabaya, Indonesia
[2]Department of Biology Education, Universitas Terbuka Surabaya, Indonesia
[3]Department of Indonesia Language, Universitas Terbuka Surabaya, Indonesia
[4]Department of Biology Education, Universitas Terbuka Jakarta, Indonesia

*Email: asnawi@unesa.ac.id

Abstract. The purpose of this community service activity is to provide knowledge and practical science skills to support the National Science Olympiad (OSN) activities for teachers at SDN I Airlangga Surabaya and its surroundings. Based on the questionnaire distributed to the teachers of the SDN I Airlangga Surabaya, it showed that most of the teachers (87%) had never carried out practicum or experimental activities to support the OSN of Science at school. One of the targets to be achieved through this activity is to change attitudes and to increase the knowledge and skills of the SDN I Airlangga Surabaya teachers who have skills in science practicum to support OSN. The training was conducted using lecture methods, slides and videos, question and answer, interactive dialogue, brainstorming and work practices. There are three things that the partners got after holding the training, namely a) Knowledge; it seems that participants have gained new knowledge about the implementation of elementary science practicums after the training, b) Skills; participants seem to have had the skills to carry out practical science activities of elementary school, c) Attitude (behaviour); according to the practicum activities carried out, scientific attitudes were formed by the participants in supporting science olympic activities. In implementing the training, the partners generally have a great willingness and interest to participate on the training.

1. Introduction

The dynamic and rapid development in the era of globalization has had a tremendous influence on the value system in the life of the nation and state. Meanwhile, the development of science, technology and art is another challenge that needs to be anticipated as well as possible [1]. These influences and challenges need to be addressed wisely and intelligently, to prevent inequality in the life of the nation and state in the future. This implies the importance of improving the quality of human resources in various fields including the world of education and the human resources. Various strategies which are improving the quality of education are directed at improving the quality of teachers in mastering science, technology, and foreign languages. To be able to achieve, the above can be done with assistance and training. This agrees with the opinion of Listyarti as Secretary General of Federation of Indonesian Teachers' Unions (FSGI) who stated "The low quality of teachers is not the fault of the teachers". This is because the capacity of educators is not built through training [2]. Likewise, the trainings that are often held for elementary school teachers in East Java are expected to make a positive contribution to teacher professionalism and need full support, follow-up and monitoring after training from the government or educational institutions which in this case are universities. The results of the field survey show that the teacher training system that has been held has not shown optimal results regarding its performance [3].

Empirically, it can be seen that the success of quality education development is influenced by the availability of various supporting components. One of them is teachers who have competences in their field of work. Therefore, it is necessary to have a teacher training system that focuses on improving teacher performance in order to support teacher professionalism. This will also have a positive impact on student achievement [4]. One of the ways to improve the quality of education related to teacher professionalism is by holding teacher training in assisting the activities of the National Science Olympic (OSN). So far, science learning is taught by classroom teachers who has unsuitable teacher education background, so that the teacher did not implement science learning accordingly with the nature or essence of science. Science learning should be oriented to the nature of science, namely as learning in the form of processes and products [5]. This means that science learning is not only enough to be carried out by delivering information about science concepts and principles but also must be balanced with real activities in the field. Students must understand the process of the occurrence of science phenomena through sensing as much as possible when they are learning science. This means that when the students are learning science, they must actively observe, try, and discuss with fellow students and teachers. The concept of science learning like this can only be implemented by teachers who are really understand the characteristics of science and its learning strategies [6].

The integration of biology, physics, and chemistry courses in science or science learning in elementary schools does not have an impact of a well-established mastery of science material with a broad range of scientific studies. By giving the incompatibility of the science teachers' qualifications with the quality of their fields of expertise, the mastery of teachers in science learning must be improved, so that teachers have more skills in learning management [7]. One alternative that is considered important in this regard is through an academic refresher (refreshing program) whose core activities include training in designing and implementing science or science practicums. It is hoped that through the implementation of this program, the teacher will gain something new and it can be used as a teacher's foothold in developing the profession which will also directly increase the productivity of their performances [8].

The urgent problems are faced by not only the teachers of SDN I Airlangga Surabaya and its surrounding s whose tenure is less than three years, but also the teachers who have already experienced in the performance period of seven years. The results of field observations related to OSN, especially for science Olympic at SDN I Airlangga Surabaya are still dominated by explanations of theoretical OSN science concepts. Teachers often complain about the classic problem such as the lack of time and facilities to implement OSN, especially science by applying the OSN Science strategy, either theory or practicum and experiment, which is the demand for the current Olympic curriculum. Likewise, according to the questionnaire distributed by community service Universitas Terbuka Surabaya team, teachers in carrying out science or its practicum activities in OSN are still low in SDN I Airlangga Surabaya and its surroundings. It shows that most of the teachers (87%) has never carried out practicum/experimental activities as a support of OSN Science activities at school. Realizing the importance of the above questions, it is seemed necessary for us to carry out community service as a tri dharma activity for higher education. We carried out this activity to related issues to increase the insight and ability of teachers in the field of science practicum, especially at the elementary school level since it is very feasible to be used as one of the focuses of community service activities at SDN I Airlangga Surabaya.

2. Method

Community service program was conducted for three meetings with activities in the form of training for teachers of SDN I Airlangga to get debriefing and understanding for training OSN to students, both theoretically and practically. It was attended by 20 participants from SDN I Airlangga. In the first stage, community service Universitas Terbuka Surabaya team conducted a survey at SDN I Airlangga Surabaya. The team conducted interviews with the school regarding the problem of teachers' development as OSN supervisors. In addition, we also offered an OSN debriefing in the form of training for teachers as OSN supervisors. In the second stage, the OSN pre-training activities were carried out, especially for science concepts and science practicums to support teachers' understanding in making OSN questions for their schools. Based on the above research, the problem-solving framework that offered as follows:

2.1. The preparation stage

At the preparation stage, a study of the science syllabus (OSN) material is carried out annually, which the provision of knowledge of OSN material both theoretically and practically is in accordance with the form, model, and characteristics of OSN. The results at this stage will provide an overview of the basic concepts and practicum of the National Science Olympiad (OSN) for elementary students. At this stage, the training participants are also able to learn examples of OSN activities in conceptually and practically as well as their scientific analysis, given by some examples of practice-based Olympic questions to provide a clear picture of OSN that has been running so far. Before the implementation stage, the participants were asked to identify OSN problems in the science practicum that had been carried out, which was done to obtain the Term of Reference (TOR). The purpose of this stage is to inform the training participants about the OSN materials and practices that will be implemented.

2.2. Implementation Stage

At the implementation stage, based on science or science practicum, an OSN implementation assistance is carried out to provide provisions for training participants in implementing OSN guidance in their respective schools [9]. Assistance for training participants will also be carried out until the analysis and completion of the practicum-based science or science Olympic are carried out, so that the training participants are expected to be able to master the concepts, types, and models of OSN natural science well.

2.3. Evaluation stage

At the end of evaluation stage, this activity is carried out by monitoring and evaluating sustainability after mentoring [10]. The results of the performance at this stage will monitor the extent of the teacher's understanding of the material that has been delivered earlier to mentoring OSN training participants both conceptually and practically about natural science based on OSN. Thus, at the end of this stage, the results of the enrichment of OSN material can be analysed either in an efficient or in a practical manner.

From the stages carried out above, the methods and the strategies used in the implementation of this activity are in the form of knowledge information, question and answer session, discussion, training, and practice as well as mentoring by using the evaluation procedures and tools to assess the success of the training participants in the form of tests. Giving tests to training participants is intended to determine the extent to which the success rate of this activity is carried out. The data were collected by testing the participant's ability to analyze questions and to practice science in supporting OSN primary school activities rather than the early activity (pre-test) and the lately activity (post-test).

3. Results and Discussion

The results of the service activities consist of: (1) the results of team discussions related to primary school science practicum materials to support OSN activities, and (2) pre-tests related to practicum materials and group discussions about practicum reports. Based on the discussion that has been carried out by community service Universitas Terbuka Surabaya team, it was decided that the primary school science practicum material is determining the optical properties of the reflection system and the results of the final image performance, magnetic system, and irregular motion of straight objects. Furthermore, the three materials are summarized into one document in the form of theoretical and practical questions to be implemented, then community service Universitas Terbuka Surabaya team members analysed the document about the practicum. The analysis is focused on the content and writing. This analysis activity is expected to produce a practical question document that is easily understood by the participants. To find out the participants' understanding regarding theories, concepts, or other supporting information, a pre-test was carried out. Questions are focused on exploring the knowledge of participants who will carry out the three practical activities mentioned above as shown in Table 1.

Table 1. Percentage of Knowledge Level of Elementary Teacher's Training Participants Using Science Practicum Materials Based on OSN.

Understanding Trainees	Percentage %			Average
	Optics	Magnetic Electricity	Motion	
Concepts of natural science (OSN)	71.5	82.3	71.2	75.0
Derivation of mathematical equations	73.7	83.3	70.5	75.8
The calculated value	77.3	82.1	73.5	77.6
Practicum implementations	70.0	85.0	73.5	76.2
Practicum data analysis	72.6	80.7	71.5	76.8
Conclude the result of practicum	74.8	81.7	73.9	76.8
Report the results of practicum	72.6	79.5	73.3	75.1

Table 2. Recapitulation of the Implementation Value of Elementary Teacher's Training Participants Science Practicum to Support OSN Activities

Aspects	Sub Aspects	Team			Score Average	Category
		I	II	III		
Science practicum preparation	Practicum Purpose	4	4	3	3.67	Very good
	Determination of tools and material	4	3	3	3.33	Good
	Practicum Step	3	3	4	3.33	Good
	Observation data	4	4	4	4.00	Very good
Natural science practicum implementation	Results of compiling practicum	3	4	4	3.67	Very good
	Operate Practicum	3	4	3	3.33	Good
	Analysing Practicum data	3	4	4	3.67	Very good
	Practicum data retrieval	3	3	3	3.00	Pretty good
	Concluding the results of the practicum	4	4	4	4.00	Very good
Submission of practicum reports	Practicum report results	3	4	4	3.67	Very good
	Submission of practicum information	4	4	4	4.00	Very good
	Report supporting data	3	3	3	3.00	Pretty good

From Table 1, it can be explained that generally the participants who are teachers of SDN I Airlangga Surabaya have been able to explain the concepts, theories, and mathematical equations related to the quantities to be measured and calculated during the natural science practicum. In addition, teacher's practicum activity is presented in question. However, most of the participants had not seen on how to process the data obtained to answer the problems and the objectives of the practicum. Given the number of participants in this activity as many as 20 teachers who become OSN guidance teachers and the number of practicums that can be done and the time availability, this activity is made into 3 groups with 6 to 7 members each. This activity is hoped that the natural science practicum activities carried out by participants as OSN supervisors can take place as planned and there will be an exchange of experiences on findings or new things during the science practicum. Therefore, the activities carried out which were observed from each group when doing practicum and reports on each group from the implementation of the practicum to reporting the results that have been obtained can be seen in Table 2.

From the results of the observations above, they are shown each aspect of the group assessment in Table 2. In the aspect of practicum preparation, the objectives and analysis of temporary observation data are in the very good category where all the groups got a score of 4 when it was only group 3 got a score of 3. Likewise, for the determination of the material tools and practicum steps to be carried out from the three groups got a score of 3.33 which is categorized good. To arrange tools in the implementation of practicum, the three groups could assemble practical tools with good categories. This means that the three groups can carry out practicum as determined by the service team. When data collection was conducted, observations of groups 2 and 3 show that the category is very good and only one group which got the good category. In other words, all groups are in good categories. Each group member contributed to data collection. The data obtained can be justified. From the results of the observational data analysed by the three groups show a score of 3 which are included as the good category. The aspect of submitting practicum reports on the sub-aspects of the results in delivering practical information with very good categories are for all groups. From the observations of community service Universitas Terbuka Surabaya team, the final activity of the practicum was carried out for each group, where the group that did the practicum with the same type delivered the results of their research in turn and was continued by other groups using the same type of practicum. The findings are based on the aspects that were assessed during group discussions for each of the following causal practices. During the group discussion, the two groups showed different abilities. However, there are several things that need to be underlined that the lack of other supporting data related to reference sources and incomplete data analysis is incomplete which is categorized as fairly good. At the end of this activity, community service Universitas Terbuka Surabaya team provided this feedback regarding the practicum activity to support OSN in schools. This feedback is needed as an evaluation material for activities in the coming year. In general, participants felt the benefits of this activity for classroom learning. Several participants suggested that this activity was not limited to theory and practicum tested in OSN but for other subjects, such as the Biology and Mathematics Olympiad as well as Social Sciences. In addition, the implementation day should not be on Sunday and the time is somewhat extended. This kind of activity should be held periodically and scheduled by moving around with the school coordinators.

4. Conclusion

From the activities that have been carried out, the understanding of the participant's activity for the science practicum material tested by OSN such as optics, magnetism and motion image of objects, magnetism-electricity system, and motion of straight objects that are irregular can be concluded. For the formulation of how the reflection system practicum and the shadow results from the practicum can be carried out, 70% could explain well, and only 15% of the participants could not explain the stages of the magnetism and electricity system practicum occurence. This was due to the fact the natural science practicum material for OSN for some of the participants who were teachers of SDN I Airlangga Surabaya was new in this case. Participants are able to compile reports and discuss the results in the class discussions. The practicum report format is in accordance with the standard form that has been determined by the service team. The presentation of practicum results by each group went quite well. In the discussion of this material, the natural sciences practicum for OSN participants describe the experiences gained during the training.

Acknowledgement
The author would like to thank the Distance Learning Program Unit – Universitas Terbuka Surabaya for the support of service facilities so that this service can be carried out well.

References

[1] Wijaya S, Arifin S, and Murnaka N P 2019 *Int. J. Sci. Technol. Res.* **8** 542
[2] Wirta M, Pujani N M, Suma K, Santyasa W and Rapi N K 1990 *Prestasi Belajar IPA Siswa Kelas VI SD Negeri se Kabupaten Buleleng tahun Ajaran 1990/1991 Sebagai Fungsi Kualitas Reinforcement dan Kualitas Guru* (Denpasar: Universitas Udayana)
[3] Asnawi 2013 *Pendampingan Eksperimen Fisika Modern Dalam Upaya Peningkatan Kompetensi Dan Keterampilan Proses Sains Guru Fisika SMA di Cluster III Kabupaten Sidoarjo* (Surabaya: Universitas Negeri Surabaya)
[4] Bimayu W, Kristiawan M, and Fitriani Y 2020 *Int. J. Progres. Sci. Technol.* **20** 6
[5] Hammond L D 2020 *Appl. Dev. Sci.* **24** 97
[6] Sutama, Narimo S, Prayitno H J, and Anif S 2021 *J. Phys. Conf. Ser.* **1836** 012046
[7] Changwong K, Sukkamart, and Sisan B 2018 *J. Int. Stud.* **11** 37
[8] Soesilowati E S 2009 *J. Indones. Soc. Sci. Humanit.* **2** 69
[9] Prastika L R, Setiawan R, Putri A A, and Triyanta 2018 *AIP Conf. Proc.* **2021** 080011
[10] Suryanto A and Djatmiko T 2016 *Evaluasi Pembelajaran di SD* (Tangerang Selatan: Universitas Terbuka)

Critical Thinking Skills on Physics Learning during COVID-19 Pandemic: A Bibliometric Analysis using VOSViewer

B Jatmiko[1], T Sunarti[1], B K Prahani[1*], E Hariyono[1], Dwikoranto[1], F C Wibowo[2], S Mahtari[3], Misbah[3], and M Asy'ari[4]

[1] Physics Education, Universitas Negeri Surabaya, Indonesia
[2] Physics Education, Universitas Negeri Jakarta, Indonesia
[3] Physics Education, Universitas Lambung Mangkurat, Indonesia
[4] Physics Education, Universitas Pendidikan Mandalika, Indonesia

*E-mail: binarprahani@unesa.ac.id

Abstract. Critical thinking skills (CTS) are still one of the research trends in this decade, including during the COVID-19 pandemic. Research will focus on generating a profile of CTS on physics learning during COVID-19 pandemic based on bibliometric analysis using VOSViewer. The findings indicated there were some parameters or interrelationships among variables to capture current and future trend of CTS on physics learning, such as analysis, implementation, effectiveness, effect, development, and technology. The research implications are supported by recent research: (1) These findings will provide an empirical basis for the development of physics education research, especially online physics learning during the COVID-19 pandemic to practice CTS; (2) CTS can still be a research trend in the latest research; (3) Digitization of physics learning is becoming a trend to be carried out gradually and continuously.

1. Introduction

The COVID-19 pandemic has still continued to be a special concern and has not been resolved. COVID-19 pandemic has had a major impact on important sectors around the world, include in education [1-10]. Based on data from worldometers.info, the global COVID-19 update until September 11, 2021, shows that the total number of COVID-19 cases currently reaches 224,599,517. 4,629,624 have died and 201,119,167 have recovered. There are still 18,850,726 active cases of COVID-19 spread across various countries. Meanwhile, the cumulative total of COVID-19 cases in Indonesia has reached 4,163,732 since the corona virus was detected in March 2020 last year. Meanwhile, recovered cases in Indonesia increased beyond the positive cases of 7,586 people. There have been 3,909,352 people who have recovered from COVID-19. Meanwhile, an additional 270 people died for a total of 138,701 people. The government continues to intensify COVID-19 vaccinations to achieve communal immunity.

The COVID-19 pandemic has had an impact on the education sector. This is shown by the results of the latest research [11-20] that education at various levels is affected by COVID-19. All levels of the education sector must adapt to deal with the COVID-19 pandemic, including still having to optimize learning to improve critical thinking skills (CTS). CTS are higher-order thinking processes that facilitate operational activities in internalizing information in analyzing and evaluating for success in life problem solving [21-30]. Recent research results [31-33] have shown critical thinking skills are still one of the research trends in physics learning.

However, in latest research there has been no research that specifically profiles research trends in online physics learning during the COVID-19 pandemic. Therefore, research will focus on generating a profile of CTS on physics learning based on bibliometric analysis using VOSViewer. Research using bibliometric was chosen because it has proven to be effective to see an overview and profile of current and future research trends. The main implication of this research is to contribute to the physics learning, in particular the profile of CTS on physics learning during COVID-19 pandemic.

2. Method

Bibliometric is the main method from this research [34-43]. Scopus document data collected in 11 September 2021. The 272 document results with keywords (TITLE-ABS-KEY (critical AND thinking AND skills) AND TITLE-ABS-KEY (physics AND learning)), then focused on 2020 and 2021 as many as 99 document results of which these 2 years are the COVID-19 pandemic period using (TITLE-ABS-KEY (critical AND thinking AND skills) AND TITLE-ABS-KEY (physics AND learning)) AND (LIMIT-TO (PUBYEAR, 2021) OR LIMIT-TO (PUBYEAR, 2020)). The VOSViewer software [43-51] was used to support and figure out the research trend of CTS on physics learning during COVID-19 pandemic.

3. Results and Discussion

Based on the results of the analysis using VOSViewer from 99 document results (Scopus), five top occurrences were obtained, including: student (90), study (71), critical thinking skill (71), research (60), physics (56), skill (48). Then the researchers visualized the research trends of CTS assisted with VOSViewer software (see Figure 1). This effort is useful for finding the novelty of the research in this domain.

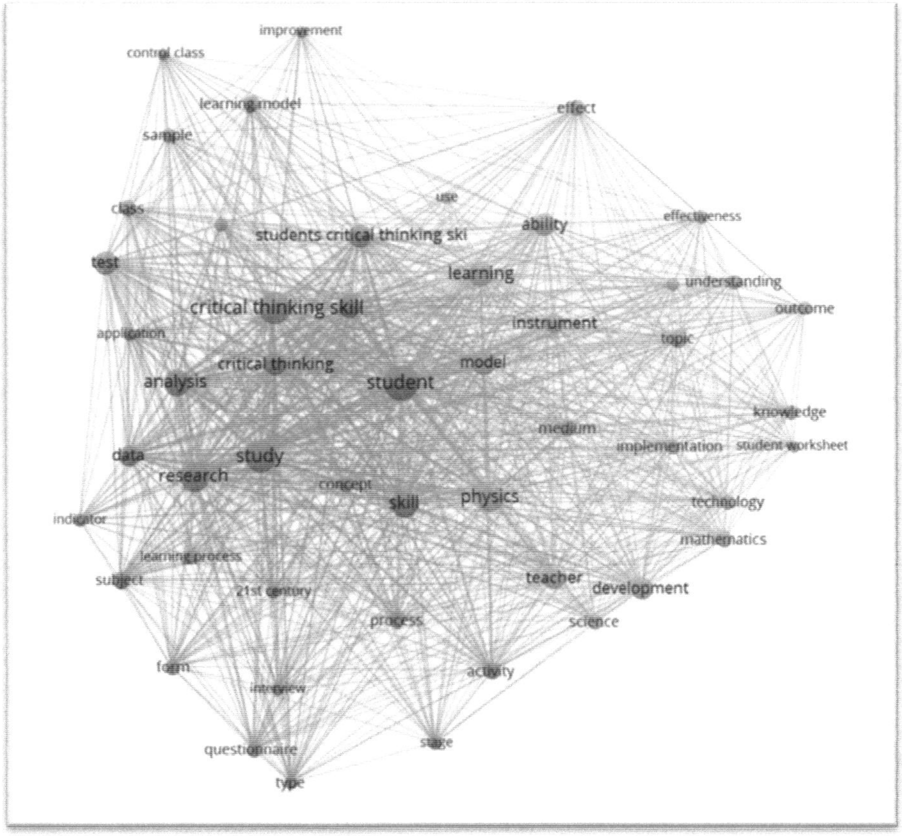

Figure 1. Network visual of CTS on physics learning during the COVID-19 pandemic

Figure 1 indicates the whole picture research of CTS on physics learning during the COVID-19 pandemic. Researchers on the world produced three main clusters (red, green, blue) and one minor cluster (yellow). The first cluster (red colour) was as a research study, analysis, learning process, indicator, form, interview, questionnaire, process, activity, type, and application. The second cluster (green) was focus on physics, model, teacher, science, mathematics, effectiveness, understanding, knowledge, ability, effect, development, instrument, topic, learning, outcome, and implementation. The third cluster (blue) was focus on student critical thinking, learning models, sample, class, test, and improvement.

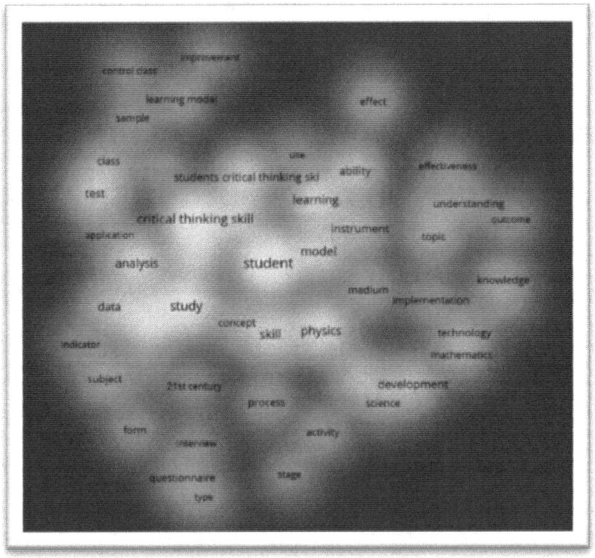

Figure 2. Density visual of CTS on physics learning during the COVID-19 pandemic

Figure 2 shows the density visual of CTS on physics learning during the COVID-19 pandemic. The yellow dominance is based on critical thinking skills, student, study, physics, analysis, test, model, learning, ability, skill, instrumnet, and development. Based on occurrence, the findings indicated there were some parameters or interrelationships among variables to capture the trend and novelty in Figure 3, 4 and 5.

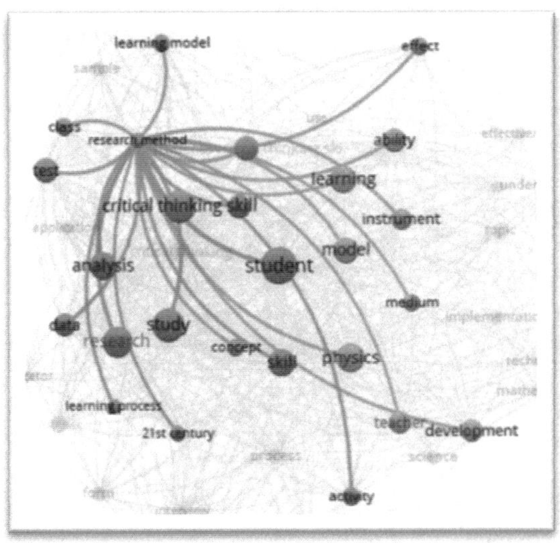

Figure 3. Research method trend on CTS on physics learning during the COVID-19 pandemic

Figure 3 shows the research method tren focus on research, study, data, analysis, teacher, development, medium, model, instrument, learning, 21st century, learning process, concept, and learning model. These results can be used by other researchers in adapting the novelty of the research to be carried out. For research trends related to collected data, it can be seen in Figure 4.

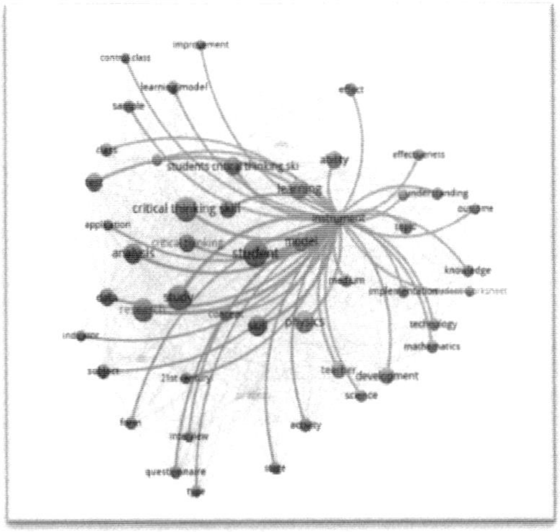

(4a) Instrument

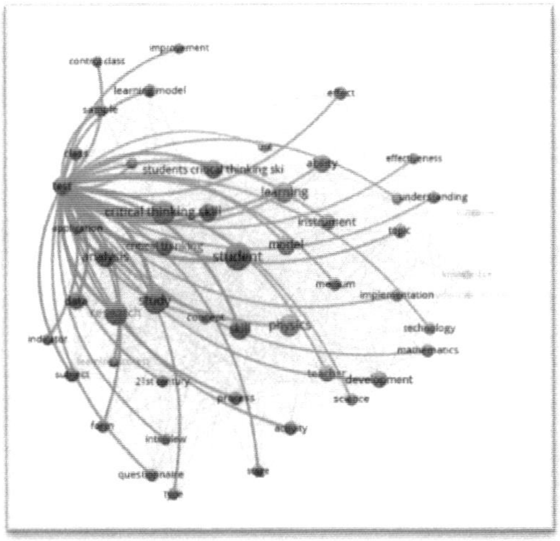

(4b) Test

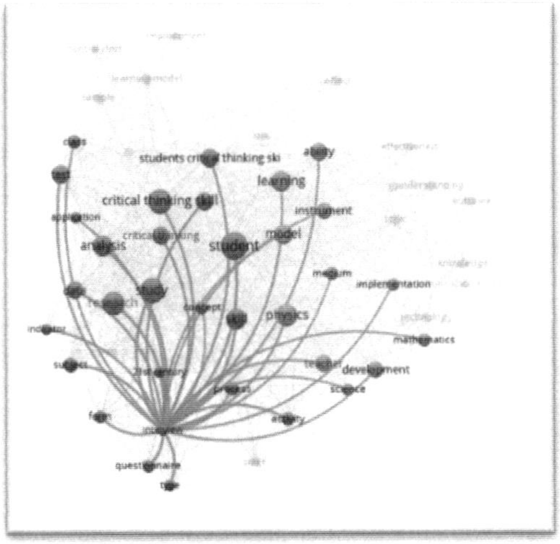

(4c) Interview

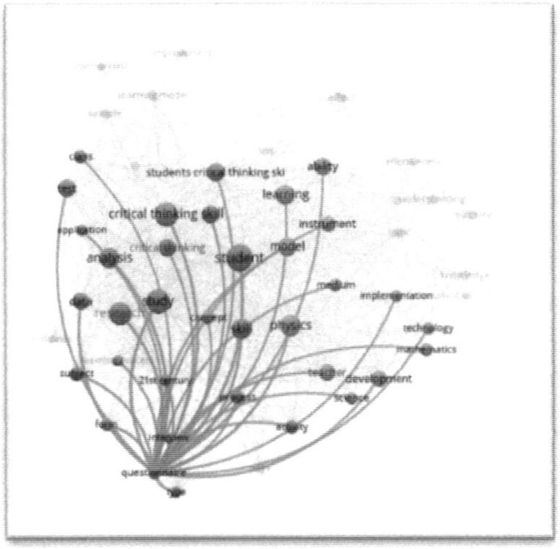

(4d) Questionnaire

Figure 4. Trends related to collected data CTS on physics learning during the COVID-19 pandemic

Figure 4 shows trends related to collected data CTS on physics learning during the COVID-19 pandemic such as in instrument, test, interview, and questionnaire. In addition to critical thinking skills, data collection is also coupled with other skills and learning outcomes such as mastery of concepts, abilities, including the learning process. More broadly, data collection interventions can also be extended to science and mathematics. This has the potential to be done related to adaptation for the undergraduate program level, modification for the master's postgraduate level, and innovation at the doctoral program level. For research trends related to current and future trend of CTS on physics learning during the COVID-19 pandemic, it can be seen in Figure 5.

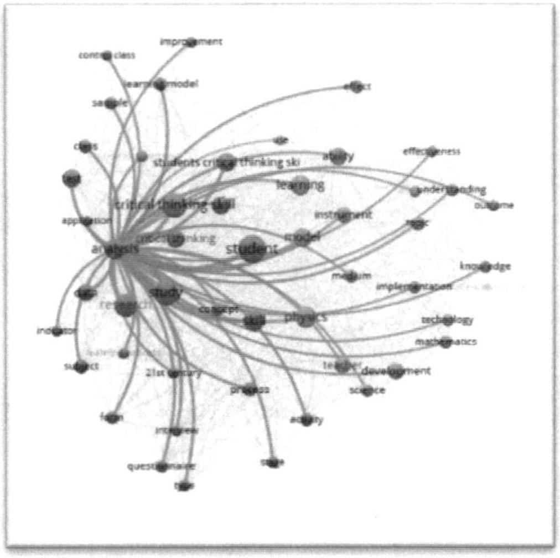

(5a) Analysis

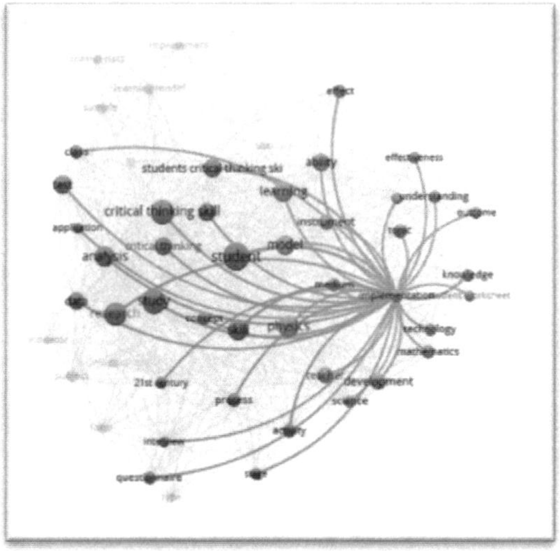

(5b) Implementation

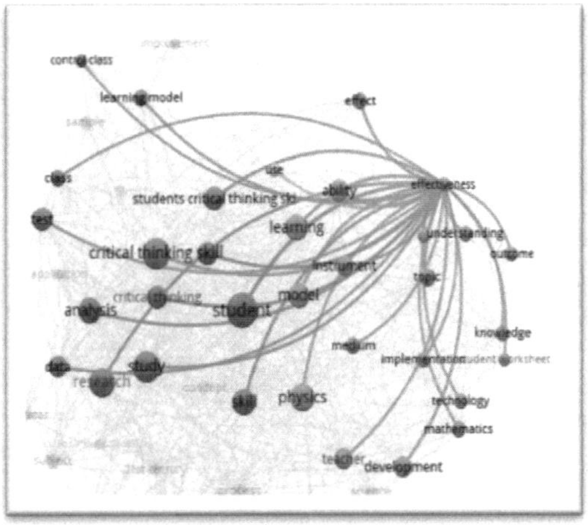

(5c) Effectiveness

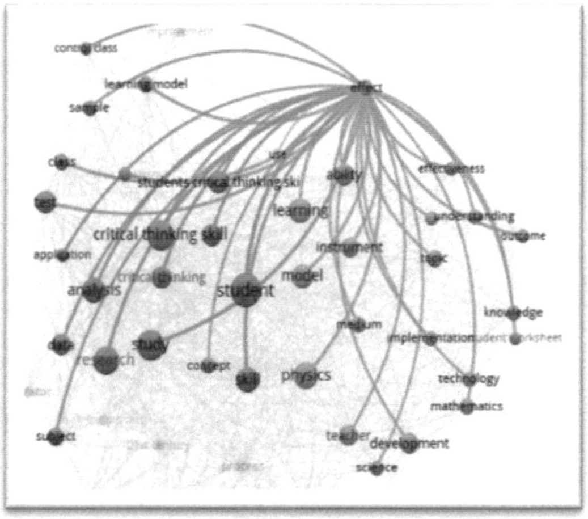

(5d) Effect

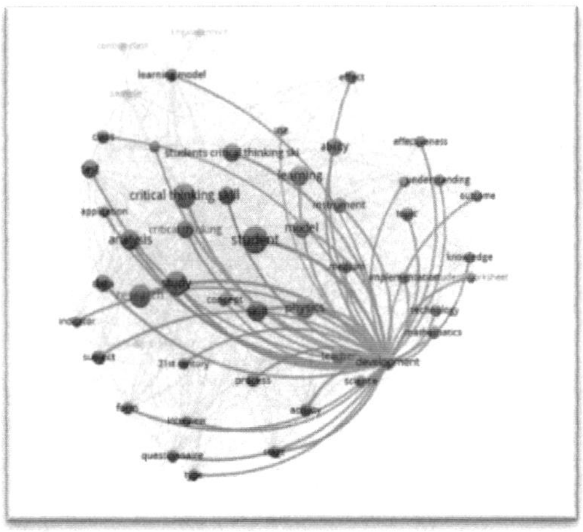

(5e) Development

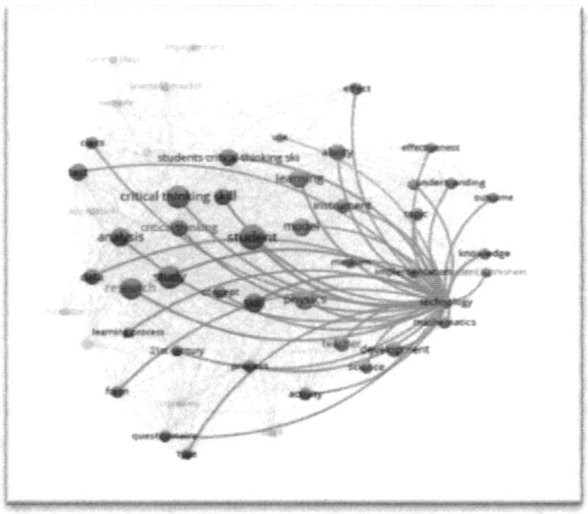

(5f) Technology

Figure 5. Current and future trend of CTS on physics learning during the COVID-19 pandemic

Figure 5 shows the findings indicated there were some parameters or interrelationships among variables to capture current and future trend of CTS on physics learning during the COVID-19 pandemic, such as Analysis (Figure 5a), Implementation (Figure 5b), Effectiveness (Figure 5c), Effect (Figure 5d), Development (Figure 5e), and Technology (Figure 5f) on physics learning during the COVID-19 pandemic. Research trends as of August 14, 2021 (Scopus Database Documents) related to experiments (6,845,363), online learning (123,697), physics teacher (16,565), and physics course (13,486). Next findings point to the top keywords trends as of September 11, 2021 (Scopus Database) connecting to CTS on physics learning during the COVID-19 pandemic, such as: Critical thinking skills (63), students (63), critical thinking (38), physics (29), learning systems (19), physics learning (18), education computing (14), surveys (11), creative thinking (9), learning process (9). Other findings point to the top source title trends as of September 11, 2021 (Scopus database) connecting to CTS on physics learning during the COVID-19 pandemic, such as: Journal of Physics Conference Series (70), IOP Conference Series Earth and Environmental Science (9), AIP Conference Proceedings (4), Universal Journal of Educational Research (4), Periodico Tche Quimica (2), Asian Journal of University Education (1), Electronic Journal of E Learning (1), International Journal of Advanced Computer Science and Applications (1), International Journal of Advanced Science and Technology (1), and International Journal of Scientific and Technology Research (1). The research implications are supported by recent research: (1) These findings will be empirical for improving the quality of physics learning research, especially on physics learning during the COVID-19 pandemic to practice CTS; (2) CTS on physics learning can still be a research trend in the latest research [52-60] and in future; (3) Technology and Digitization of CTS on physics learning is becoming a trend to be carried out gradually and continuously.

4. Conclusion

The findings indicated there were some parameters or interrelationships among variables to capture current and future trend of CTS on physics learning during the COVID-19 pandemic, such as analysis, implementation, effectiveness, effect, development, and technology on physics learning during the COVID-19 pandemic. Fundamental implication is Technology and Digitization of CTS on physics learning become a trend to be carried out gradually and continuously. The research limitation is that it only involves Scopus documents. Further research can be done by involving document data from the Web of Science (WoS), Google Scholar (GS), and DOAJ.

5. Acknowledgment

Thanks for the financial supporting and conducting research in 2021 from the LPPM, Universitas Negeri Surabaya; Number: 855/UN38/HK/PM/2021 dated June 24, 2021.

6. References

[1] Liang X, Li C, Kong, H, and Zhao, Z. 2021 *J. Phys. Conf. Ser.* **1992** 042074
[2] Haryani F, and Ayuningtyas, N. 2021 *J. Phys. Conf. Ser.* **1957** 012006
[3] Yunita H 2021 *J. Phys. Conf. Ser.* **1957** 012038
[4] Payadnya I P A A, and Wibawa K A 2021 *J. Phys. Conf. Ser.* **1957** 012012
[5] Karlimah K, Hamdu G, Pratiwi V, Herdiansah H, Kurniawan D 2021 *J. Phys. Conf. Ser.* **1987** 012026
[6] Putri H E, Sasqia A S, Abdulloh A, Muqodas I, and Majid N W A 2021 *J. Phys. Conf. Ser.* **1987** 012025
[7] Cheng Y, Liu J, and Sriboonchitta S 2021 *J. Phys. Conf. Ser.* **1978** 012043
[8] Taufiq M, Dewi N R , Savitri E N, Parmin, and Darmawan M S 2021 *J. Phys. Conf. Ser.* **1968** 012021
[9] Maheswari R, Ganesan R, and Venusamy K 2021 *J. Phys. Conf. Ser.* **1964** 062112
[10] Hakim M A R, Serasi R, Efrizal D, and Kurniawan D 2021 *J. Phys. Conf. Ser.* **1933** 012082
[11] Avila E C, Abin G J, Bien G A, Acasamoso D M, and Arenque, D D 2021 *J. Phys. Conf. Ser.* **1933** 012130
[12] Oktafiani R, Widiatningrum T, and Retnoningsih A 2021 *J. Phys. Conf. Ser.* **1918** 052026

[13] Rudyatmi E, Iswari R S, Rusilowati A, and Alimah S 2021 *J. Phys. Conf. Ser.* **1918** 052093
[14] Wiradinata R, Jaja, Rosita C D, and Amalia S 2021 *J. Phys. Conf. Ser.* **1918** 042096
[15] Suyitno A, Suyitno H, and Sugiharti E 2021 *J. Phys. Conf. Ser.* **1918** 042083
[16] Puspa R D, As'ari A R, and Sukoriyanto 2021 *J. Phys. Conf. Ser.* **1918** 042078
[17] Hidayah I, Tatirah, Latiana L, and Widowati 2021 *J. Phys. Conf. Ser.* **1918** 042156
[18] Sulistyo L, and Kartono 2021 *J. Phys. Conf. Ser.* **1918** 042052
[19] Fiantika F R, Sugianto F A, Jannah H N, Junior F A, and Perdani C W 2021 *J. Phys. Conf. Ser.* **1918** 042094
[20] Sholikhah B U, and Cahyono A N 2021 *J. Phys. Conf. Ser.* **1918** 042063
[21] Bağ H K, and Gürsoy E 2021 *Think. Ski. Creat.* **41** 100910
[22] Dykhne M, Hsu S Y, McBane S, Rosenberg E, and Taheri R 2021 *Curr. Pharm. Teach. Learn.* **13** 659
[23] Hart C, Da Costa C, D'Souza D, Kimpton A, and Ljbusic J 2021 *Think. Skills Creat.* **41** 100877
[24] Hursen C 2021 Technol. Knowl. Learn. **26** 515
[25] Jatmiko B, Prahani B K, Munasir, Supardi Z A I, Wicaksono I, Erlina N, Pandiangan P, Althaf R, and Zainuddin 2018 *J. Baltic Sci. Educ.* **17** 1.
[26] Patonah S, Sajidan, Cari and Rahardjo S B 2021 *Int. J. Inst.* **14** 39
[27] Subekti M A S, and Prahmana R C I 2021 *Math. Educ. Res. J.* **13** 137
[28] Supratman, Zubaidah S, Corebima A D, and Ibrohim 2021 *Int. J. Inst.* **14** 187
[29] Taylor T Z, Prescott R, and Harrup K, Lee C 2021 *J. Transp. Secur.* **14** 107
[30] Varenina L, Vecherinina E, Shchedrina E, Valiev I, and Islamov A 2021 *Think. Ski. Creat.* **41** 100906
[31] Dwijananti P, Ruwaida A, Mindyarto B N, and Wiyanto 2021 *J. Phys. Conf. Ser.* **1918** 052076
[32] Marfu'i L N R, Ilfiandra, and Nurhudaya 2019 *J. Phys. Conf. Ser.* **1280** 052012
[33] Ferty Z N, Wilujeng I, Jumadi, and Kuswanto H 2019 *J. Phys. Conf. Ser.* **1233** 012062
[34] Ferreira J J, Fernandes C I, Schiavone F, and Mahto R V 2021 *Technol. Forecast. Soc. Change* **173** 121077
[35] Aalirezaei A, and Kabir G 2021 *Innov. Infrastruct. Solut.* **6** 230
[36] Ojeda-Pereira I, and Campos-Medina F 2021 *Resour. Policy* **74** 102272
[37] Yurtcu M, and Güzeller C O 2021 *Participatory Educ. Res.* **8** 426
[38] Marcal J, Bishop T, Hofman J, and Shen J 2021 *Chemosphere* **284** 131267
[39] Zhang J Z, Srivastava P R, Sharma D, and Eachempati P 2021 *Expert Syst. Appl.* **184** 115561
[40] Yin J, Wan J, Zhu J, Pan Y, and Zhou H 2021 *Chinese Med.* **16** 53
[41] Morin C, and Gaboury I 2021 *BMC Complement. Med. Ther.* **21** 196
[42] Milán-García J, Caparrós-Martínez J L, and Rueda-López N, de Pablo Valenciano J 2021 Glob. Health **17** 74
[43] Marín-Marín J -A, Moreno-Guerrero A -J, Dúo-Terrón P, and López-Belmonte J 2021 *Int. J. STEM Educ.* **8** 41
[44] Rahmatika H, Lestari S R, and Sari M S 2021 *AIP Conf. Proc.* **2330** 030046
[45] Yennita Y, and Zukmadini A Y 2021 *J. Phys. Conf. Ser.* **1731** 012007
[46] Medriati R, Hamdani D, and Harjilah N 2021 *J. Phys. Conf. Ser.* **1731** 012074
[47] Sari Nurza, U M, Suyanti, R D, and Dewi R 2021 *J. Phys. Conf. Ser.* **1811** 012097
[48] Ratna S D, Simanjuntak M P, and Sahyar 2021 *J. Phys. Conf. Ser.* **1811** 012004
[49] Sari Simatupang, T I, Siregar, N, and Sinulingga K. 2021 *J. Phys. Conf. Ser.* **1811** 012131
[50] Kurniawan, R., Syafriani 2021 *J. Phys. Conf. Ser.* **1876** 012067
[51] Sutiani A, Situmorang M, and Silalahi A 2021 *Int. J. Instr.* **14** 117
[52] Waluyo A A, Hartono, and Sulhadi 2021 *J. Phys. Conf. Ser.* **1918** 022011
[53] Dewanthikumala, Jasruddin, and Abdullah H 2021 *J. Phys. Conf. Ser.* **1805** 012004
[54] Firmansyah J, and Suhandi A 2021 *J. Phys. Conf. Ser.* **1806** 012047
[55] Sa'Diyah L H, Siahaan P, Samsudin A, Riani V R, and Fatima, W O 2021 *J. Phys. Conf. Ser.* **1806** 012032
[56] Styawan A, and Arty I S 2021 *J. Phys. Conf. Ser.* **1806** 012177
[57] Febriana R, and Sinaga P 2021 *J. Phys. Conf. Ser.* **1806** 012012

[58] Medriati R, Hamdani D, and Harjilah N 2021 *J. Phys. Conf. Ser.* **1731** 012074
[59] Hanicza Y, Putri D H, and Hamdani D 2021 *J. Phys. Conf. Ser.* **1731** 012069
[60] Yusal Y, Suhandi A, Setiawan W, and Kaniawati I 2021 *J. Phys. Conf. Ser.* **1731** 012075

The Effect of PjBLL Online Platform on Student Collaboration Skills and Basic Science Process Skills During the Covid-19 Pandemic

Dwikoranto[1]*, R Setiani[2], and Widiasih[3]
[1]Physics Education, Universitas Negeri Surabaya, Indonesia
[2]Universitas Bhinneka PGRI, Indonesia
[3]Physics Department, Faculty of Teacher and Training, Universitas Terbuka, Indonesia

*Email: dwikoranto@unesa.ac.id

Abstract: The purpose of this research to describe the effect of Project Based Laboratory Learning (PjBLL) on collaboration skills and basic science process skills of students. This type of research is quasi-experimental with one group pretest-posttest design. The subjects of this study were 60 first year students from two different classes who were enrolled in the Basic Physics course at the Department of Chemistry, Universitas Negeri Surabaya Indonesia. The collaboration ability test on kinematics and particle dynamics was used to measure students' collaboration skills, the basic science process skills test was used to measure students' science process skills as a pre-test and post-test. The analysis of the results showed that students' collaboration skills were complete on all indicators with gain scores in the high and sensitive categories, while basic science process skills were completed on all indicators with gain scores in the medium and high criteria. PjBLL Online platform can be used for lectures during the COVID-19 pandemic.

1. Introduction

Project Based Laboratory Learning (PjBLL) as one of the innovative learning models, accommodates the needs of 21st century skills and basic literacy and enters the industrial revolution era 4.0 [1] towards a society era 5.0 which must be able to solve various challenges and social problems by take advantage of various innovations. Innovation in terms of internet on things, artificial intelligence, robots to improve the quality of human life, and big data. Society 5.0 can also be interpreted as a concept of a human-centered and technology-based society in solving various problems to improve the quality of human life. The 21st century education era considers every student to be a learning community, learning is obtained from various sources, not only from books, but also from the internet, various technology and information platforms and global curriculum developments.

Internet on things and mobile phones to improve the quality of human life have changed habits and brought a big influence in student life, changing the mindset, personality, and behavior of students [2]. The negative effect causes introvert, antisocial and difficult nature to join the real world [2]. The influence of these behaviors will have an impact later when they are in matters related to work, where students are required to interact with other people competently and respect each other [3,4]. To deal with these problems is to foster collaboration skills in students. The importance of having these skills so that students are able to socialize, are sensitive to the surrounding environment, and control ego and emotions [5,6].

Collaboration skills as the ability to participate in every activity to build relationships with other people, respect each other's relationships and teamwork to achieve the same goal play an important role in the industrial revolution era 4.0 and society era 5.0 [7,8]. The indicators of collaboration skills in this study are (1) actively contributing, (2) working productively, (3) showing flexibility and compromise, (4) showing responsibility, and (5) showing respect. In order for students to have these abilities, they can be trained through lectures by providing challenging problems that exist in real life [3-9].

In physics courses on college level, these skills are important to give students supporting information from observations and experiences. Students also can develop their skills [10]. When taught using constructivist strategies physics learning becomes more effective combined with development of process skills and conceptual understanding will facilitate students [11,12]. Some education experts state that its can be classified into integrated process skills and basic process skills [13-16]. The two process skills, namely integrated science process and basic skills, should taught and trained for students [7].

For students from the Department of Chemistry the basic ability aspects used in this study are: (1) observing real objects or events; (2) classifying, there is an opportunity to look for or find similarities, differences or certain criteria are given to do the grouping or determine the number of groups that must be formed; (3) communicating there is a certain form of statement to be changed to another form of presentation, for example the form of a description to a chart form, or a table to a graphic form; (4) inferring draw conclusions based on data, tables obtained; (5) predicting must have clear patterns or tendencies to be able to make predictions or predictions [13-16]. These basic ability are not hands-on skills using tools but process thinking used scientific processes so that the use of the Online platform during the COVID-19 pandemic is suitable to be applied in lectures [17].

The action given to first-year students whose effect will be analyzed on collaboration skills and basic science process skills is in the form of implementing Project Based Laboratory Learning Online platforms using Google Meet combined with WhatsApp and Google Form. The purpose of this research is to describe the effect of implementing PjBLL online platform using Google Meet combined with WhatsApp and google form on collaboration skills and basic science process skills.

2. Method

This study used a quasi-experimental one group pretest-posttest design which is presented in Figure 1. The effect of PjBLL (Project Based Laboratory Learning) on students' collaboration skills and basic science process skills was replicated with the available classes, namely class B* and class C*. The replication was carried out to see the consistency of the effect of using PjBLL.

Class B*	O_1	X	O_2
Class C*	O_1	X	O_2

Figure 1. Replication research design [18]

Information:

O_1 = Pretest conducted before the implementation of PjBLL Online platform

O_2 = Final test is carried out after the implementation of PjBLL Online platform

X = Treatment using PjBLL Online platform

In this class, learning begins with an initial test [O_1] in which class B* and class C* students take tests of basic science process skills and tests of collaboration skills. With the PjBLL Online platform lecturers carry out the learning process (X) the lecturer carries out learning for 30 class B* students and 30 class C* students. Activity of learning ends with final test [O_2] students in both classes take two test, basic science process skills and collaboration tests.

The subjects were students of class B* and class C* of the chemistry department at the Universitas Negeri Surabaya who took general physics courses as many as 60 students. Research implementation in the 2019/2020 school year. Because the online lecture platform, the instruments used are observation sheets, basic science process and collaboration skills tests. Collaboration skills, basic science process skills from students are assessed through videos made by students as a means of control that the process is really carried out by students by doing the test entries provided. The data obtained through the test instrument was then analyzed descriptively and N-gain analysis [19].

3. Results and Discussion

3.1. Student collaboration skills

Based on the research data obtained and the analysis of two (2) class, the results of student collaborative skills are shown in Table 1 for class B and Table 2 for class C below. Completeness of learning outcomes from class B and class C as the replication class is calculated and described based on the specified benchmark.

Table 1. Collaborative skills class B

Indicators*	Description	Pre-test			Post-test		
		ΣN	Completeness of Indicators	x̄	ΣN	Completeness of Indicators	x̄
Actively contribute	C	12	40.00	56.40	26	86.67	80.50
	NC	18			4		
Work productively	C	10	33.33	68.60	27	90.00	89.60
	NC	20			3		
Demonstrate flexibility and compromise,	C	14	46.70	68.50	24	80.00	80.20
	NC	16			6		
Show responsibility	C	12	40.00	64.12	26	86.67	83.12
	NC	18			4		
Show respect	C	12	40.00	66.40	25	83.33	82.50
	NC	18			5		

Information: x̄= Average score, ΣN = Number of students, C = Complete, NC = Not Complete [20]

Based on Table 1, the pre-test class B students have not achieved completeness for all indicators of collaboration skills. Each of the five indicators is still below 76%. Collaboration skills as the ability to participate in every activity to build relationships with other people, respect each other's relationships and teamwork to achieve the same goal play an important role in the industrial revolution era 4.0 and society era 5.0 [7-8]. After the PjBLL lectures on the Online platform, the achievement of the collaboration skill indicators (1) actively contributes, (2) works productively, (3) shows flexibility and compromise, (4) shows responsibility, and (5) shows an attitude of respect for getting everything done [20].

Table 2. Collaborative skills class C

Indicators*	Description	Pre-test			Post-test		
		ΣN	Completeness of Indicators	x̄	ΣN	Completeness of Indicators	x̄
Actively contribute	C	12	40.00	56.40	26	86.67	80.50
	NC	18			4		
Work productively	C	10	33.33	68.60	27	90.00	89.60
	NC	20			3		
Demonstrate flexibility and compromise,	C	14	46.70	68.50	24	80.00	80.20
	NC	16			6		
Show responsibility	C	12	40.00	64.12	26	86.67	83.12
	NC	18			4		
Show respect	C	12	40.00	66.40	25	83.33	82.50
	NC	18			5		

Information: x̄= Average score, ΣN = Number of students, C = Complete, NC = Not Complete [20]

For class C, the results are not much different from class C. Where at the time of the pre-test, all indicators of collaborative skills have not been completed and after studying with PjBLL online platforms, all indicators are completed. This indicates that the treatment given to the two classes has an effect on students' collaboration skills. These abilities can be trained through lectures by providing challenging problems that exist in real life [2-9].

Table 3. N-gain score and sensitivity class B and class C

Indicators*	**Clas B**				**Class C**			
	Gain score		**Sensitivity**		**Gain score**		**Sensitivity**	
	Coef.	**Descrip.**	**Coef.**	**Descrip.**	**Coef.**	**Descrip**	**Coef.**	**Descrip.**
Actively contribute	0.87	High	0.58	Sensitive	0.88	High	0.66	Sensitive
Work productively	0.89	High	0.68	Sensitive	0.88	High	0.69	Sensitive
Demonstrate flexibility and compromise,	0.80	High	0.69	Sensitive	0.82	High	0.68	Sensitive
Show responsibility	0.88	High	0.66	Sensitive	0.87	High	0.69	Sensitive
Show respect	0.85	High	0.57	Sensitive	0.88	High	0.68	Sensitive

Table 3 underlies the conclusion that there is an influence on the implementation of PjBLL online platform which in its implementation uses google meet combined with whatsapp and google form in class B* and class C*. The N-gain scores of the five indicators of collaborative skills are in the high and sensitive category. It is in accordance with and supports the research that has been done on the same topic as the PjBLL model [8-21].

The use of WhatsApp in learning is very helpful when there are problems in the synchronous mode. Lecturers can check assignments and implementation of learning via WhatsApp and students can also collaborate with fellow students or ask lecturers to overcome urgent difficulties. The use of the online PjBLL platform supports students' collaborative abilities [11].

3.2. Basic science process skills

The results of the study after implementation PjBLL Online platform for students from the Department of Chemistry who programmed basic physics courses in physics education study programs with aspects: (1) observing; (2) classifying; (3) communicating; (4) inferring; (5) predicting as shown in Table 4 [13,15,16].

Table 4. Basic science process skills class B*

No	**Pre test**			**Post test**			***N-Gain***	
	Score	**value***	**Completeness Indicator Individual**	**Score**	**value***	**Completeness Indicator Individual**	**<g>**	**Information**
1*	41.63	D	NC	71.99	B	C	0.54	Medium
2	44.75	D	NC	78.83	B+	C	0.67	Medium
3	38.50	E	NC	78.93	B+	C	0.68	Medium
4	51.00	D	NC	78.53	B+	C	0.59	Medium
5	35.38	E	NC	84.78	A-	C	0.77	High
6	45.75	D	NC	84.88	A-	C	0.79	High
7	46.78	D	NC	79.33	B+	C	0.58	Medium

No	Pre test			Post test			*N-Gain*	
	Score	value*	Completeness Indicator Individual	Score	value*	Completeness Indicator Individual	<g>	Information
8	51.12	D	NC	73.78	B	C	0.46	Medium
9	30.01	E	NC	78.85	B+	C	0.73	High
10	35.38	E	NC	82.65	A-	C	0.76	High
11	52.00	D	NC	76.40	B+	C	0.52	Medium
12	34.38	E	NC	85.88	A-	C	0.76	High
13	46.63	D	NC	76.30	B+	C	0.59	Medium
14	67.63	B-	C	73.48	B	C	0.38	Low
15	43.75	D	NC	76.40	B+	C	0.59	Medium
16	33.25	E	NC	79.63	B+	C	0.69	Medium
17	65.66	B-	C	71.66	B	C	0.17	Low
18	56.65	C	NC	81.57	A-	C	0.57	Medium
19	34.48	E	NC	81.75	A-	C	0.71	High
20	37.37	E	NC	72.89	B	C	0.56	Medium
21	34.38	E	NC	81.65	A-	C	0.71	High
22	33.33	E	NC	66.73	B-	C	0.53	Medium
23	66.71	B-	C	72.21	B	C	0.28	Low
24	45.86	D	NC	77.73	B+	C	0.62	Medium
25	47.68	D	NC	78.11	B+	C	0.54	Medium
26	66.73	B-	C	81.24	A-	C	0.46	Medium
27	36.60	E	NC	69.65	B-	C	0.53	Medium
28	24.87	E	NC	65.73	B-	C	0.55	Medium
29	43.64	D	NC	68.85	B-	C	0.46	Medium
30	33.35	E	NC	56.85	C	NC	0.37	Medium
Completeness Classical Indicator: 13% *(not achieved)*				Completeness Classical indicators: 97% *(achieved)*				

Information: C = Complete, NC = Not Complete

Based on Table 4, 30 students of class B for the pre-test achieved classical completeness of 13% in the incomplete category. Getting an A is not there yet. Obtaining B grade of 4 students, obtaining C value of 1 student, obtaining D value of 12 students and the rest obtaining E score. Only 4 students completed the individual indicators.

The results of the post-test showed that classically the completeness reached 97% in the category of complete completion and there was a large increase as the effect of the application of PjBLL Online platform in basic physics lectures. There are no more students who get D and E grades (do not pass). These skills are not hands-on skills using tools but thinking skills process using scientific processes so that the use of the Online platform during the COVID-19 pandemic is suitable to be applied in lectures guided using WhatsApp and google forms [21].

While the results of the basic science process skills in class C, the results are not much different from those obtained in class B, the results basic science process skills from students are presented at Table 5.

Table 5. Basic science process skills class C

No	Pre-test			Post-test			*N-Gain*	
	Score	Inf.	Completeness Indicator	Score	Inf.	Completeness Indicator	<g>	Information
1*	40.63	D	NC	65.63	B-	C	0.42	Medium
2	40.63	D	NC	78.13	B+	C	0.63	Sedang
3	26.12	E	NC	79.28	B+	C	0.76	High
4	35.57	E	NC	76.26	B+	C	0.68	Medium

No	Pre-test			Post-test			*N-Gain*	
	Score	Inf.	Completeness Indicator	Score	Inf.	Completeness Indicator	<g>	Information
5	40.63	D	NC	76.10	B+	C	0.59	Medium
6	34.38	E	NC	84.38	A-	C	0.76	High
7	43.75	D	NC	72.89	B	C	0.53	Medium
8	43.75	D	NC	73.18	B	C	0.54	Medium
9	25.00	E	NC	44.85	D	NC	0.30	Low
10	34.38	E	NC	81.25	A-	C	0.71	High
11	53.13	D	NC	81.25	A-	C	0.60	Medium
12	40.63	D	NC	87.50	A	C	0.79	Tinggi
13	37.50	E	NC	81.25	A-	C	0.71	Medium
14	68.75	B-	C	72.78	B	C	0.19	Low
15	46.88	D	NC	69.85	B-	C	0.45	Medium
16	34.38	E	NC	79.12	B+	C	0.68	Medium
17	59.38	C	NC	79.53	B+	C	0.47	Medium
18	46.88	D	NC	88.51	A	C	0.77	Tinggi
19	21.88	E	NC	90.63	A	C	0.84	Tinggi
20	29.34	E	NC	76.12	B+	C	0.65	Med.
21	40.63	D	NC	84.38	A-	C	0.74	Tinggi
22	40.63	D	NC	81.25	A-	C	0.68	Medium
23	51.00	D	NC	76.00	B+	C	0.53	Medium
24	38.51	E	NC	75.87	B+	C	0.66	Medium
25	44.65	D	NC	72.85	B	C	0.52	Medium
26	47.18	D	NC	69.76	B-	C	0.43	Medium
27	42.64	D	NC	78.45	B+	C	0.67	Medium
28	35.38	E	NC	69.85	B-	C	0.51	Medium
29	41.63	D	NC	67.58	C+	C	0.47	Medium
30	32.25	E	NC	58.98	C	NC	0.45	Medium
Classical Indicator Completeness 3.3% (not achieved)				Completeness Classical indicators 93.3% (achieved)				

Description: T = Completed, TT = Not Complete

Based on Table 5, 30 students of class C for the pre-test achieved classical completeness of 3.3% in the incomplete category. Getting an A is not there yet. Obtaining a B- value of 1 student, obtaining a C value of 1 student, obtaining a D value of 16 students and the rest obtaining an E score. Only 1 student who completed the indicators individually.

The results of the post-test showed that classically the completeness reached 93.3% in the category of complete completion and there was a large increase as the effect of the application of PjBLL Online platform in basic physics lectures. There are still students who get a D as much as 1 person and pass all of them. These basic skills can be taught and trained for students even through online lectures because they are not hands-on skills using tools but process thinking skills using scientific processes so that the use of the online platform during the covid-19 pandemic is suitable to be applied in lectures. guided using the WhastApp and google form [21]. The increase in the N-gain of basicprocess skills obtained by class B* and class C* students due to the influence of the PjBLL Online platform with five indicators is shown in Table 6.

Based Table 6 shows that implementation PjBLL can increase completeness process skill indicators in two classes namely class B* and class C* which was originally not completed to 86% completed; because all indicators in research have been completed. This shows that there is an effect of using the PjBLL Online platform combined with WhatsApp and google forms. The results of this research are in accordance with and support the research conducted by [13,14]. The N-gain value shows the level improvement each indicator basic science process skills in the medium criteria.

Table 6. N-Gain Process Skills Class B* and Class C*

Class	Indicator Basic Science Process Skills	Pre-test				Post-test				*N-Gain*	
		Score	Completenes			Score	Completenes				
			Σ	%	info		Σ	%	Info	<g>	Info
B*	Observing	46.88	8	25.00	NC	78.13	28	87.50	C	0.59	Med.
	Classifying	40.63	5	15.63	NC	80.47	27	84.38	C	0.67	Med.
	Communicating	42.97	4	12.50	NC	74.22	25	86.13	C	0.64	Med.
	Inferring	27.34	3	9.38	NC	72.66	26	86.25	C	0.62	Med.
	Predicting	30.47	4	12.50	NC	73.44	25	88.13	C	0.62	Med.
C*	Observing	44.53	2	6.25	NC	78.91	27	84.38	C	0.62	Med.
	Classifying	40.63	0	0.00	NC	80.47	28	87.50	C	0.67	Med.
	Communicating	39.06	2	6.25	NC	75.78	27	84.38	C	0.61	Med.
	Inferring	25.00	3	9.38	NC	74.22	28	87.50	C	0.67	Med.
	Predicting	25.78	3	9.38	NC	74.22	27	84.38	C	0.65	Med.

Description: C = Completed, NC = not complete

4. Conclusion

The effect of implementing PjBLL online platform using Google Meet combined with WhatsApp and google form on collaboration skills can complete all indicators with gain scores in the high and sensitive category, while on basic science process skills can complete all indicators with gain scores in medium and high criteria.

References

[1] Dwikoranto, Erman, and Madlazim 2019 *J. Phys. Conf. Ser.* **1387** 012074
[2] Nursyifa A 2018 *Proc. Community Dev.* **2** 649
[3] Zubaidah S 2016 *Semin. Nas. Pendidik.* **2** 1
[4] Redhana I 2015 *Pros. Semin. Nas. FMIPA UNDIKSHA* **5** 141
[5] Kusumadewi 2018 *J. Pendidik.* **5** 24
[6] Tama D M 2108 *Proses Pembuatan Lagu Anak Melalui Metode Tadasa Sesuai dengan Kecakapan Abad-21* (Bandung: Universitas Pasundan)
[7] Sari K A, Prasetyo Z K, and Wibowo W S 2017 *J. Pendidik. Ilmu Pengetah. Alam* **6** 461
[8] Le H, Jeroen J, and Theo W 2017 *Camb. J. Educ.* **48** 110
[9] Trinova Z 2012 *J. Al-Ta'lim* **1** 209
[10] Usta E and Akkanat C 2015 *Procedia-Soc. Behav. Sci.* **191** 1408
[11] D Dwikoranto, Setiani R, Prahani B K, and Mubarok H 2019 *J. Penelit. Pengkaj. Ilmu Pendidik. e-Saintika* **4** 259
[12] Espinosa A A, Monterola S L C, and Punzalan A E 2013 *Cypriot J. Educ. Sci.* **14** 211
[13] Dogan I and Kunt H 2016 *J. Europ. Educ.* **6** 32
[14] Abdullah C, Parris J, Lie R, Guzdar A, and Tour E 2015 *CBE-Life Sci. Educ.* **14** Ar34
[15] Kefi S and Uslu M 2015 *Middle-East J. Sci. Res.* **23** 2619
[16] Zeidan A F and Jayosi M R 2015 *World J. Educ.* **5** 13
[17] Dwikoranto, Hariyono E, Lestari N A, Jatmiko B, Prahani B K, and Suyidno 2021 *J. Phys. Conf. Ser.* **1805** 012006
[18] Sugiono 2014 *Statistika untuk Penelitian* (Bandung: Alfabeta)
[19] Plomp T J and Nieveen N 2013 *Educational Design Research, part A: An introduction* (Netherland: SLO)
[20] Greenstein L 2012 *Assessing 21st Century Skills: A Guide to Evaluating Mastery and Authentic Learning* (California: Corwin)
[21] Dwikoranto, Setiani R, Munasir, Surasmi W A, Tresnaningsih S, Suyidno, and Pramonoadi 2020 *J. Phys. Conf. Ser.* **1491** 012006

A Bibliometric Analysis of Minimum Competency Assessment Research with VOSViewer Related to the Impact in Physics Education on 2019-2020

Suliyanah[1], B D Adelia[1], M N R Jauhariyah[1], Misbah[2], S Mahtari[2], A Saregar[3], and U A Deta[1]*

[1]Department of Physics, Faculty of Mathematics and Natural Sciences, Universitas Negeri Surabaya, Indonesia
[2]Department of Mathematics and Science Education, Faculty of Teacher Training and Education, Lambung Mangkurat University, Indonesia
[3] Department of Physics Education, Universitas Islam Negeri Raden Intan Lampung, Indonesia

*Email: utamadeta@unesa.ac.id

Abstract. This study discusses the *Asesmen Kompetensi Minimum* (AKM) or Minimum Competency Assessment using bibliometrics, the top 5 articles and authors. The method used is the literature review using Harzing's Publish or Perish application on March 9 and 16, 2021, with keywords "*Asesmen Kompetensi Minimum*" and "Minimum Competency Assessment" from the Google Scholar database and then processed by the VOSViewer application. The data obtained were 593 articles in Indonesian and 507 articles in English. The results of the bibliometric study obtained five clusters in Indonesian and English. Minimum competency assessment in Indonesian relates to eliminating the national exam, while English relates to student readiness. Top authors and articles discuss science learning, learning activities and mathematical literacy analysis skills. The top five articles in Indonesian and English, there is one study on applying minimum competency assessment and literacy skills in minimum competency assessment, which can also increase knowledge of physics in static and dynamic electricity materials. The conclusion obtained is that there is still minimum research on minimum competency assessment. The impact of this research on physics learning can be applied to static and dynamic electricity, and the top articles and authors discuss the discussion of science learning and PISA design in the pandemic era.

1. Introduction

Education in Indonesia has not run optimally. Especially in learning physics because of the low learning outcomes of students at school [1]. Many students are not interested in learning physics so that it has an impact on learning achievement [2]. In this case, only a few students took physics in the national exam.

Evaluation of student achievement results, namely by providing an assessment that students can pass or not. According to Khamim [3], the national exam is an evaluation system used for traditional primary and secondary education in Indonesia. The national exam is defined as a measuring tool for evaluating the learning process. There are evaluation of learning activities and controversy in the community. The issuance of the National Education System Law No. 20 of 2003, namely education, is a conscious, planned effort to create an atmosphere of learning and the learning process.

To achieve national education standards, a continuous monitoring process is necessary. The results obtained from the evaluation of education standards can be used as a milestone for improving the quality of education. The drawback of the national exam is that it is not effective in improving the quality of education. The implementation of the national exam is not carried out transparently so that the related parties do not understand the assessment process. The value of the national exam is used as a determining factor for graduation [4], which results in students cheating in working on questions [5].

The advantage of the national exam is that it can be used as an educational evaluation by applying educational standards and can be considered as a mapping of the quality of education in academic units. National exams can also develop existing learning [6] to stimulate students to become better at taking educational programs. The National Examination is given only in certain subjects, while Permendikbud will hold the exam is conducted for all subjects. Nadiem Makarim made a breakthrough called the National Assessment. The National Assessment consists of a Minimum Competency Assessment and a character survey.

Several studies on Minimum Competency Assessment have been carried out, such as student readiness [7,8], online learning [9], pedagogy [10], critical thinking [11], design PISA [12], perceptions of prospective teachers [13] and on children with special needs [14]. An article discusses the Minimum Competency Assessment from the Perdana [7] regarding SMA and SMK students' readiness to face the Minimum Competency Assessment questions. However, there are no articles that discuss Minimum Competency Assessment using bibliometric maps. In this study, we will explore the Minimum Competency Assessment (AKM) using a bibliometric map included in the new topic, find the top 5 articles based on the Minimum Competency Assessment (AKM), and find the top authors.

2. Method

The method used is by using a literature review which is a literature review method by summarizing several articles on a particular topic. The search for research articles was carried out using Harzing's Publish or Perish application on March 9 and 16, 2021, using the keyword "Asesmen Kompetensi Minimum" in Indonesian language articles and "Minimum Competency Assessment" in English articles. Then look for the top 5 articles related to keywords and the article year ranges from 2020-2021 as the article on Minimum Competency Assessment is a new topic. Then the data is stored in .ris form, which is then analyzed using the VOSviewer application to get a visualization in a bibliometric map. There are 593 articles in Indonesian and 507 articles in English taken from the Google scholar data used in the study, then the articles obtained are analyzed in the form of a bibliometric map. The research procedures can be seen in Figure 1.

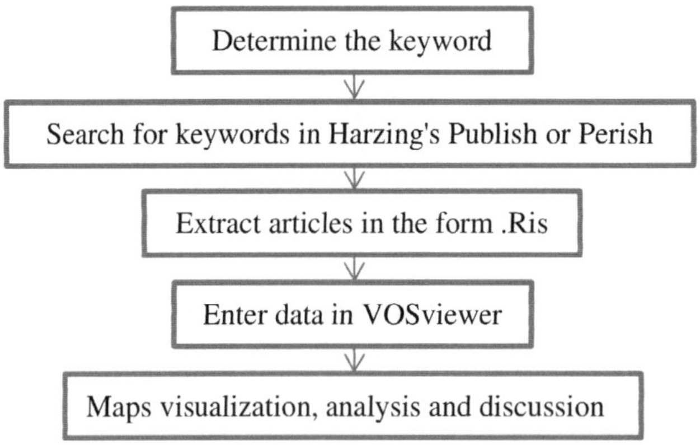

Figure 1. Research flowchart

3. Results and Discussion

The results obtained from Harzing's Publish or Perish application in Indonesian and English language articles are the year of publication, year of citation, paper, citation, citations per year, citations per paper, author per paper, and h-index in Table 1.

Table 1. Results of Harzing's Publish or Perish application on Indonesian and English language articles.

Result	Indonesian language	English language
Publication year	2020-2021	2020-2021
Paper	593	507
Citations	217	789
Citations per paper	0.37	1.56
Author per paper	1.80	2.25
h-index	6	9
g-index	11	25
hI-norm	5	4
hI-yearly	1.67	4

The data of the articles in Table 1 was restricted in year 2020-2021. The number of papers on the Minimum Competency Assessment in Indonesian is more than in English, insignificant. However, in terms of citations, citations per paper, and authors per paper, articles in Indonesian are less than in English. The h-index (the article referred to) in Indonesian is also less than in English.

After the h-index will appear the g-index (suggested authors based on the publication). The articles in Indonesian are less than in English. The hI-norm (index h- on individual/author) in Indonesian language articles is more numerous than English. However, in hI-annually (the number of single authors who write the same articles per year) in Indonesian is less than in English.

After analysing Indonesian and English articles using VOS viewer by mapping the title and abstract to determine the minimum number, there are several colours, and each colour has a relationship between one topic and another. There are two visualisations when analysing using VOS viewer: network visualisation and density visualisation with different functions. In-network visualisation explains the attachment of the topic to the title and abstract. In contrast to network visualisation, density visualisation describes frequently used topics marked in bright yellow and the larger the text of the bibliometric map visualisation.

Bibliometric maps in Indonesian language articles are obtained 15 words in 5 clusters, namely red, yellow, green, blue and purple that showed in Figure 2. Figure 2 shows the studied clusters. The first is the red cluster consisting of the national exam keywords, minimum competency and character survey, replaced and future. In the yellow cluster, there are assessment keywords and Indonesia. In the green cluster, there are the words AKM, teacher, improvement, and outcome. In the blue cluster, there are keywords PISA, aspect and literacy. In the purple cluster, there are students' keywords and skills. This shows that there are several keywords in each cluster.

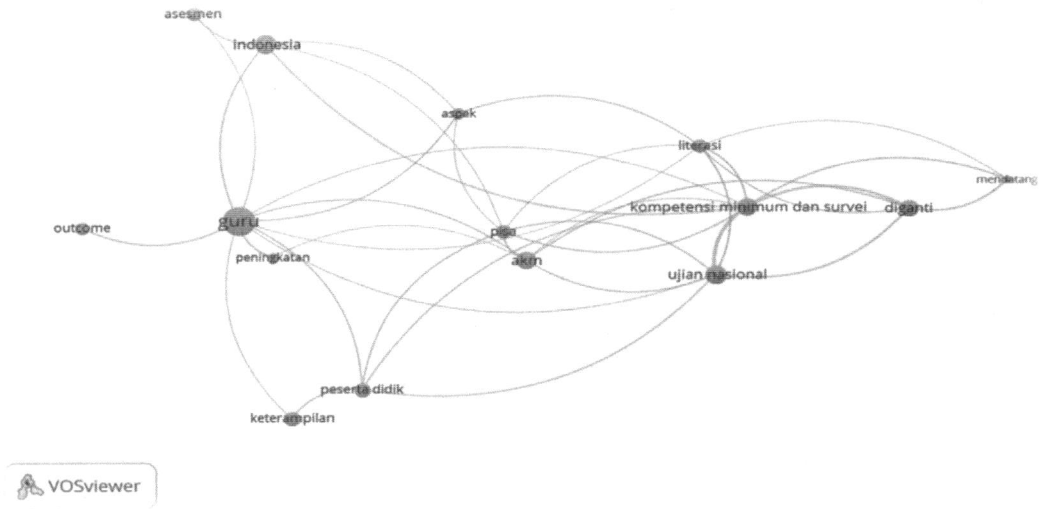

Figure 2. Visualization network of Indonesian language article

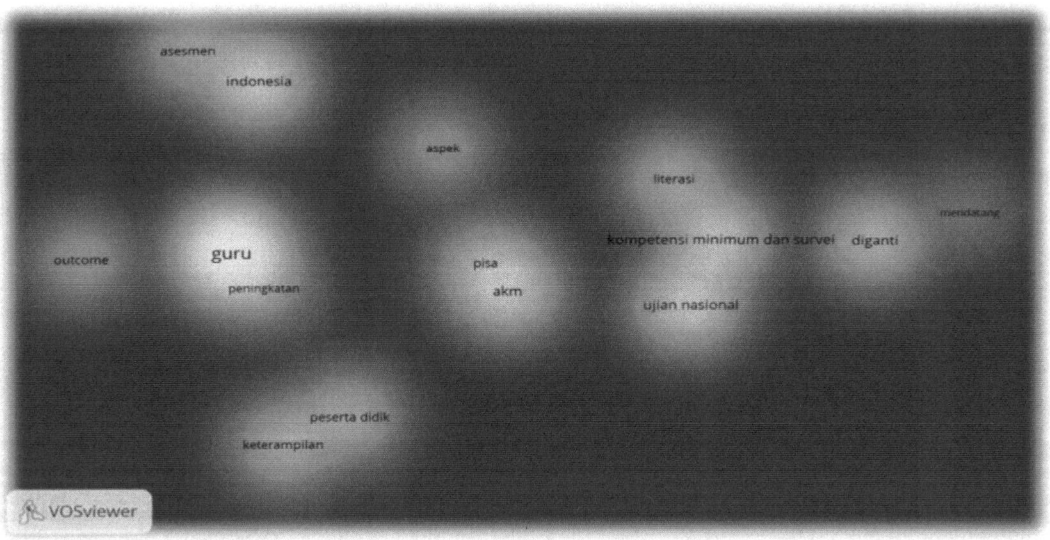

Figure 3. Visualisation density of article in Indonesian

Figure 3 shows the frequently used keywords, which are written in more giant letters and lighter colours. This article has the keywords teacher, improvement, Indonesia, AKM, PISA, national exams, and minimum competencies and surveys. These data can search for novelty AKM and outcomes. For example, the AKM already running in Indonesia aims to develop students' potential by involving literacy and numeracy.

The implementation of AKM activities is carried out for students at the primary and secondary school levels, as has been done by Widianti & Hidayati [16]. Students can memorise theory and be able to participate with the community, such as research done by Sidiq & Muqowin [15]. This AKM learning is also carried out for children with special needs through independent learning, as has been done by Asfiati & Mahdi [14]. The results of AKM can hone thinking skills to improve students' skills, as in the research of Lestari, Ambarsari, Prahani, Jauhariyah, Yantidewi and Deta [17].

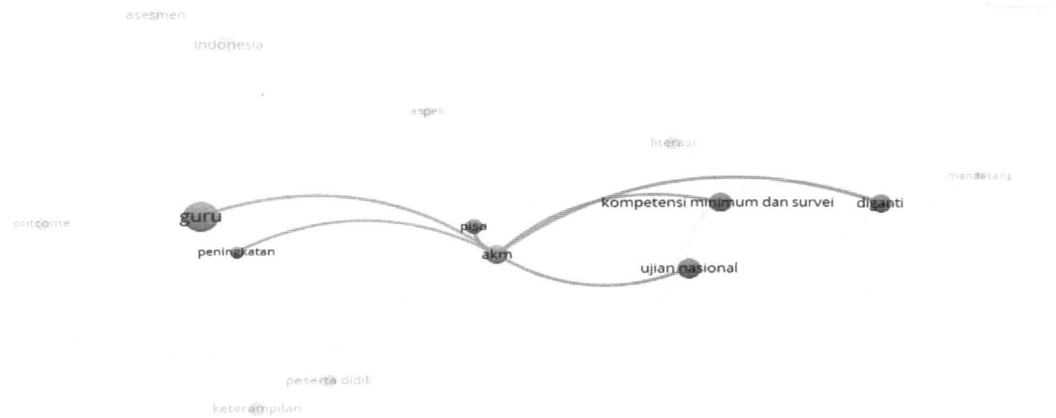

Figure 4. Visualization of keywords relating to AKM in Indonesian articles

In Figure 4, it can be seen that the AKM keywords relate to PISA, national exams, minimum competencies, and surveys, replaced, improvement, and teachers. From the keyword results, it can be seen that AKM is related to the minimum competency and the survey. In this case, the minimum competence consists of two competencies: literacy (reading) and numeracy (mathematics). According to the Nudiati & Sudiapermana research [18], there is six basic literacy knowledge: reading and writing, numeracy, science, digital, finance, culture, and citizenship. Then this basic literacy is shortened to only two aspects, according to the research of Ayuningtyas & Sukriyah [19].

AKM is implemented in 2021 in Indonesia to improve the quality of education at the education level unit, as in Muntholi'ah research [20]. This was done because he saw no improvement in the quality of education in Indonesia when determining the national exam as a graduation competency, and there were pros and cons. The new system, which the Minister of National Education created, will also impact universities, especially in the teacher training department, to organize the teaching and learning process. The learning process is used only to use theory, causing misconceptions in students. Therefore, there are various ways to reduce misconceptions in students, such as one in Hidayatulloh, Humairoh, Wachidah, Iswati and Suliyanah research [21], but there is AKM because the questions given are related to everyday life.

Bibliometric maps of English articles obtained 19 words in 5 clusters, namely red, yellow, green, blue and purple that showed in Figure 5.

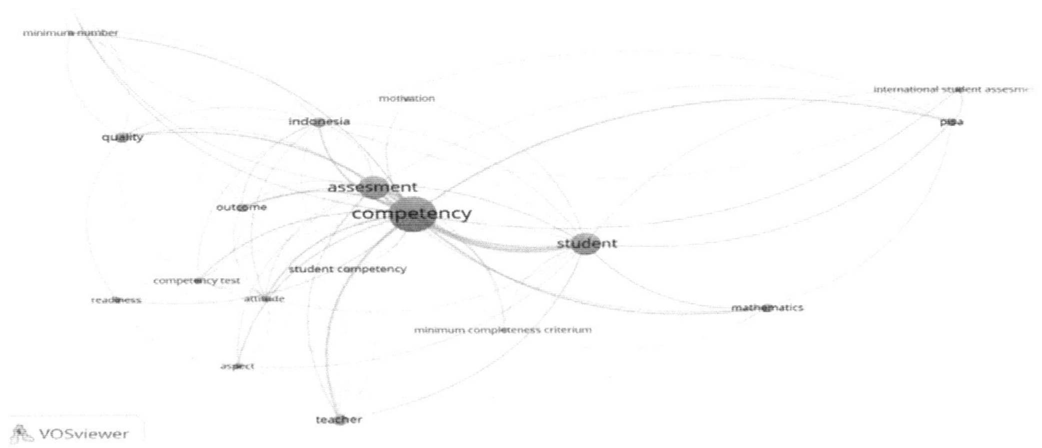

Figure 5. Visualization network of english article

Figure 5 shows the studied clusters, and it can be seen that the English articles found more keywords. The first is the red cluster which consists of the keywords competency, competency test, attitude, and readiness. In the yellow cluster, there are keywords, minimum completeness criterium and motivation. The words Indonesia, issue, minimum number, quality, and outcome are in the green cluster. In the blue cluster, there are the keywords student, mathematics, PISA and international student assessment. In the purple cluster, the keyword is assessment and teacher. This shows that there are several keywords in each cluster.

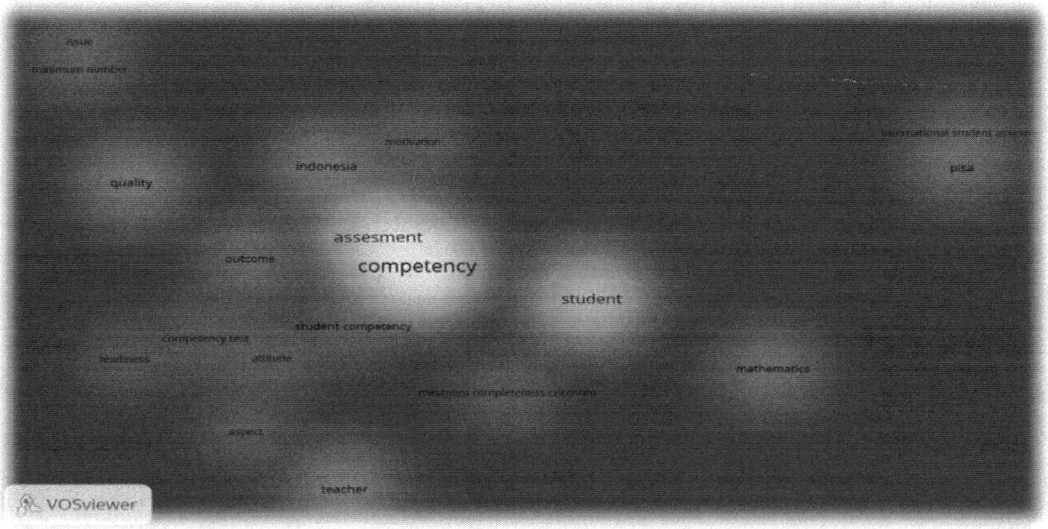

Figure 6. Visualization density of english articles

Figure 6 shows the frequently used keywords, for example, the competence used by students in learning. Competence in students is divided into two, namely literacy (reading) and numeracy (mathematics) competencies, as in the research of Mauizdati [22]. This competency is carried out to replace the national exam as an existing education system, like Alawiyah research [23]. Students now prefer to memorize formulas so that literacy (reading) is needed to add insight to students, such as research by Rafiqoh [24]. The results of the application of this competency will foster students' enjoyment of literacy (reading), such as research by Fadila, Suliyanah and Deta [25]. This competence is contained in a new system called independent learning, as in the research of Prahani, Deta, Yasir, Astutik, Pandiangan, Mahtari and Mubarok [26].

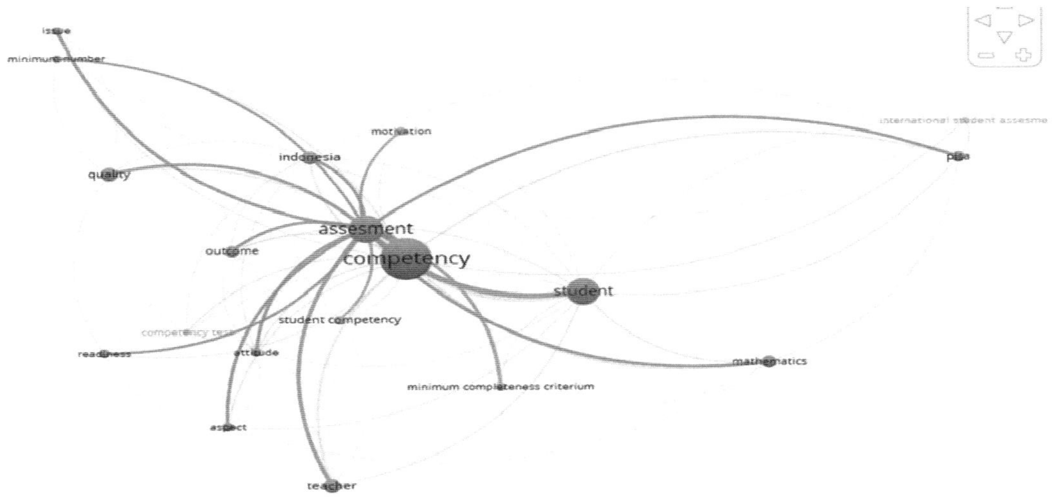

Figure 7. Visualization of keywords relating to assessment of english articles

In Figure 7, it can be seen that the assessment keywords relate to issue, minimum number, quality, Indonesia, motivation outcome, competency, PISA, student, mathematics, minimum completeness criterium, student competency, teacher, aspect, attitude, readiness. In this case, the keyword assessment is related to competency. This competency assessment improves the quality of education by using online learning, as in Chainchumpa and Temdee [9].

In Indonesia, a competency assessment has been prepared, which is named AKM. Currently, Indonesia is implementing independent learning, as in Yamin & Syahrir research [27]. The government makes competency assessments to support learning with independent learning, such as research by Setiawan, Sa'dijah, and Akbar [28]. Competency assessment is carried out to assess the course of learning as in the research of Coates, Karen, Wilson, and Sadberry [29]. The competency assessment was also carried out to assess students' interest in reading, not only memorizing as in the Perdana Research [7].

Table 2. Harzing's publish or perish in Indonesian language articles

Rank	Author	Title	Year
1.	Sari, Daulay, Yuliani and Epriani [30]	*Penghapusan ujian nasional tahun 2021 dalam perspektif guru SMA di kota Tebing Tinggi.*	2021
2.	Novita, Mellyzar and Herizal [13]	*Asesmen nasional (an): pengetahuan dan persepsi calon guru.*	2021
3.	Aisah, Zaqiah and Supiana [31]	*Implementasi kebijakan asesmen kemampuan minimum (AKM): analisis implementasi kebijakan AKM.*	2021
4.	Mustofa [32]	*Kompetensi numerasi siswa SMK ditinjau dari gender dan berbagai kesulitannya.*	2020
5.	Hidayah [33]	*Asesmen ujian nasional dari masa ke masa dan solusinya.*	2021

Table 2 describes the top 5 articles in Indonesian related to the Minimum Competency Assessment in Harzing's Publish or Perish application. The article was written by Sari, Daulay, Yuliani, and Epriani

[30]; Novita, Mellyzar, and Herizal [13]; Aisah, Zaqiah, and Supiana [31]; Mustofa [32]; and Hidayah [33].

The first research from Sari, Daulay, Yuliani and Epriani [30]. This research wants the opinion of high school teachers in Tebing Tinggi to find pros and cons in eliminating the national exam into a minimum competency assessment and character survey. Teachers who refuse to change the national exam are deemed not ready to face the minimum competency assessment and character survey, and there is a lack of socialization from the government regarding this matter. However, teachers who accept this change feel that the change can improve the quality of education in Indonesia. With the new government policy, all teachers should be ready to accept the replacement of this system.

The second research from Novita, Mellyzar, and Herizal [13]. This study asks prospective teachers from the mathematics education, physics education and chemistry education courses at the University of Malikussaleh. It turns out that only 49% of prospective teachers understand the national assessment correctly. Many people do not understand the implementation of the national assessment, including the form of questions, the parties involved, the reasons for the national assessment's formation, and the national assessment's output. Therefore, the government should provide socialization to the public before implementing new policies.

The third research from Aisah, Zaqiah, and Supiana [31]. This study discusses the implementation of the minimum competency assessment in the community. Researchers disseminate information to the community, including the objectives, benefits, people involved, and the minimum competency assessment program mechanism. After conducting the research, all people's new policies implemented by the Ministry of Education and Culture were accepted. This study shows that the community has started to recognize the minimum competency assessment, but this study does not discuss the content or components of the minimum competency assessment.

The fourth research from Mustofa [32] discusses the difficulties of numeracy competence in the Minimum Competency Assessment by differentiating gender in vocational students. It turns out that there is no difference in numeracy abilities between men and women. However, students still have difficulty working on numeracy problems. Therefore, schools need to provide learning that can train students in working on numeracy questions to become skilled and accustomed to working on numeracy questions.

The fifth research from Hidayah [33], according to PISA, this study discusses changes to the national exam from 2005-2020, later changed to the Minimum Competency Assessment. Changes to the national exam were made to improve the quality of education. The national exam is considered burdensome for students because the material submitted is too much so that students are more dominant in memorizing. Therefore, the national exam will be replaced with a Minimum Competency Assessment which involves literacy and numeracy skills.

Table 3 describes the top 5 articles in English related to the Minimum Competency Assessment. The article was written by Perdana [7]; Aunurrahman [8]; Ismail, Mudjiran, Neviyarni, and Nirwana [34]; Nusantara, Zulkardi, and Putri [12] and Wigati, Wardono, and Purwanto [35].

The first research from Perdana [7] that discusses high school and vocational high school student's readiness to face the Minimum Competency Assessment questions. It turns out that the questions given by the teacher are different from the questions in the Minimum Competency Assessment, so that students are still not ready to face the Minimum Competency Assessment. Before the government carries out a Minimum Competency Assessment, examples of questions should be given about it. Therefore, students can prepare it.

The second research from Aunurrahman [8] discusses the readiness related to implementing the minimum competency assessment and character survey in 2021 for elementary school teachers. However, some teachers think that curriculum changes are too fast so that they cannot take advantage of existing technology as a communication and information system and the principal is not maximally involved in engaging teachers in exploring new policies so that many teachers do not understand the minimum competency assessment and character survey. Socialization should be carried out to parties who play a role in the Minimum Competency Assessment to prepare adequately.

Table 3. Harzing's Publish or Perish in English Articles

Rank	Author	Title	Year
1.	Perdana [7]	Analysis of student readiness in facing minimum competency assessment.	2021
2.	Aunurrahman [8]	Taxonomy of educational problems in support of readiness for the implementation of minimum competency assessment and character survey in elementary schools	2020
3.	Ismail, Mudjiran, Neviyarni, and Nirwana [34]	Creative approach guidance and counseling facing independence learning policy minimum competency assessment and survey characters in the industrial revolution 4.0	2020
4.	Nusantara, Zulkardi and Putri [12]	Designing PISA-like mathematics problem in covid-19 pandemic (PISAComat)	2020
5.	Wigati, Wardono and Purwanto [35]	Analysis of Mathematical Literacy Skills through PMRI Approaches of Elementary School Students	2020

The third research from Ismail, Mudjiran, Neviyarni and Nirwana [34]. This research analyzes the era of 4.0 counselling guidance needs to be done to determine students' intelligence, interests, personality, creativity, attitudes, and achievements using the minimum competency assessment approach and character surveys. The application of the Minimum Competency Assessment impacts the field of counselling, and the results of this application can increase HOTS thinking in students.

The fourth research from Nusantara, Zulkardi and Putri [12]. This study discusses mathematical problems similar to PISA questions using literacy and numeracy knowledge methods in pandemic conditions called PISAComat. Students are grouped into small groups, then given PISA questions changed using language that is easy to understand. From the results received, students can understand and solve PISAComat questions. In addition, this method can involve the ability of reasoning and students. With the PISAComat method, students can freely express their opinions and not stick to existing formulas.

The fifth research from Wigati, Wardono, and Purwanto [35]. This study discusses the analysis of mathematical literacy in elementary students using the PMRI approach. This PMRI approach applies mathematics learning that is linked to student experiences. Mathematical literacy skills in elementary students are still lacking. This can be seen from solving math problems associated with everyday life. After being given the PMRI approach, students understand how to process math problems. So this PMRI approach can be one of the methods that can be used in discussing the Minimum Competency Assessment.

Figure 8. Visualization of top author in Indonesian language articles using minimum competency assessment keywords

Figure 8 explains that the authors are not tied to each other, namely Rahmawati, Agung and Hidayati. In the research written by Rahmawati & Rahmawati [36] discusses that science learning at elementary schools in the pandemic era is oriented towards science and technology literacy skills and builds students' character by making the learning atmosphere comfortable. Susanti, Agung & Wulandari research [37] discusses the activeness of learning mathematics using the Inside-Outside Circle model. It turns out that this research can be used as an improvement in students' learning skills and creativity because the Inside-Outside Circle model can generate mathematical learning suitable for use in the ongoing Minimum Competency Assessment. Widianti and Hidayanti research [16] discusses analyzing mathematical literacy skills in junior high school students. This research has been described in Figure 3 (visualization of density articles in Indonesian).

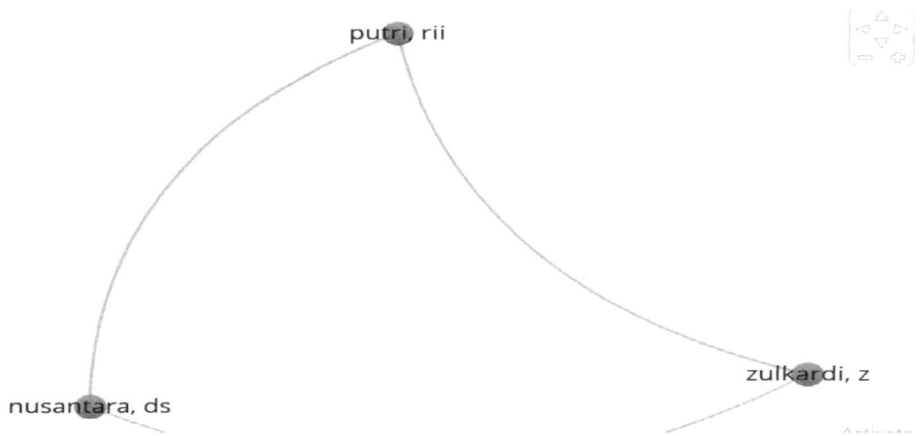

Figure 9. Visualization of top author in english language articles using minimum competency assessment keywords

Figure 9 above explains the top author who wrote three articles about the Minimum Competency Assessment in English. Three authors are linked to each other, namely Nusantara, Zulkardi, and Putri. Nusantara writes that research, Zulkardi, and Putri [12], are included in the top 5 articles, which have been described in Table 3.

Minimum Competency Assessment Research that impacts physics education, namely literacy skills, can be applied to physics material associated with everyday life. Students only read at the exam time, but this literacy needs to be done before doing a practicum in the laboratory. However, as a teacher, the teacher must design activities in the laboratory that are fun for students, such as Admoko & Supriyono research [38]. In addition, literacy skills can also increase knowledge of physics on static and dynamic electricity material [39]. Numerical abilities can also be applied to physics education as the calculation of formulas in physics problems.

Research on the Minimum Competency Assessment has limitations in conducting research. The limitations of this research on Minimum Competency Assessment are only found in 2020-2021 because it is still a new topic. 2021 is not over yet, so many new studies have not been included in this research. With increasing days, there must be many people researching the Minimum Competency Assessment. Therefore, it is necessary to do further research on Minimum Competency Assessment.

4. Conclusion

The Minimum Competency Assessment is divided into literacy (reading) and numeracy (mathematics) competencies. Literacy skills in the minimum competency assessment can be applied to physics material associated with everyday life and can also increase knowledge of physics on static and dynamic electricity material. Numerical abilities can also be applied to physics education as the calculation of formulas in physics problems. The Minimum Competency Assessment (AKM) is used to develop student's abilities to reason and think logically using the concepts under control related to their surroundings. The red colour cluster has the exact keywords as the top article. The Minimum Competency Assessment (AKM) relates to PISA, national examinations, minimum competencies and surveys, upgrades and teachers.

The top article of the Indonesian Language Minimum Competency Assessment (AKM) deals with eliminating national exams, perception and implementation, while the top article in English deals with student readiness, implementation, PISA, and abilities. The top author in Indonesian language articles discusses science learning associated with independent learning, active learning mathematics uses the Inside-Outside Circle model and analyzes the literacy skills of junior high school students. In contrast, the top author in the English article discusses PISA design in the pandemic era. Many people do not understand the need to carry out a Minimum Competency Assessment and the aspects contained in the Minimum Competency Assessment, and it is necessary to follow up on the Minimum Competency Assessment. There is still little research on minimum competency assessment, so there is a need for further development related to this research.

References

[1] Supardi, Leonard, Suhendri H, and Rismurdiyati 2015 *J. Ilm. Pendidik. MIPA* **2** 71
[2] Astuti S P 2015 *J. Ilm. Pendidik. MIPA* **5** 68
[3] Khamim 2019 *Misykat: J. Ilmu-Ilmu Al-Quran, Hadist, Syari'ah Tarb.* **4** 125
[4] Indahri Y 2014 *Kaji. Singk. Terhadap Isu-Isu Terkini* **7** 13
[5] Raharjayanti Y 2019 *J. Promkes. Indones. J. Health Promot. Health Educ.* **7** 133
[6] Erlina 2013 *UG J.* **7** 1
[7] Perdana N S 2021 *J. Pendidik. Sej. Ilmu-ilmu Sos.* **5** 15
[8] Aunurrahman 2020 *J. Educ. Teach. Learn.* **5** 296
[9] Chaichumpa S and Temdee P 2018 *Int. J. Innov. Learn.* **23** 19
[10] Makasau R 2020 *J. Masalah Pastor.* **8** 107
[11] Marunung A S, Hasanah M and Siswanto W 2020 *J. Pendidik. Teor. Penelit. Pengemb.* **5** 730
[12] Nusantara D S, Zulkardi and Putri R I I 2020 *J. Phys. Conf. Ser.* **1657** 012057
[13] Novita N, Mellyzar M and Herizal H 2021 *J. Ilmu Sos.Pendidik.* **5** 172
[14] Asfiati and Mahdi N I 2020 *J. Islamic Early Child. Educ.* **3** 59
[15] Sidiq A M and Muqowim 2020 *J. Program Studi PGRA* **6** 146

[16] Widianti W and Hidayati N 2021 *J. Pembelajaran Mat. Inov.* **4** 27
[17] Lestari N A, Ambarsari R, Prahani B K, Jauhariyah M N R, Yantidewi M and Deta U A 2021 *J. Phys. Conf. Ser.* **1805** 012033
[18] Nudiati D and Sudiapermana E 2020 *Indones. J. Learn. Educ. Couns.* **3** 34
[19] Ayuningtyas N and Sukriyah D 2020 *J. Mat. Pendidik.* **9** 237
[20] Muntholi'ah 2020 *J. Pendidik. Islam* **7** 161
[21] Hidayatulloh M, Humairoh F, Wachidah U, Iswati D A, and Suliyanah 2015 *J. Penelit. Fis. Apl.* **5** 28
[22] Mauizdati N 2020 *J. Rev. Pendidik. Pengajaran* **3** 315
[23] Alawiyah F 2015 *J. Masalah Sos.* **6** 189
[24] Rafiqoh S 2020 *J. MathEduct. Nusant.* **3** 58
[25] Fadila D, Suliyanah, and Deta U A 2020 *J. Kependidikan Fis.* **8** 39
[26] Prahani B K, Deta U A, Yasir M, Astutik S, Pandiangan P, Mahtari S, and Mubarok H 2020 *Stud. Philos. Sci. Educ.* **1** 21
[27] Yamin M and Syahrir 2020 *J. Ilm. Mandala Educ.* **6** 126
[28] Setiawan H, Sa'dijah C, and Akbar S 2017 *Pros. TEP & PDs* **1** 593
[29] Coates R D, Karen R, Wilson, and Sadberry 2015 *Sociol. Focus* **27** 173
[30] Sari A, Daulay S, Yuliani P, and Epriani P 2021 *Proc. Semin. Nas. Pembelajaran Bhs. Sastra Indones.* **3** 213
[31] Aisah H, Zaqiah Q Y, and Supiana A 2020 *J. Pendidik. Islam Al-Affan* **1** 128
[32] Mustofa Z 2020 *J. Pendidik. Mat.* **8** 227
[33] Hidayah L N 2020 *J. Pendidik. Guru Madrasah Ibtidaiyah* **4** 11
[34] Ismail R N, Mudjiran, Neviyarni, and Nirwana H 2020 *J. Ilm. Teknol. Pendidik.* **8** 1
[35] Wigati T, Wardono W, and Purwanto E 2020 *J. Prim. Educ.* **9** 303
[36] Rahmawati D N U and Rahmawati R 2020 *Glob. Conf. Ser. Soc. Sci. Educ. Humanit.* **6** 131
[37] Susanti P I, Agung A A G, and Wulandari I G A A 2020 *J. Ilm. Pendidik. Profesi Guru* **3** 22
[38] Admoko S and Supriyono 2016 *J. Penelit. Fis. Apl.* **6** 34
[39] Solihudin J H T 2018 *J. Wahana Pendidik. Fis.* **3** 51

Development of STEM Model Student Worksheets with PhET Simulation on Hooke's Law Material to Improve the Ability Students' Critical Thinking

F R Agustina and Dwikoranto*
Physics Education, Faculty of Science and Mathematics, Universitas Negeri Surabaya, Indonesia.

*Email: dwikoranto@unesa.ac.id

Abstract. This research aims to describe the feasibility of STEM model student worksheets in guided inquiry learning with PhET Simulation media on Hooke's Law material in terms of validity, effectiveness and practicality, due to the lack of critical thinking skills of students. This development research uses a four-D model with the following steps: design, define, develop and disseminate with high school students as research targets. Data were analyzed descriptively quantitatively. The results showed that: (1) The level of validity of student worksheets get a score of 86.36% with a very valid category; (2) Practicality level of student worksheets get through observation sheets and questionnaire responses from students get a practicality score of 88% with a very practical category; and (3) The level of effectiveness of the worksheets obtained from the N-gain calculation get a score of 0.56 in the medium category. Based on these things, it can be concluded that the developed student worksheets are feasible to be used to improve the ability students' critical thinking.

1. Introduction

The current era is the industrial revolution era 4.0 or can be said that the current era has entered the 21st century. The characteristics of industrial revolution era are digitization, optimization production, customization production, automation and adaptation, interaction of human with machines, value-added services and businesses, automatic data exchange and communication, as well as using information technology [1].

In line of the times, the curricula of educational is also developing. Currently using the curriculum of 2013 which emphasizes the learning process with a scientific approach means providing understanding to students in recognizing, understanding various materials using a scientific approach.

The STEM model is one of the learning that is in accordance with the curriculum of 2013 which emphasizes the learning process so that students can understand anyvarious materials using a scientific approach. This model of STEM is an integration of the four disciplines of science, technology, engineering and mathematics learning which are suggested to be able to help the successful development of 21st century skills [2]. One of the efforts that can be taken to increase the quality of education is the development of student worksheets. Student worksheets are generally a learning tool as a complement or a means of supporting the implementation of the Lesson Plan [3]. Student worksheets are a tool used to assist the learning process so that interactions between students and teachers are formed and can increase learning activities and achievements [4].

The development of STEM-based student worksheets can solve problems for students and teachers. The developed student worksheets contain four disciplines, they are science, technology, engineering and mathematics. Student worksheets can be used to increase science process skills and higher order thinking skills. The use of STEM in learning activities that are applied in the form of student worksheets can have a good impact, including being able to improve conceptual understanding and students' critical thinking skills [5]. STEM can train critical thinking skills, as evidenced by an increase in thinking skills on each critical thinking indicator [6]. Critical thinking skills referred to in this study are students' analytical and evaluative abilities to process information and experiences obtained in the learning process, with indicators of skills to interpretazion, analyze, evaluate, explain, and draw conclusions. Critical thinking skills will make students able to make decisions or actions on the problems they face. This critical thinking skill will equip students in dealing with every problem in everyday life [7].

In the implementation in the classroom, it is combined with guided inquiry which places students in solving a problem by formulating hypotheses, collecting data, testing hypotheses to drawing conclusions by students [8]. There was an enhancement in the ability critical thinking after the application of guided inquiry learning in the class [9]. Guided inquiry is able to motivate students to become individuals with a high level of intelligence with their critical and rational thinking and have broad insight [10].

STEM is suitable when combined with guided inquiry so that it can develop critical thinking skills [11]. STEM is a learning and strategy that is seen as an approach that can make changes in 21st century skills [12]. STEM is able to create a modern learning system in the midst of technological developments because it incorporates four disciplines that are really needed, namely science, technology, engineering and mathematics [13]. Continuous application will produce students who have scientific qualifications, are critical and innovative, are skilled in solving mathematical equations, and are able to take advantage of increasingly sophisticated technology along with the times [11].

Aspects of science and technology are two things that are interrelated, these two things are indispensable factors in the development of education in the digital era. Natural sciences including physics are subjects that are closely related to experiments, so with increasingly sophisticated technology we can conduct virtual experiments [13]. One of the virtual physics experimental media that can explain physical phenomena through experiments so that students are able to explore certain concepts and materials through virtual media simulations. Students' critical thinking skills increase after learning that refers to the guided inquiry syntax with PhET simulation experimental media [11].

With the explanation of the educational problems above, it is necessary to conduct research to produce innovative student worksheet material teaching that can be used in learning by linking the disciplines of science, technology, engineering and mathematics, namely the development of STEM-based student worksheets in guided inquiry learning with PhET Simulation on Hooke's law material to improve the ability students' critical thinking.

2. Method

The type of research used Research and Development with a 4-D model (Define, Design, Develop, and Disseminate [1]. The subjects of research are the level of validity, practicality and effectiveness of a STEM model student worksheet in guided inquiry learning with using PhET Simulation on the Hooke's Law Material to increase the students' critical thinking. The object used for limited dissemination in this research was class XI students of State Senior High School 1 Sumberrejo Bojonegoro.

The student worksheets integrated STEM were tested using a one group pretest-posttest research design [14]. In this research design, a pretest was given at the beginning of learning, then students were given treatment using a student worksheet developed on the material of Hooke's Law which was then given a posttest at the end of the lesson. The response questionnaire given to students contained questions about student worksheets developed in physics learning.

The Define stage is carried out by determining and defining learning requirements by conducting interviews with physics teachers, analyzing physics learning syllabus, analyzing teaching materials, literature reviews about teaching materials similar to student worksheets. The design stage is carried out by designing student worksheets that will be developed so that they are in accordance with the planned objectives, selecting media, selecting formats, designing student worksheets which include the formulation of indicators and learning objectives, making lesson plans, syllabus, drafts of developed student worksheets, and making learning instruments. The Develop stage is carried out by compiling learning tools, compiling and making student worksheets, observation sheets and questionnaires which will later be validated by expert validators. This stage consists of the validation stage, effectiveness stage, and practicality stage. The last research stage, namely Disseminate, is the stage of implementing the learning process using student worksheets that have been made.

The instrument used is the instrument of validity, practicality and effectiveness of student worksheets. Student worksheets are declared valid if they reach a minimum of 61 validity criteria. The results are then matched with the validity score criteria which can be observed in Table 1.

Table 1. Criteria for validity scores [14]

No	Score Percentage	Criteria
1	0% - 20%	Not very valid
2	21% - 40%	Not valid
3	41% - 60%	Quite valid
4	61% - 80%	Valid
5	81% - 100%	Very valid

The practical instrument of a student worksheet is viewed from the observation sheet on the implementation of learning and the results of the questionnaire responses given by students. The assessment data is in the form of a checklist on the response questionnaire that has been distributed. Assessment Student response analysis is based on the Guttman Scale in the form of statements with scores as shown in Table 2.

Table 2. Guttman Scale Criteria [14]

No	Response	Score
1	Yes	1
2	No	0

Instrument effectiveness of student worksheets in terms of whether or not there is an increase in students' critical thinking skills. The effectiveness of student worksheets can be done through the provision of pretest and posttest questions. Then the score from the question is used to calculate the N-Gain value. Student worksheets are declared effective if they meet a minimum of medium or high criteria. The score criteria can be observed in Table 3.

Table 3. Interpretation of N-gain Value [15]

No	Score N-Gain	Normalized Gain Criteria
1	0.70 > N-Gain	High
2	$0.30 \leq$ N-Gain ≤ 0.70	Medium
3	N-Gain < 0.30	Low

The use of N-gain is used to determine the category of process performance improvement and product performance, which is classified as a high improvement category if the N-Gain value is >0.70, medium category if the N-Gain value is 0.30-0.70, or low category if the value is N-Gain <0.30 as suggested by Hake [15].

3. Results and Discussion

3.1 Student worksheet validity

Before the student worksheets are implemented in learning, the student worksheets must go through a validation stage to test the level of validity [6]. Validation was carried out by 2 lecturers of physics education experts. The validation sheet is calculated based on the given score. The validation score by the validator on the development of student worksheets can be seen in Table 4.

Table 4. Validation score by validator

Aspect	Validator 1	Validator 2	Total	Percentage	Category
1	5	4	9	90%	Very valid
2	5	4	9	90%	Very valid

3	4	5	9	90%	Very valid
4	4	4	8	80%	Valid
5	4	4	8	80%	Valid
6	5	4	9	90%	Very valid
7	4	5	9	90%	Very valid
8	4	4	8	80%	Valid
9	4	4	8	80%	Valid
10	4	5	9	90%	Very valid
11	5	4	9	90%	Very valid
Average : 86.36 %					Very valid

Based on Table 4, there are 11 aspects of validity studied. The first aspect is that the material refers to the curriculum. In this aspect, a validation score of 90% is obtained and is included in the very valid category. In the second aspect, it covers some of the main concepts on the suitability of content. In this aspect, a validation score of 90% is obtained and is included in the very valid category. The third aspect, namely the activities carried out to support the understanding of the concept, obtained a validation score of 90% and was included in the very valid category. In the next aspect, namely the suitability of the display that contains aspects of the student activity sheet format, it makes it easier for students to learn to get a validation score of 80% and fall into the valid category. In the fifth aspect, namely the harmony of colors, text, images and tables, the validation score is 80% and is included in the valid category. Furthermore, the aspect containing STEM elements obtained a validation score of 90% and was included in the very valid category according to research [3-6]. Furthermore, the suitability of the presentation consists of several aspects in the seventh aspect, namely the aspect of clearly stating the learning objectives and obtaining a validation score of 90% and is included in the very valid category. In the eighth aspect, namely the aspect of including instructions or procedures for students clearly obtaining a validation score of 80% and included in the valid category. In the ninth aspect of the suitability of the presentation, namely critical thinking skills, it gets a validation score of 80% and is included in the valid category. In the last assessment aspect, namely linguistic suitability, consisting of the tenth aspect, namely conformity with standard and correct Indonesian language rules, it obtained a validation score of 90% and was included in the very valid category. Then in the eleventh aspect, namely the use of language effectively and efficiently, it obtained a validation score of 90% and was included in the very valid category.

Based on Table 4 which is described, it can be said that all aspects of the validation instrument are categorized as very valid, meaning that the developed student worksheets are suitable for use in learning [2-3].

3.2 Student Worksheets Practicality

Table 5. Practicality questionnaire based on assessment aspects

Aspect	Score	Category
1	100%	Very practical
2	100%	Very practical
3	80%	Practical
4	80%	Practical
5	80%	Practical
Average	88%	Very practical

The feasibility of the practical aspect was assessed from the results of the student response questionnaires given after the learning activities using the developed student worksheets [16]. The response questionnaire consists of 5 questions that will be answered by students with "Yes" and "No" answers. Assessment of the response questionnaire using the Guttman scale obtained a percentage value of 88% with a very practical category. This shows that the developed student worksheets are declared feasible from the practical aspect.

3.3 Student Worksheet Effectiveness
The effectiveness of student worksheets can be seen and measured from the learning outcomes before doing the learning (pretest) and the learning outcomes after doing the learning (posttest). The learning outcomes can be used to determine the improvement of students' critical thinking skills, if there is an increase, the student worksheets are proven to be effective in learning.

Table 6. Calculation results of N-Gain

Abs	Test Student Score			
	Pre-test	**Post-test**	**N-gain**	**Category**
1	30	70	0.57	Medium
2	30	80	0,71	High
3	70	90	0.67	Medium
4	30	70	0.57	Medium
5	40	70	0.50	Medium
6	50	70	0.40	Medium
7	60	90	0.75	High
8	30	70	0.57	Medium
9	50	70	0.40	Medium
10	60	80	0.50	Medium
Average	**45**	**76**	**0.56**	**Medium**

From the N-gain calculation above, an average score of 0.56 is obtained in the medium category. Based on the use of STEM-based student worksheets using PhET media on Hooke's Law material which is applied to guided inquiry learning is able to optimize students' critical thinking skills [2]. Table 6 shows that STEM-based student worksheets in guided inquiry learning with help of PhET Simulation media on Hooke's Law material obtained effective results in improving critical thinking skills of the State Senior High School 1 Sumberrejo Bojonegoro students.

4. Conclusion
Based on research that has been carried out the State Senior High School 1 Sumberrejo Bojonegoro by conducting learning using STEM-based student worksheets in guided inquiry learning with the help of PhET Simulation media on Hooke's Law materials to improve students' critical thinking skills, the results obtained in the form of the level of validity, practicality and effectiveness of the developed student worksheets. At the student worksheet validation stage, a score of 86.36% was obtained and was included in the very valid category. In the aspect of practicality, a score of 88% is obtained and is included in the very practical category. At the implementation stage using the one group pretest posttest design, the N-gain was 0.56 and was included in the medium category. Thus, it can be concluded that the developed student worksheets meet the criteria of being valid, practical and effective in improving the ability critical thinking.

References
[1] Ramli R, Yohandri Y, Sari, Y S, and Selisne M 2020 *J. Eksakta Pendidik.* **4** 10
[2] Aldila C, Abdurrahman A, and Sesunan F 2017 *J. Pembelajaran Fis.* **5** 85
[3] Syamsurizal S, Epinur E, and Marzelina D 2014 *J. Indones. Soc. Integr. Chem.* **6** 35
[4] Umbaryati U 2016 *PRISMA Pros. Semin. Nas. Mat.* 217
[5] Pangesti K I D, Yulianti, and Sugianto 2017 *J. Literasi Pendidik. Fis.* **2** 44
[6] Pertiwi R S, Abdurrahman A, and Rosidin U 2017 *J. Pembelajaran Fis.* **5** 11

[7] Facione P A 2013 *Critical Thinking: What it is and Why it Counts* Retrieved from: https://www.nyack.edu/files/CT_What_Why_2013.pdf
[8] Nur'Azizah H, Jayadinata A K, and Gusrayani D 2016 *J. Pena Ilm.* **1** 51
[9] Puspita A T and Jatmiko B 2013 *Inov. Pendidik. Fis.* **2** 121
[10] Nurmayani L and Nurmayani L 2018 *J. Pendidik. Fis. Teknol.* **4** 98
[11] Fitriyah L and Madlazim 2021 *J. Inov. Pendidik. Fis.* **10** 99
[12] Khairiyah Nida'ul, 2019 *Pendekatan Science, Technology, Engineering and Mathematics (STEM)* (Medan: Guepedia)
[13] Sukmana R W 2018 *Pendas: J. Ilm. Pendidik. Dasar* **2** 189
[14] Riduwan 2015 *Skala Pengukuran Variabel-Variabel Penelitian* (Bandung: Alfabeta)
[15] Hake R R 1998 *J. Phys Am. Assoc. Phys. Teach.* **66** 64
[16] Simatupang H, Sianturi A, and Alwardah N 2019 *J. Pelita Pendidik.* **7** 170

Improving the Ability to Develop Scientific Articles among Physics Teachers

T Sunarti*, Wasis, N Suprapto, and S Admoko

Physics Education Study Program, Universitas Negeri Surabaya, Indonesia

*Email: titinsunarti@unesa.ac.id

Abstract. There have been training scientific articles online for physics teachers. The purpose of this activity is to (1) improve Physics teachers in the ability to create scientific articles and (2) increase the number of scientific articles written by Physics teachers. Seventy-five physics teachers in Mojokerto attended the training. They implemented online by forming a synchronous meeting zoom class and continued with asynchronous assistance using WhatsApp group facilities. The success of this training workshop is evaluated using an assessment of the resulting article and a questionnaire of the trainee's response to the implementation of training activities. Scientific articles that have been obtained are included in the review process and have been revised to meet the criteria to be published in scientific journals. The article have been prepared aspects of the assessment of articles compiled by Physics teachers include titles and identities, abstracts, introductions, theoretical foundations, results and discussions, conclusions, and bibliography. The articles compiled have met the criteria well, with an average score of 87. The results of the trainee response questionnaire showed that 67 % strongly agreed and 33 % agreed. It shows that there is still a need to increase the quantity and quality of articles compiled by physics teachers.

1. Introduction

The learning from home policy began to be implemented on March 9, 2020, after the minister of education and culture issued circular number 2 of 2020 and number 3 of 2020 on online learning and working from home to prevent Corona Virus Disease (COVID-19). When the COVID-19 outbreak appeared, all human activities were restricted, including learning activities both at the elementary school level to the lecture level began to implement learning activities from home [1]. This is done to limit the spread of massive viruses [1].

Related to the position of teachers and lecturers, the Government has also implemented the Minister of Utilization of State Apparatus and Bureaucratic Reform No. 16 of 2009 concerning the Functional Position of Teachers and Their Credit Numbers. This regulation contains a change that the promotion and the class of teachers into higher class must make scientific papers or scientific articles [2,3]. This requires that all teachers, including high school teachers, must be able to compile a paper or scientific article [2,3]. Writing scientific work expresses ideas and conveys information effectively by using language so that others know and understand the information in descriptions, expositions, arguments, or persuasions [4]. Scientific articles are scientific studies in the form of ideas or thoughts based on empirical evidence or theoretical studies that can be proven true [5].

Related to scientific articles, many senior high school teachers who have not filled out articles in various media publications of articles or not many who make scientific journals [6] The low ability to write senior high school teachers is also seen from the rareness of senior high school teachers who participate in seminar activities by becoming a companion in national or regional seminars [7,10]. The results of the observation of the community service team in the field at the time of community service in 2019 obtained information that most (almost all) senior high school teachers in Mojokerto Regency had conducted Class Action Research (PTK) and made their reports. However, from senior high school teachers who had carried out the PTK, there were still few who published their research into scientific seminars or scientific journals. The observations also show that teachers still have difficulty conducting *literature reviews* to complete article creation and compile literature-based *articles* review. A literature-based *review* article is an article to analyze topics using secondary data in the form of journals or other references to produce descriptive data about the results of an analysis that has been done [8]. It led to many senior high school teachers who stopped their promotions due to bumps on the requirements of proof of writing scientific papers [7]. On this basis, the Community Service team of the Department of Physics UNESA, together with partners, namely MGMP Physics Mojokerto Regency, agreed to hold an Online *Literature* Review Based Article Making Training activity for MGMP Physics High School in Mojokerto Regency. The main purpose of this effort is to increase high school Physics teachers' ability to make literature review-based scientific articles in the Covid 19 pandemic and increase the number of scientific articles based on literature reviews or based on research which written by Physics teachers.

2. Method
Research-based "Training of Writing Scientific Article based literature review ore based on research" to improve the skills of teachers in writing scientific articles was conducted in October - November 2020 online using the Zoom Meeting platform with links https://zoom.us/j/96792119046?pwd=aHlITG1SY0NUWTF3ckk1eXJ2OFRJdz09 and WA group. The subjects used in this study were 75 trainees who were teachers of MGMP Physics members from high school and vocational school, the junior high school both public and private located in the District and Mojokerto City.

The components of the article writing skills are developed through three stages of training. They are the delivery of suitable writing materials, the implementation of scientific articles examples (or modelling) based on literature review or research, and the implementation of theories obtained into workshops to create literature-based or review-based scientific articles conducted in groups where group members come from several different schools within a month with technical coordination of online mentoring through WhatsApp. The instruments used in this study are article assessment instruments and the questionnaire response of trainees to implement training activities. The data was analyze by providing a maximum score of three an aspect. The assessment according to Tanjung and Ardial, includes title and identity, abstract, preliminary, theoretical foundation, results and discussion, conclusions, and bibliography [9]. The article assessment instrument concludes the feasibility of scientific articles and analysis of response questionnaires with categories strongly agree (SS), agree (S), disagree (TS), and strongly disagree (STS) on each indicator assessed to conclude the response to the conductibility of training.

3. Results and Discussion
3.1. Training activities
The performance of scientific article writing materials aims to understand trainees' benefits and technicalities of writing scientific papers. In this training, the materials presented include (1) continuous professional development, (2) theory literature review, (3) modelling literature review, (4) paraphrase engineering, and (5) scientific articles. then, practicing aspects of writing scientific articles include (1) collecting references of journals and other relevant references, (2) constructing abstract, (3) compiling introductions, (4) compiling literature studies, (5) conducting analysis preparations, and (6)

Concluding [11]. In general, the presentation of material and discussion activities went very well. Participants and keynote speakers showed a positive response with the number of questions that emerged from participants regarding the training material. This also showed that there were still many things that still needed to be known regarding the technical preparation of scientific writing articles to be published in online scientific journals. The process of conducting the training can be seen in Figure 1.

Figure 1. Screenshot of training activities

The implementation of training activities in Figure 1 indicates that the initial knowledge of participants related to the technical preparation of scientific papers to be published in online scientific journals is relatively lacking. However, after being given training, the understanding level of trainees showed promising results. This statement is reinforced by the results of the questionnaire on 30 participants shown in Table 1.

Based on the exposure of the results of the questionnaire in Table 4, the two appear that the average percentages of positive responses (Strongly Agree and Agree) is 64.75 % and 37.25 %, respectively. No one gives a negative response (Disagree and Strongly Disagree). This shows that all participants rate positively on the implementation of training activity. High assessments were given to participants for the response that the training supported the profession's development (86 %) and that facilitators mastered the training well (86 %). The benefits of training for participants (81 %) were relevant to the needs of teachers in schools (81 %). Slightly low assessments on delivery methods that were less attractive were done online (48 %), and participants provided less effective training responses because they could not be provided offline mentoring.

Table 1. Results of response training participants

No	Indicators	Response			
		SA	A	D	SD
1	This training is useful for me	81 %	19 %	0	0
2	Training materials are well organized	57 %	43 %	0	0
3	Have relevance to the needs of teachers in school	81 %	19 %	0	0
4	The delivery of training materials is communicative	58 %	42 %	0	0
5	Delivery of interesting training materials	48 %	52 %	0	0
6	The speaker mastered the training materials well	86 %	14 %	0	0
7	Training materials helped me in developing skills in literature review-based article writing	58 %	42 %	0	0
8	Through this training, I gained a clear picture of literature review-based article writing	58 %	42 %	0	0
9	The implementation of the training takes place effectively	48 %	52 %	0	0
10	Implementation time in accordance with training activities (not too long and not too short)	63 %	38 %	0	0
11	Support the improvement of the development of the teacher profession	86 %	14 %	0	0
12	If the FMIPA Unesa Community Service team holds another training, I will gladly take the training	53 %	47 %	0	0
	Average	64.75	37.25	0	0

Information: SA = Strongly Agree, A = Agree, D = Disagree, SD = Strongly Disagree

3.2. Results of Articles

Table 2 shows the implementation of the training activities in the form of writing literature review articles, scientific articles that have obtained input in the review process and have been revised so that they meet the criteria to be published in online scientific journals that have been prepared based on components assessed on scientific article assessment instruments.

Based on Table 2, participants' ability to compile scientific articles is excellent by getting an average of 87. The input and advice given, among others, are that the abstract's content includes a brief background, goals, methods, results, and conclusions in abstract writing. The background writing of the problem lists the references used.

Table 2. Average assessment of 2 articles of training participants

No	Rated Components	Maximal Score	Average Score
Article Identity			
1	The title of the article is written in the form of a phrase that does not mean double, describing the content of the article	3	3
2	Register the author's name, email address, and institution	2	3
Abstract			
3	Abstracts are arranged in the form of a single paragraph containing goals, methods, and results	3	3
4	Abstracts are arranged with a maximum number of 250 words and pay attention to conformity with keywords	3	3
Introduction			
5	There is rationalization and urgency of the problem/purpose studied supported by relevant references	4	3
6	The theoretical foundation is formulated in-depth based on theoretical and empirical studies relevant to the problem/purpose studied	4	3
7	Research ideas and ideas are formulated by prioritizing *state of art*	2	2
Theoretical Foundation			
8	Relevance of theoretical studies to research topics	6	5
9	Adequacy of supporting theories for use in the proof or resolution of the problems discussed	7	5
10	Validity/truth of the theory written	10	8
Results and Discussions			
11	Discussion of research results is carried out clearly and quickly, relevantly with the formula/ purpose, and the theoretical foundation used	10	9
12	Quality of argument or proof provided	10	8
13	Breadth and depth studied	10	8
Conclusion			
14	The formulation of the conclusion describes the full results of the study	7	7
References			
15	Libraries/references used are sufficient	3	3
16	Current reference sources used	3	3
17	Libraries/references used in accordance with the topic	3	3
Language			
18	The writing of symbols, formulas, notations, and terms is used appropriately	5	5
19	Articles are written in Indonesian/English that is good, correct, and following scientific rules	5	5
Number of Scores		**100**	87

Figure 2 illustrates the sample of scientific article produced through the training process. Accordingly, the paper aims to determine the effectiveness of simple KIT-based media in improving students learning achievement in dynamic electricity material as an example of the training results.

Studies in Learning and Teaching
Homepage: https://scie-journal.com/index.php/SiLeT
Email: silet@scie-journal.com

p-ISSN: 2722-399X; e-ISSN: 2722-1857
SiLeT, Vol. 1, No. 3, December 2020: 209-217
©2020 Studies in Learning and Teaching

Media Innovation Based on Simple Kit for Dynamic Electricity Learning

*Subkan Nina[1]
[1] Ministry of Religion Kab. Mojokerto, Madrasah Aliyah Negeri 2 Mojokerto, Mojokerto 61361, East Java, Indonesia

Article Info

Article history:

Received October, 31 2020
Revised December 7, 2020
Accepted December 29, 2020
Available Online December 30, 2020

Keywords:

Simple KIT
Dynamic Electricity
learning achievement
increase
learning outcomes

ABSTRACT

This study aims to determine the effectiveness of simple KIT-based media in improving student learning achievement in dynamic electricity material. The sample used was the students of class grade 12 of IPA-3 Madrasah Aliyah Negeri 2 Mojokerto, totaling 35 students. The type of research used is research development (Research and Development). The research data were obtained from questionnaire responses during learning and the results of the pre-test and post-test. The questionnaire's responses were as many as 85.14% of students who thought that the development of KIT-based learning media was very effective in using dynamic electricity learning. Moreover, as many as 14.85% think that KIT-based learning media is creative and interactive learning and is an additional supplement in the teaching and learning process. From the test results given to students, there was a significant increase in students' achievements. This indicator can be seen from the t count value is smaller than the t table (t count <t table), then H1 (learning with media treatment) can be applied in learning. Another indicator is that the value of p-value is smaller than alpha 5% (0.05) Ho (learning without treatment) is rejected, so it is concluded that there is a significant increase between learning outcomes before using Simple KIT-based media after implementing learning media using simple KIT.

Figure 2. Sample of scientific article as a product of the physics training

Figure 3 illustrates the sample of scientific article produces through the training process. Accordingly, the paper aims to determine the effectiveness of simple KIT-based media in improving student learning achievement in dynamic electricity material as an example of the training results.

ABSTRAK

Salianto : PENGGUNAAN QUIPPER SCHOOL DALAM PEMBELAJARAN JARAK JAUH PADA MASA PANDEMI COVID-19 DI SMAN 1 BANGSAL KABUPATEN MOJOKERTO

Kata Kunci : Penggunaan Quipper School, Pembelajaran Jarak Jauh.

Pemerintah Indonesia menerapkan kebijakan untuk memutus penyebaran Covid-19 memunculkan berbagai dampak pada kehidupan masyarakat, termasuk bidang pendidikan dengan mengubah metode pembelajaran dari mayoritas tatap muka menjadi Pembelajaran Jarak Jauh (PJJ). Pembelajaran Jarak Jauh (PJJ) digunakan untuk mengajar kelas maupun berkomunikasi dengan siswa. Untuk itu bermunculan begitu banyak platform yang menawarkan kemudahan, isi, dan harga yang bervaraiasi.

Penelitian ini untuk mendeskripsikan hasil belajar sisiwa materi Teori Kinetik Gas dengan menggunakan platform Quipper School. Sampel penelitian adalah sisiwa Klas XI MIPA. Lebih dari 20% waktu guru bekerja dihabiskan untuk membuat tugas, mengoreksi, dan menilaihasil kerja siswa. Dengan memanfaatkan teknologi, Quipper School membantu meringankan beban guru dalam :1) Mengelola kelas, 2) Mengirim tugas ke siswa, 3) Mengoreksi, menganalisa, dan menilai hasil pekerjaan siswa secara efisien, 4) Merangkum performa individual siswa.

Dengan strategi menggunakan Quipper School, diperoleh hasil rata-rata nilai mata pelajaran Fisika materi Teori Kinetik Gas pada siswa kelas XI MIPA 82,5. Nilai rata-rata ini memenuhi ketuntasan pembelajaran klasikal di SMAN 1 Bangsal Kabupaten Mojokerto. Sehingga dapat disimpulkan bahwa penggunaan Quipper School sangat bermanfaat dalam peningkatan kemampuan *memahami konsep Fisika* dalam materi *Teori Kinetik Gas* .

Figure 3. Another sample of scientific article as a product of the physics training

The purpose of this study was to describe the effectiveness of using Quipper School in distance learning for the Kinetic Theory of Gas. So it can be concluded that the use of Quipper School is very useful in improving the ability to understand the concepts of Physics in the Kinetic Theory of Gases

Both articles have complied with the rules of article writing, in terms of titles and identities, abstracts, introductions, theoretical foundations, results and discussions, conclusions, and bibliography.

4. Conclusion

Based on study results, it can be concluded that the article compiled by physics teachers in Mojokerto has met the criteria both with an average score of 87 with aspects of assessment in the form of titles and identities, abstracts, introductions, theoretical foundations, results and discussions, conclusions and bibliography. The results of the trainee response questionnaire showed that 67 % strongly agreed and 33 % agreed. This shows that there is still a need to increase the quantity and quality of articles compiled by physics teachers.

References

[1] Abidah A, Hidaayatullaah H N, Simamora R M, Fehabutar D, and Mutakinati L 2020 *Stud. Philos. Sci. Educ. (SiPoSE)* **1** 38
[2] Redhana W I 2012 *J. Pengabdi. Kpd. Masy. Widya Laksana* **2** 74
[3] Undang-Undang Nomor 20 Tahun 2003 *Sistem Pendidikan Nasional* (Jakarta: Depdiknas)
[4] Syafi'i I 1988 *Retorika dalam Menulis* (Jakarta: Depdikbud)
[5] Suparno and Yunus 2007 *Keterampilan Dasar Menulis* (Jakarta. Universitas Terbuka)
[6] Udil P A 2021 *J. Nas. Pengabdi. Masy.* **2** 21
[7] Wahyuningtyas N and Ratnawati N 2018 *J. Praksis Dedik. Sos.* **1** 40
[8] IKIP Malang 1996 *Pedoman Penulisan Karya Ilmiah: Skripsi, Tesis, Desertasi, Artikel, Makalah, Laporan Penelitian* (Malang: BAAKPSI)
[9] Tanjung H B N and Ardial H 2010 *Pedoman Penulisan Karya Ilmiah dan Mempersiapkan Diri Menjadi Penulis Artikel Ilmiah* (Jakarta: Prenada Media Group)
[10] Suprapto N 2020 *Stud. Philos. Sci. Educ. (SiPoSE)* **1** 50
[11] Jatmiko W, Santoso H B, Purbarani S, Syulistyo A R, Firmansyah D, Yusuf M, A'yunina Q, and Laili N A 2015 *Panduan Penulisam Artikel Ilmiah* (Jakarta: Fakultas Ilmu Komputer Universitas Indonesia)
[12] Pusat Bahasa Diknas 2003 *Kamus Besar Bahasa Indonesia* (Jakarta: Balai Pustaka)

E-LS on The Subject of Temperature: The First Work of a Science-Physics Teacher at Siti Aminah Junior High School Surabaya

A Kholiq*, I Sucahyo, T Sunarti, D H Kusumawati, and M N R Jauhariyah

Physics Department, Faculty of Mathematics and Natural Sciences, Universitas Negeri Surabaya, Indonesia

*E-mail: kholiq@unesa.ac.id

Abstract. The effect of COVID-19 pandemic in Indonesia requires learning that is usually done face-to-face has turned into online learning. For teachers, the implementation of online learning requires engaging media so that learning becomes more various and exciting. This study aims to analyse the validity of the science literacy-based e-book, which will later be called E-LS. The first work of the Science-Physics teacher at Siti Aminah Junior High School Surabaya can train students' scientific literacy skills. This study using a research method with the DDD-E design (Decide, Design, Develop, Evaluate). Based on the data analysis that has been done, we obtained descriptive-quantitatively using the Likert scale percentage score criteria. The results show that the percentage of validity in the aspect of Curriculum Conformity of 91.30 % (very valid). The material aspect is 93.45 % (very valid). The language aspect is 91.67 % (very valid). The media design aspect of 91.91 % (very valid). Scientific literacy aspect of 87.50 % (very valid). Overall the level of validity of the E-LS has an average of 91.17 % (very valid). It can be concluded that the E-LS media (temperature topic) is very suitable for use as a science-physics learning medium to train science literacy for junior high school students.

1. Introduction

The world of education in Indonesia has changed due to the coronavirus (COVID-19) pandemic, which lasted 1.5 years. Learning that used to be face-to-face now has to be done online. In implementing online learning, teachers need skills in making media with an adequate IT base, such as making scientific literacy-based e-book media (E-LS), so that online learning can take place better, interestingly and more various. To improve teachers' ability to design science literacy-based e-book, science-physics teachers participate in training to make science literacy-based e-book independently. In line with the research by Sudarisman that science learning contributes to technological development, science is the fundamental science that underlines technological development [1]. Based on ICASE, students need adequate scientific literacy skills to live productively and obtain the best quality of life as the goal of science education itself [2]. It is also in line with the research results [3], which states that science is fundamental in all aspects of life. Therefore, it is necessary to learn and teach to achieve scientific literacy (science literacy community) and retain national character. The fact that the 2018 PISA test results obtained from the scientific literacy test of Indonesian students showed that they were ranked 72 out of 78 countries with a scientific literacy score of 396 [4,5]. It showed that the scientific literacy ability of Indonesian students is still low. PISA defines scientific literacy as the capacity to use scientific knowledge, identify questions, and draw conclusions based on facts and data to understand the universe and make decisions about changes due to human activities [6]. One indicator is that an individual with high scientific literacy skills can master the concept and understand its application in everyday life and technology [7].

The teaching and learning process is communication between teachers and students in conveying a message or material delivered by the teacher or a learning resource into a visual and verbal communication symbol. Communication of teachers and students in the delivery of material should be arranged as attractive as possible so that students can be motivated and learning can be meaningful for students. Fitriani et al state that learning-science, meaningfulness can be obtained if the students' scientific literacy skills are promising [8]. Meaningful learning can be carried out if students connect the newly acquired knowledge with previous knowledge [9].

In implementing learning by performing the 2013 curriculum, which leads to 21st-century competencies by utilizing advances in science and technology, a media that supports science and technology-based learning is needed, creating e-book (digital books). The application of e-book in the learning process is one alternative that teachers can use to integrate technology to make learning more exciting and varied [10]. E-Book is an application program that a teacher should develop in the 21st century to provide a fun learning process by prioritizing students' understanding of concepts. E-books have been understood only in the text that can be opened with a PC or gadget.

The development of E-LS is different from the e-book that other researchers have developed. The difference is that features are integrated to train students' scientific literacy in E-LS and provide examples of physics concepts in everyday life related to the concept of temperature for junior high school students. So it can be seen that this study aims to analyze the feasibility level of the E-LS as seen from the level of validity of the E-LS on the temperature topic developed by the science teacher at SMP Siti Aminah as a support for online learning during the covid-19 pandemic.

2. Method

E-LS development uses DDD-E (Decide, Design, Develop, Evaluate) research design according to Figure 1. Analysis uses the technique-quantitative descriptive with E-LS content validity instruments in terms of Curriculum Conformity Aspects, Material Aspects, Media Design Aspects and Scientific Literacy Aspects, and for further analysis using quantitative descriptive analysis with Likert scale according to Table 1. The criteria for scoring on the E-LS content validity instrument using a Likert scale are shown in Table 1.

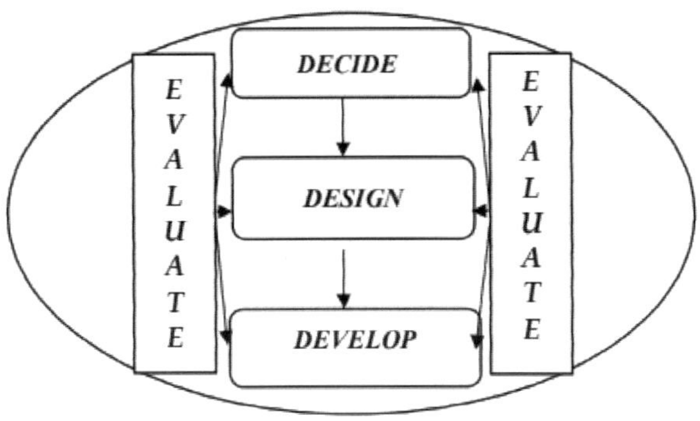

Figure 1. DDD-E research design [11]

Table 1. The criteria of the E-LS content validity instrument [12]

Score	Criteria
4	Very Valid (VV)
3	Valid (V)
2	Invalid (IV)
1	Very Invalid (VIV)

The percentage of student response sheets is calculated using the formula:

$$\text{Score percentage} = \frac{\Sigma\ \text{Score}}{\Sigma\ \text{Maximum Score}} \times 100\ \% \quad (1)$$

The criteria for the percentage score of the E-LS validity instrument can be seen in Table 2, and the developed E-LS is said to be feasible if the percentage of E-LS validity is ≥ 61 %.

Table 2. E-LS validity interpretation criteria [12]

Percentage	Information
$0\ \% < x \leq 20\ \%$	Very Invalid
$21\ \% < x \leq 40\ \%$	Less Valid
$41\ \% < x \leq 60\ \%$	Quite Valid
$61\ \% < x \leq 80\ \%$	Valid
$81\ \% < x \leq 100\ \%$	Very Valid

3. Results and Discussion

3.1 E-LS Product that has been Developed by the Teacher of SMP Siti Aminah Surabaya

The science literacy-based e-book (E-LS) developed by the science teacher at SMP Siti Aminah on temperature is different from the e-book developed by previous researchers. This E-LS contains temperature material and is accompanied by a video associated with scientific literacy indicators. It can motivate and train students' scientific literacy skills, thus making online learning more exciting and various. The following is a display of the E-LS that has been developed by a science teacher at SMP Siti Aminah Surabaya.

Figure 2. Display cover E-LS

Figure 2 show the display cover of E-LS. It represents the physics-science content material, that is, temperature at 7 grade of junior high school. It is validated by the expert judgment that is the resource persons in training on making E-LS from experts in the field of physics learning media. Besides, there are experts on the physics content and experts on developing physics learning based on scientific literacy.

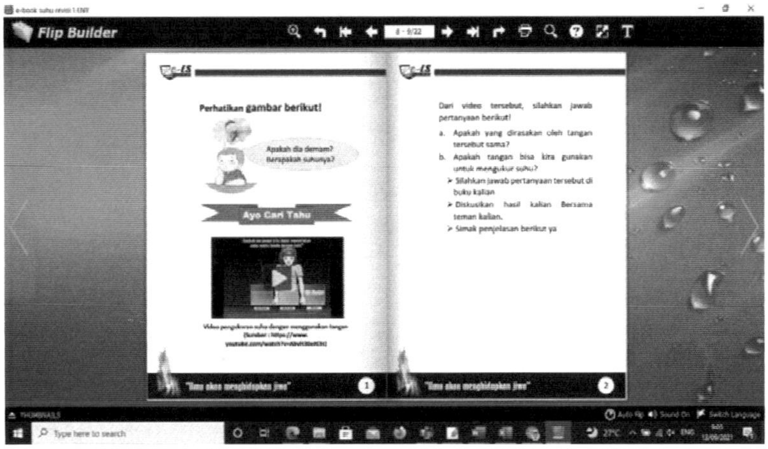

Figure 3. E-LS content display (motivation video)

There are three competencies of scientific literacy: explaining phenomena scientifically, evaluating and designing scientific inquiry, and interpreting data and evidence scientifically [5,6]. E-LS was developed based on that competencies. Figure 3 shows the E-LS content display for the motivation video. This part stimulates the students to explain phenomena scientifically through contextual phenomena that can be explained scientifically. Figure 4 shows the role of E-LS to serve the competence of evaluating and designing scientific inquiry through practice or experiment activity. Based on the data and evidence obtained in experimental activities, students are trained to interpret data and evidence scientifically.

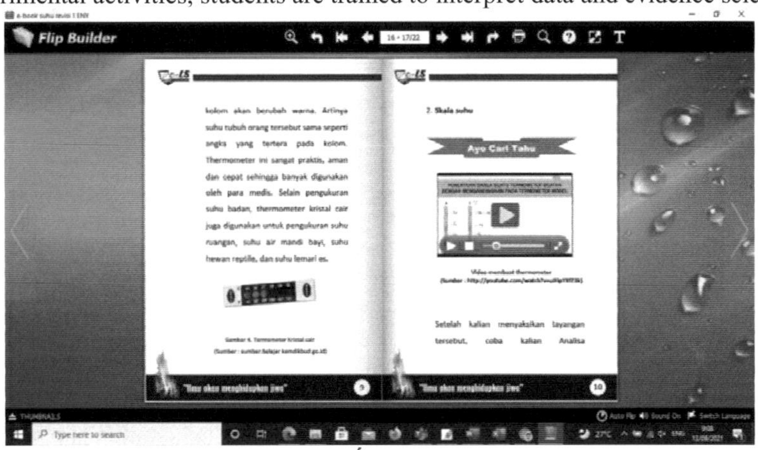

Figure 4. Content display of E-LS (video practicing science literacy)

Based on the data analysis that has been carried out from the validity process, obtained overall validity is shown in Figure 5. The results show that the percentage of validity in the aspect of Curriculum Conformity of 91.30 % (very valid). The material aspect is 93.45 % (very valid). The language aspect is 91.67 % (very valid). The media design aspect of 91.91 % (very valid). Scientific literacy aspect of 87.50 % (very valid). Overall the level of validity of the E-LS has an average of 91.17 % (very valid).

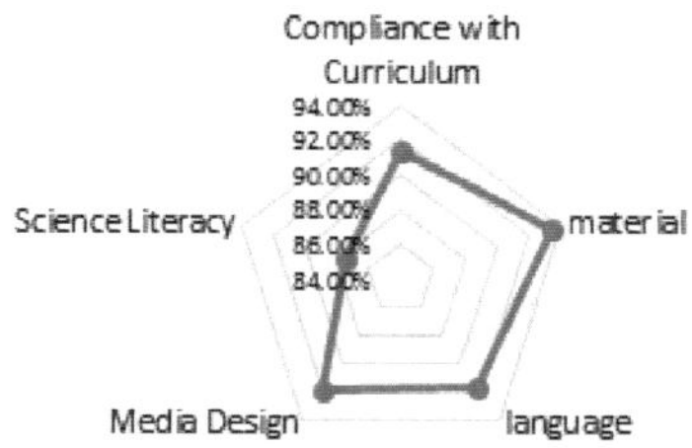

Figure 5. E-LS validity of all aspects

The validity of each aspect is shown in Figure 6 until Figure 8. Figure 6 shows the E-LS validity aspects of conformity with the 2013 Curriculum. The components of curriculum validation are compliance with the 2013 curriculum, content standard, learning process standard, and evaluation standard. Figure 7 shows the material aspect of E-LS Validity. The validity components are basic competence, adequacy, the need for teaching materials, the material substance, and consistency.

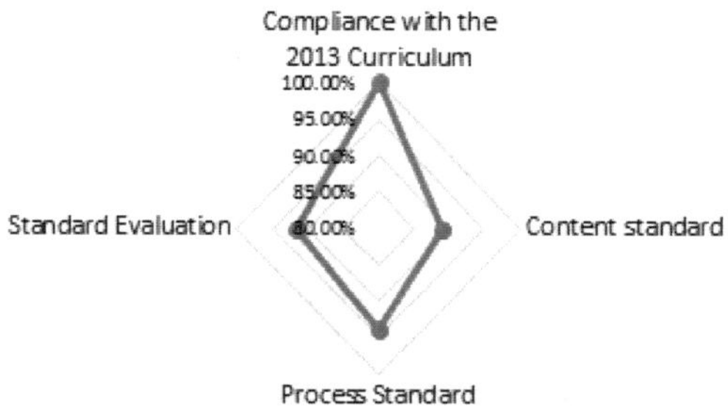

Figure 6. Material aspect E-LS validity of compliance with curriculum

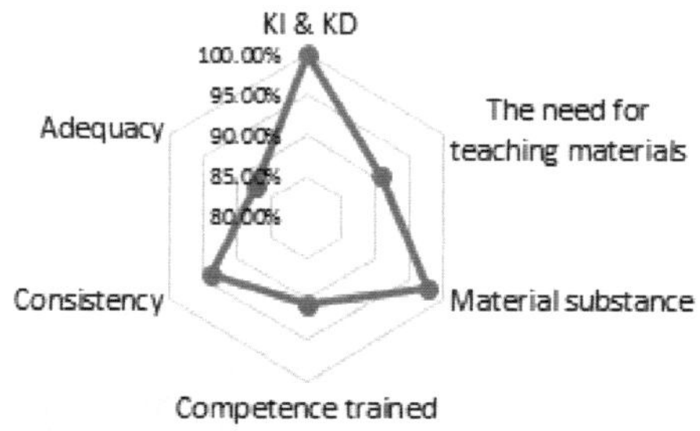

Figure 7. Material aspect E-LS validity

Figure 8 shows the result of E-LS media design aspect validity. The validity of the media design is based on the principle of learning media, the media principle and curriculum, also writing systematics and media design.

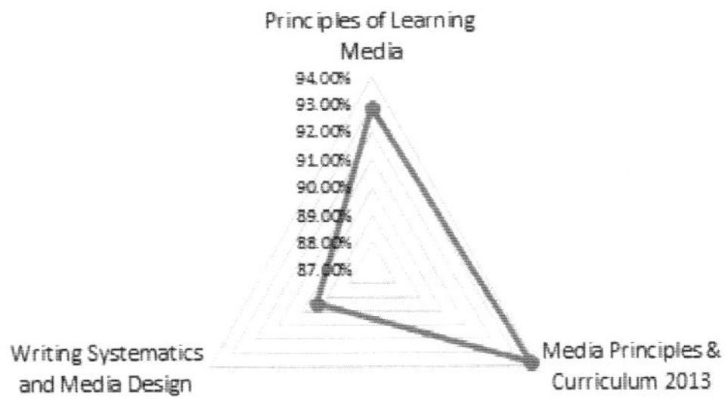

Figure 8. Media aspect E-LS validity

Figure 9 shows the result of the E-LS validity for the science literacy aspect. It is based on the competencies of scientific literacy, that is, explaining phenomena scientifically, evaluating and designing scientific inquiry, and interpreting data and evidence scientifically [5,6].

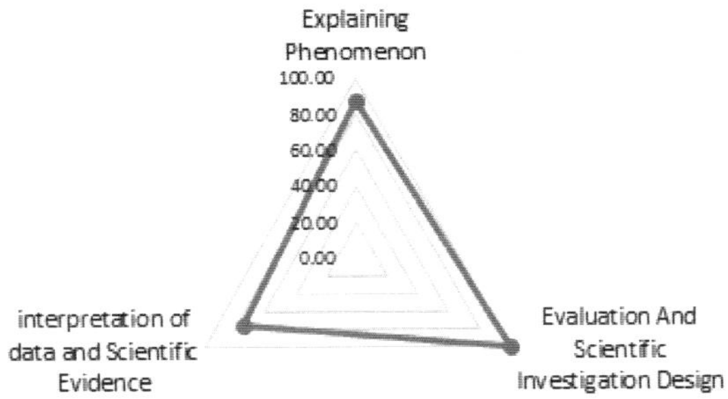

Figure 9. Validity of E-LS aspects of scientific literacy

Figure 5 illustrates that the results of the validity of the E-LS developed by the Physics Science teacher of SMP Siti Aminah, which is the first work, have an average validity of 91.17 % with very valid criteria. But suppose you look at the E-LS Validity diagram. All aspects can be seen that the validity of the scientific literacy aspect has the lowest validity compared to other aspects. In that case, the teacher as the maker of the E-LS work is not familiar with science literacy competencies which is something new for teachers of SMP Siti Aminah, which incidentally is a private junior high school in Surabaya. Also in the 2013 curriculum for junior high school level does not include scientific literacy skills. In the junior high school curriculum, it is stated that basic competence (KD) for temperature only has understanding, applying and conducting experiments on a concept. Teachers tend to focus on core competence (KI) and KD in the curriculum. As results, teachers are less likely to provide examples of scientific literacy on the temperature topic. For the validity of conformity with the 2013 curriculum, materials, language, and media design have balanced validity.

3.2 The Temperature Concept in E-LS

Temperature is the degree of hotness and coldness of an object. To determine the degree of hotness of an object, requires a device called a thermometer [13]. In the E-LS, it is described that measuring the temperature of an object cannot be done by hand but must use a thermometer to produce accuracy in measuring the object temperature, according to the video contained in the E-LS (Figure 10).

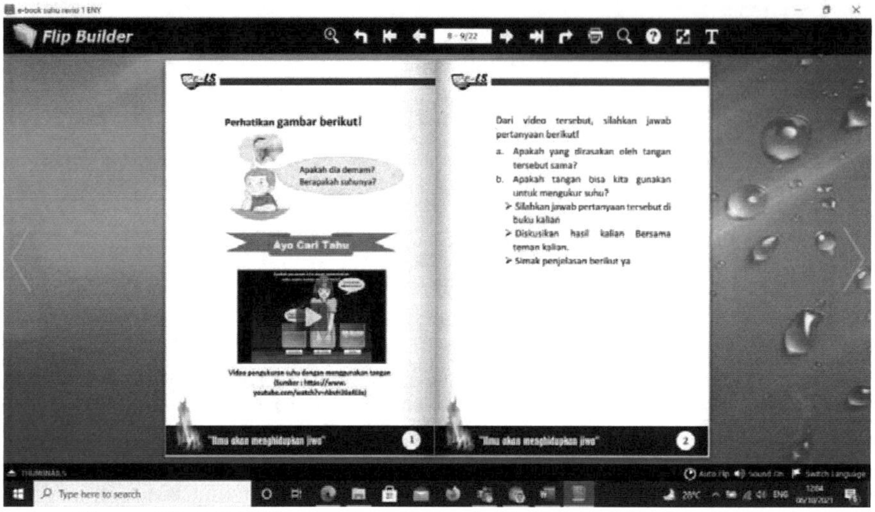

Figure 10. E-LS: Temperature determination cannot be done by hand

In learning, when explaining the concept of temperature, students are invited to think and be literate that measuring the temperature of an object cannot be done by hand but must use a tool called a thermometer.

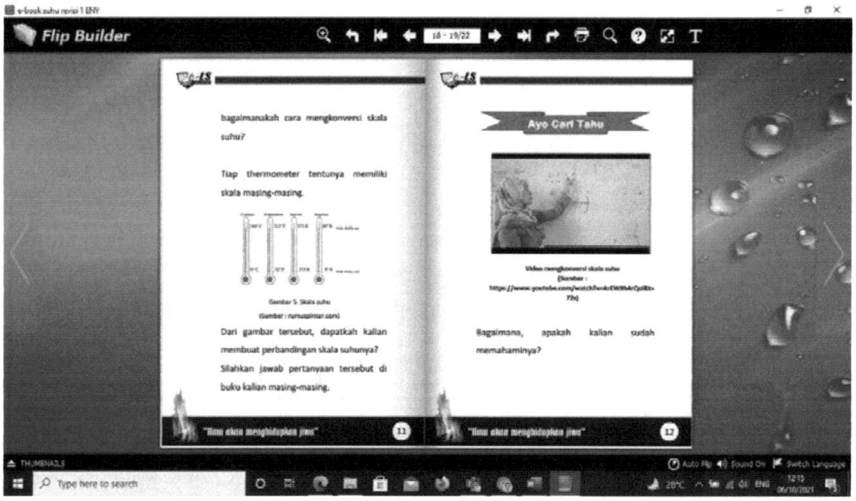

Figure 11. E-LS: Temperature scale conversion

Temperature is a physical quantity that has units. There are several units of temperature, namely: degrees Fahrenheit, degrees Celsius, degrees Reaumur and Kelvin. The temperature unit requires conversion. For example, it can be done using equation [13] from degrees Celsius in the conversion to degrees Fahrenheit. In E-LS, a teacher is given a video explaining how to convert temperature units according to the concepts in physics literature books. It aims to make distance learning students feel like a teacher is explaining to them in class the concept of temperature unit conversion.

The research results on E-LS development are in line with research by Ravenscroft, which states that 21st-century learning is process-oriented learning, solving a problem, and the ability to adapt and work effectively, and efficiently in different situations [14]. The purpose of learning in the 21st century in the 2013 curriculum is to prepare students to enter the world of work with a broader scope. The more comprehensive range of work is caused by high mobility in times like today. It can be taken as a common thread that students must adapt and survive outside their environment. In other words, the goals of 21st-century learning are shifting to be global. The demands of individuals in the 21st century should have hard skills and soft skills to enter the world of work and be ready to compete with other countries. The core subjects and interdisciplinary 21st-century themes are surrounded by three sets of skills most in-demand in the 21st century said that among the 21st-century skills are (1) learning and innovation skills; (2) information, media, and technology skills; and (3) life and career skills [15].

This research is also in line with the results of a study related to Assessment and Teaching for 21st Century Skills (ATCS), which concludes that four main things are related to 21st-century life skills: ways of thinking, working, work tools, and life skills. Ways of thinking include creativity, critical thinking, problem-solving, decision making and learning. The form of working provides communication and collaboration with others. Tools for work include Information and Communication Technology (ICT) and information literacy. Life skills include living as a citizen, life and career, personal responsibility, and social responsibility [16]. In implementing learning in the 21st-century, 21st-century skills are needed to be trained on students. Twenty first century skills include critical thinking skills.

In the E-LS, there are several examples of how scientific literacy is related to the concept of temperature. The E-LS adopts the concept of scientific literacy, which defines scientific literacy as a person's understanding of science and can apply that science in social life [18]. The scientific literacy indicators used in the E-LS are based on the OECD's scientific literacy indicators, which include (a) explaining phenomena scientifically, (b) evaluating and designing scientific investigations, and (c) interpreting data and evidence scientifically [6,18]. Scientific literacy skills are also in direct contact with higher-order thinking skills (HOTS), which have indicators of analyzing, evaluating, solving problems, thinking logically, creatively, and systematically [19]. Scientific literacy skills, which are also related to higher-order thinking, require someone to apply new information or knowledge and manipulate data to reach possible answers in new situations [20].

Online-based learning has the advantage of reducing paper usage and being more communicative. In this case, e-book, IT media can be used to practice indicators of these science process skills such as observing, classifying, analyzing, predicting, and communicating. In addition, online learning media can be used to develop several elements in critical thinking and literacy skills, such as analyzing, evaluating, applying, generating ideas, and expressing ideas [18,21-23].

Concerning the results of the E-LS validity, the results of this study are in line with the research conducted by AT Kusumawati, Wasis, IGM Sanjaya, and Abd. Kholiq. Their results found that Elite (E-Book Literacy) for Junior High School Student's Scientific Literacy in Solar System Materials obtained valid results with very valid criteria and suitable for junior high school students [24].

4. Conclusion
Based on the data analysis, the E-LS (science literacy e-book) on the topic of temperature (developed and is the first work of a teacher at SMP Siti Aminah Surabaya) that has very valid criteria. The E-LS media is suitable for being used as an online learning medium and can train students' scientific literacy skills.

Acknowledgement
This acknowledgement for Universitas Negeri Surabaya. It is part of the Faculty of Mathematics and Natural Sciences policy programme of community service at Universitas Negeri Surabaya based on Decree No. 823/UN38/HK/PM/2021 date June 21st 2021, with the research contract number 35459/UN38.3/PM.06.03/2021.

References
[1] Sudarisman S 2015 *Florea: J. Biol. Pembelajaranya* **2** 29
[2] International Council of Associations for Science Education (ICASE) 2008 *Covid-19 Phenomenon-based Learning (PhBL) Engineering Challenge* Retrieved from: http://www.icaseonline.net/

[3] Liliasari and Fitriana 2014 *Persiapan Literasi Sains Generasi Muda Indonesia Menjelang ASEAN Community Prosiding Seminar Nasional Pendidikan IPA 2014* (Surabaya: Universitas Negeri Surabaya) 1.
[4] Ministry of Education and Culture 2017 *Guidelines for the National Literacy Movement* (Jakarta: Ministry Education and Culture)
[5] OECD 2019 *PISA 2018 Assessment and Analytical Framework* (Paris: OECD Publishing)
[6] OECD 2017 *PISA 2015 Science Framework* (Paris: OECD Publishing)
[7] Yager R E 2000 *Sch. Sci. Math.* **100** 327
[8] Fitriani W and Lestari I 2014 *J. Pendidik. Pembelajaran Khatulistiwa* **3** 4432
[9] Haristy D R, Enawaty E, and Lestari I 2013 *J. Pendidik. Pembelajaran Khatulistiwa* **2** 4002
[10] Gueval J, Tarnow K, and Kumm S 2015 *Teach. Learn. Nurs.* **10** 181
[11] Tegeh I M and Kirna I M 2012 *Research Methods for Educational Development* (Singaraja: Graha Ilmu)
[12] Riduwan 2015 *Measurement Scale of Research Variables* (Bandung: Alpha Beta)
[13] Serway R A and Jewett J W 2004 *Physics for Scientists and Engineers 6th Edition* (California USA: Thomson Brooks/Cole)
[14] Ravenscroft A 2012 *J. Comput. Assist. Learn.* **28** 235
[15] Trilling B and Fadel C 2009 *21st Century Skills: Learning for Life in Our Times: The Benefits of Teachers' Workshops on Their Social and Emotional Intelligence in Four Countries* (San Francisco: John Wiley & Sons)
[16] Trisdiono H 2013 *21st Century Learning Strategy* (Yogyakarta: Young Widyaiswara Provincial Education Quality Assurance Institute)
[17] Ananiadou K and Claro M 2009 *21st Century Skills and Competences for New Millennium Learners in OECD Countries* (Paris: OECD Publishing)
[18] Wasis, Rahayu Y S, Sunarti T, and Indana S 2020 *HOTS and Science Literacy: Concepts, Learning and Assessment* (Jombang: Kun Fayakun)
[19] Shidiq A S, Masykuri M, and Susanti V H E 2014 *J. Pendidik. Kim.* **3** 83
[20] Brookhart S M 2010 *How to Assess Higher-Order Thinking Skills in Your Classroom* (Virginia: ASCD)
[21] Kustijono R and Zuhri F 2018 *IOP Conf. Ser. Mater. Sci. Eng.* **296** 012025
[22] Martono K T and Nurhayati O D 2014 *Int. J. Comput. Sci.* **11** 168
[23] Irma Z U and Kustijono R 2017 *J. Inov. Pendidik. Fis.* **6** 224
[24] Kusumawati A T, Wasis, Sanjaya I G M, and Kholiq A 2020 *J. Phys. Conf. Ser.* **1491** 012070

Research Trends on Ethnoscience based Learning through Bibliometric Analysis: Contributed to Physics Learning

H N Hidaayatullaah*, **N Suprapto, E Hariyono, B K Prahani, and D Wulandari**
Science Education Postgraduate Program, Universitas Negeri Surabaya, Indonesia

*Email: hasan.20011@mhs.unesa.ac.id

Abstract. The research purposes are 1) Analyse the results of bibliometric mapping visualization on research trends ethnoscience-based learning and its contribution to physics learning, 2) Analyse author collaboration on research trends ethnoscience-based learning, and 3) Recommendations for future ethnoscience research in physics learning. In this research, the descriptive method is applied with a bibliometric approach. Data was obtained through the Scopus and Google Scholar databases published in the period 2011-2020. The results showed that ethnoscience-based learning made a major contribution to science learning at the elementary to junior high school level and physics at the high school to university level. Ethnoscience can be integrated with learning innovations in schools and can also train/improve students' thinking skills. Ethnoscience also dominates the social sciences. The recommended research trends in the Scopus and Google Scholar databases are physics learning about ethno-stem, development of ethnoscience-based physics teaching materials, ethnoscience-based physics learning to train students' scientific literacy and problem-solving skills.

1. Introduction

Indonesia is a country with a wide variety of natural resources and cultural traditions [1]. Globalization in the 21st century has transformed the Indonesian original cultural values, causing local culture to decrease and be forgotten. In the current era of globalization, students are more familiar with foreign cultures and do not understand the culture and local wisdom of the Indonesian, so that the sense of nationalism of students begins to fade. For the existence of culture and local wisdom to remain strong, students as the nation's next-generation need to be instilled with a sense of love for culture and local wisdom by integrating cultural knowledge in the learning process [2–4]. Therefore, an educational breakthrough is needed, especially in learning that combines culture with science or commonly called ethnoscience.

Learning science, particularly physics, is an important aspect of preparing students to respond to societal problems socially, critically, and imaginatively [5–7]. Physics does not only memorize its products in the form of concepts, principles, laws, and theories but also conducts a re-examination to create a scientific attitude in students. Physics is the study of the natural environment [8,9]. In learning physics, a pleasant atmosphere and conditions are needed. According to the facts on the ground, students' ability to integrate physics study into daily life is still lacking [10]. For this reason, teachers must be able to incorporate local cultural values into the science and physics learning process. Teaching science involves a curriculum and approach that goes beyond theory and facts to boost the relevance of education [11,12]. Students will be able to study facts and phenomena that occur in society more easily using ethnoscience-based learning, which may be combined with scientific knowledge [13–15].

Authors all over the world have conducted ethnoscience research at the primary, intermediate, and tertiary levels. For example, research with the title Top 100 Cited Publications in Physics Education in The last Thirty Years: A Bibliometric Analysis [16], about students' innovative and creative thinking skill profile in designing chemical batik after experiencing ethnoscience integrated science technology engineering mathematic integrated ethnoscience (ethno-stem) learnings [17], anything about The validity of e-module based on guided inquiry integrated ethnoscience in high school physics learning to improve students' critical thinking [18], there is also about Local and scientific knowledge in the school context: Characterization and content of published works, etc [19]. Several researchers focus on literature studies on ethnoscience-based learning using bibliometric analysis [19].

Bibliometrics has 2 roots, namely (i) Biblio and (ii) Metrics [20]. The term "metrics" denotes knowledge of meters or measurements, derived from the Latin or Greek word metrics meaning measure [21]. Bibliometrics is a term that uses mathematical and statistical methods to study and identify patterns in the use of literature/publications as analysis material to determine the development of specific literature, especially authorship, publication, and use [22]. The purpose of bibliometrics is to explain the process of written communication and the nature and direction of developing descriptive means of calculating and analyzing the various facets of communication [23–25].

Ethnoscience study is an attempt to integrate culture into the existing science curriculum, based on past research. Studies on bibliometric literature review, on the other hand, have not been considered. So, this study focuses on the trend of ethnoscience based learning research in the last 10 years interval with three research questions:

a. How the results of bibliometric mapping visualization on research trends ethnoscience-based learning and its contribution to physics learning?
b. How does the author collaborate on research trends in ethnoscience-based learning?
c. What are the recommendations for future ethnoscience research in physics learning?

2. Method

A descriptive research method with a bibliometric approach was used in this study. Bibliometrics aims to measure the progress of the publication of scientific articles and scientific contributions [26]. Bibliometric analysis was carried out by looking at the publications distribution to evaluate the contribution of articles to the progress of knowledge of various literature using a statistical approach [27]. Bibliometric reviews are a sort of review that focuses on quantitative research in journals, articles, books, and other forms of written communication [28]. This study was conducted by searching the Scopus and Google Scholar databases. The article metadata search was based on parameters that matched the study emphasis with the keyword ethnoscience in learning in the search results. The stage is to collect/harvest metadata in the 2011-2020 range. Exploring data on the Scopus and Google Scholar databases. Publish or Perish (PoP) is software to make it easier to search metadata in the Google Scholar database. Furthermore, to visualize metadata about ethnoscience-based learning, VOSviewer software is used. Making visualizations in the form of maps is by exporting the search results from the database into RIS format, then inputting them into the VOSviewer. The results of data visualization shown by VOSviewer were used to carry out an analysis of ethnoscience-based learning and its relationship to physics learning.

3. Results and Discussion

3.1 The results of bibliometric mapping visualization on research trends ethnoscience-based learning and its contribution to physics learning

Based on the search results of metadata about ethnoscience-based learning from the Scopus database in the last ten years (2011-2020) obtained as many as 230 documents from 434 throughout the year. From the Google Scholar database with the help of the Publish or Perish (PoP) search software, 386 documents were obtained from a maximum number of results of 500 documents. Figure 1 shows a network visualization of the VOSviewer software on ethnoscience-based learning from the Scopus database.

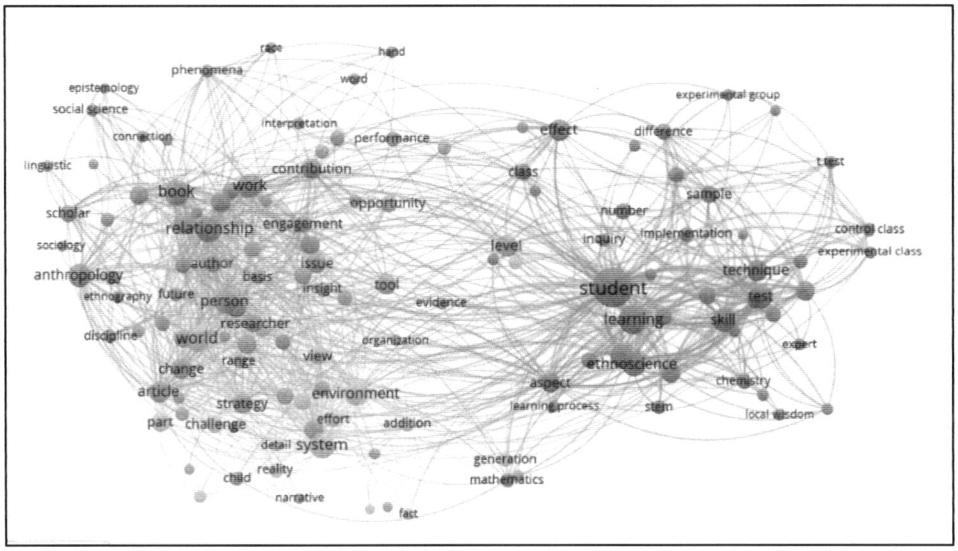

Figure 1. Network visualization from term ethnoscience-based learning in the last 10 years (2011-2020) on the Scopus database

The display in Figure 1 is a network visualization of the ethnoscience-based learning with a minimum number of occurrences of 5 of the 5,200 terms, 227 meet the threshold resulting in 136 items composed of 4 clusters with 3,910 links. In cluster 1, red is related to the terms of student, learning, ethnoscience, local wisdom, implementation, learning process, etc. Cluster 2 in green is related to the terms of strategy, opportunity, level, issue, person, etc. In Cluster 3, blue is related to the terms of relationship, interpretation, contribution, work, social science, etc. The last cluster is cluster 4 in yellow with related terms system, environment, reality, fact, etc.

If we look at the relationship between the term ethnoscience and learning itself, from the visualization of VOSviewer it is clear that the two have a very close and close relationship. The most dominant visualization display of ethnoscience-based learning is the term/keyword student and the second is learning (Figure 2), where the student itself has 125 links with a total link strength of 588 and occurrences of 70. The term student is connected with learning, such as learning process, implementation, strategy, skill, inquiry, stem, etc.

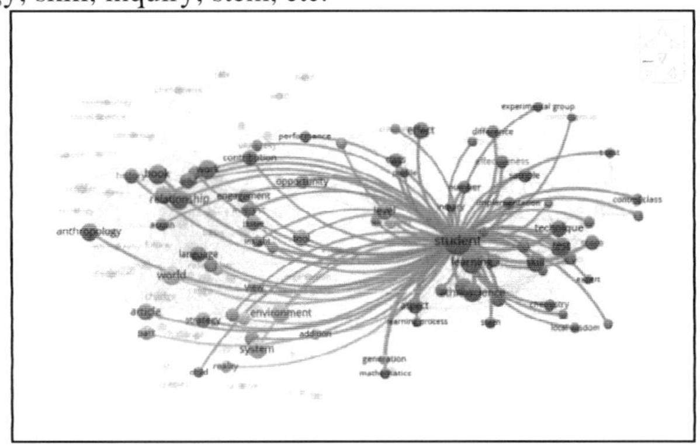

Figure 2. The dominant term in the visualization of ethnoscience-based learning outcomes in the Scopus database

Figure 2 also describes the various data collections carried out by the author in researching ethnoscience-based learning such as samples in research, experimental groups, control classes, experimental classes, test data collection methods, questionnaires, surveys, and so on. This means that the authors researched ethnoscience-based learning using quantitative and qualitative research methods. In addition, it is also seen that ethnoscience-based learning over the past 10 years has dominated the learning of science (physics) and social science (Figure 3).

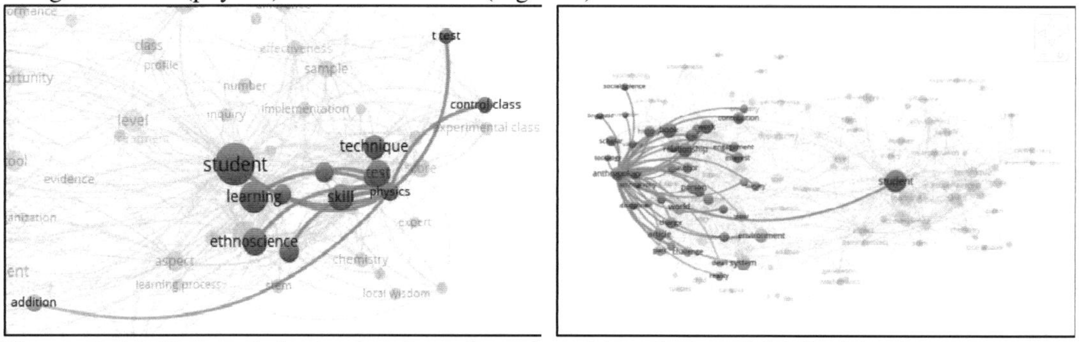

Figure 3. (a) Terms of physics in ethnoscience-based learning) and (b) Social science terms on ethnoscience-based learning

In Figure 3a, it can be seen that physics is related to learning, students, and skills. This means that ethnoscience-based physics learning can train/improve students' skills during the learning process. This is in line with research that the effectiveness of learning physics can improve students' creative skills [29]. The term ethnoscience-based learning does not only focus on learning about science which consists of the physical sciences and life sciences but also dominates the branches of social sciences such as sociology, history, anthropology, language, and culture (Figure 3b).

Furthermore, based on the metadata search results about ethnoscience-based learning from the Google Scholar database with the help of the Publish or Perish (PoP) search software, 386 documents were obtained from a maximum number of results of 500 documents.

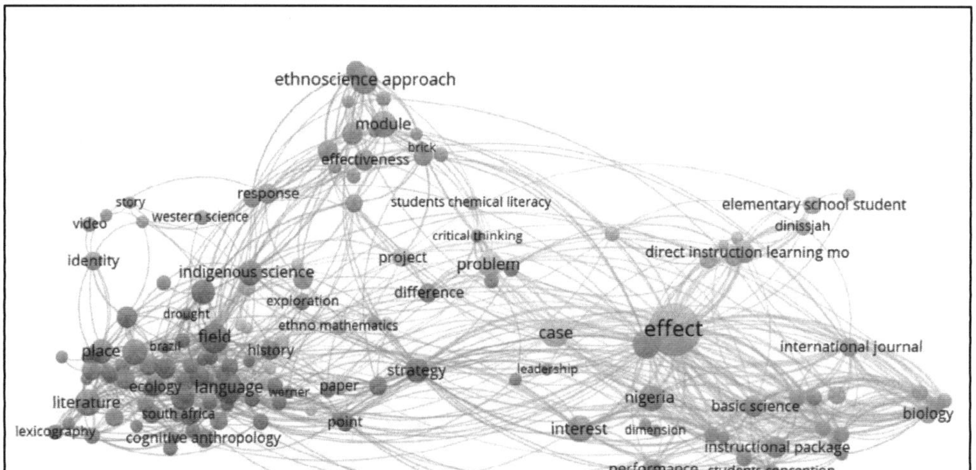

Figure 4. Visualization of the ethnoscience-based learning network in the last 10 years (2011-2020) on the Google Scholar database

The network visualization in Figure 4 about ethnoscience-based learning on the Google Scholar database with a minimum number of occurrences of a term 3 of the 3180 terms, 291 meet the threshold resulted in 156 items composed of 8 clusters. Cluster 1 (red) is related to the terms of language, cognitive, history, etc. Cluster 2 (green) is related to the terms ethnoscience approach, response, effectiveness, etc. Cluster 3 (dark blue) is related to the terms place, literature, indigenous science, etc. Cluster 4 (yellow) is related to the terms problem, project, ethnomathematics, etc. Cluster 5 (purple) is related to the terms of students' conception, basic science, etc., Cluster 6 (light blue) is related to the terms effect, physics learning, etc. Cluster 7 (orange) is related to the terms of innovations, scientific, etc. The last cluster is cluster 8 (brown) related to the terms critical thinking skills, scientific science, etc. Where each cluster consists of terms that describe the keyword ethnoscience-based learning.

From the Google Scholar database, it can be seen that during the 2011-2020 period there were more researchers on the ethnoscience approach, which can be seen in Figure 5a that the ethnoscience approach has a total link strength of 33 with 14 occurrences. Based on the visualization of the google scholar database, it can be seen that the application of ethnoscience has been implemented in physics learning (Figure 5b).

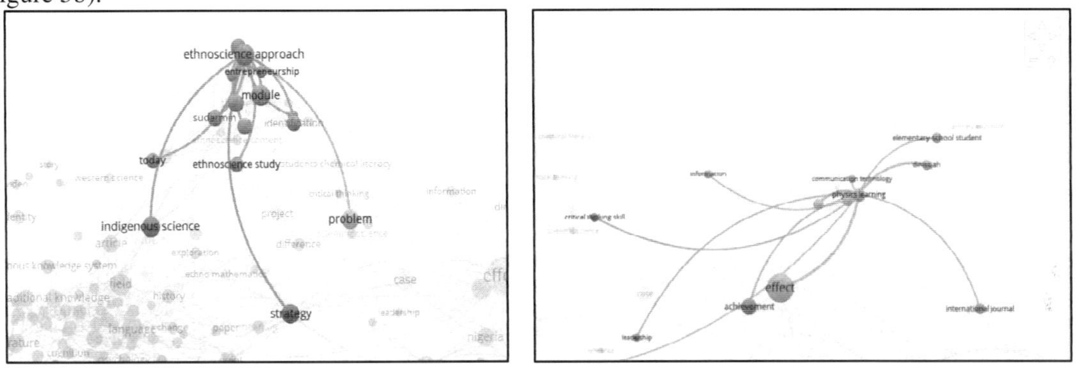

Figure 5. (a) Network visualization in terms of ethnoscience approach and (b) Network visualization in terms of physics learning

In Figure 5b, it can be seen that physics learning related to ethnoscience-based learning is connected with terms in the same cluster, namely elementary school student, effect, communication technology. In addition, physics learning is also related to leadership, achievement, and critical thinking skills. This means that ethnoscience-based learning researchers in physics learning are not only studied at the high school level but also the elementary school (science) level by applying learning innovations to improve students' thinking skills. In line with research that ethnoscience can be integrated into learning at the high school level [9]. In addition, ethnoscience in learning is also closely related to critical thinking, problems, ethnomathematics, and student chemical literacy (Figure 6). This means that ethnoscience in learning in addition to improving cognitive can also train students' skills.

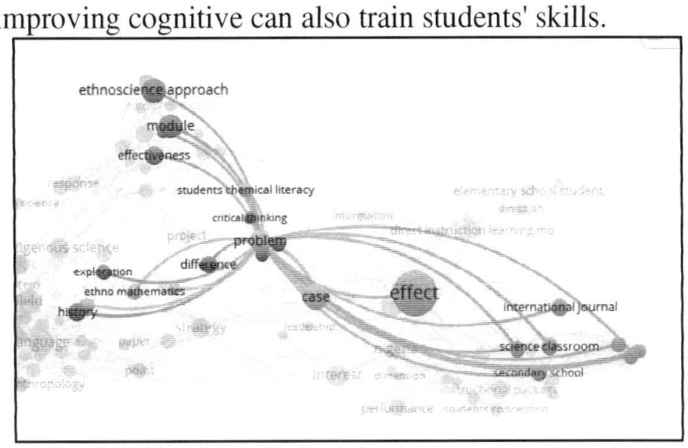

Figure 6. Network visualization in terms of critical thinking, literacy, and other thinking skills

Furthermore, the VOSviewer display also displays metadata in the form of overlay visualization and density visualization. Overlay visualization can show traces of history in ethnoscience research in learning and density visualization which shows the density/emphasis on research groups on ethnoscience in learning over the last 10 years (Figure 7).

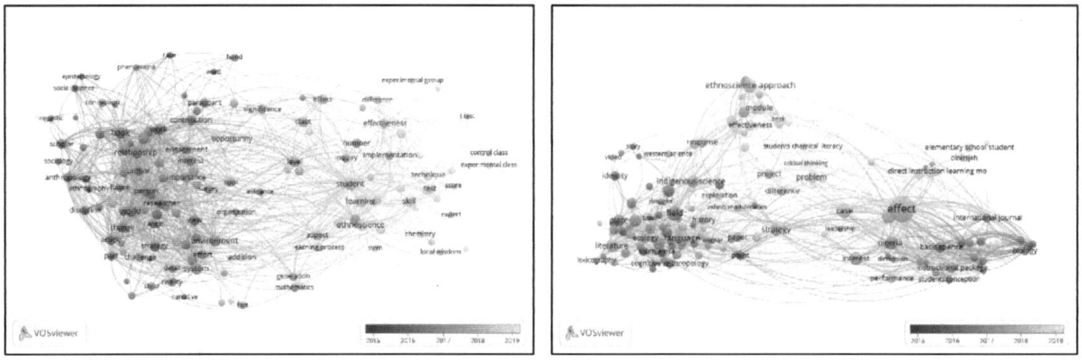

Figure 7. (a) Overlay visualization of ethnoscience-based learning on the Scopus database and (b) Overlay visualization of ethnoscience-based learning on the google scholar database

Each colour in the variable represents the last 10 years (2011-2020), the yellow colour indicates the most recent research conducted in 2020. While dark blue represents research that was conducted several years ago. It can be seen that each cluster is connected with the title, keywords, abstract, and words in the abstract about ethnoscience-based learning. Researchers can also detect future trends and crucial changes in clusters, identify study gaps and research limits where they can contribute to future research fields [30].

3.2 Visualization of the author's collaboration on the ethnoscience-based learning research trend

The results of the Co-authorship visualization on ethnoscience-based learning with a minimum number of documents of an author 1 of the 490 authors in the Scopus database and 656 authors in the Google Scholar database obtained visualization results as shown in Figure 8. Based on the Scopus and Google Scholar databases, it can be seen that the top three researchers during the 2011-2020 period were Sumarni et al., Sudarmin et al., and Wiyanto, et al. This shows that the authors who have researched about ethnoscience based learning for the last 10 years are Indonesian authors.

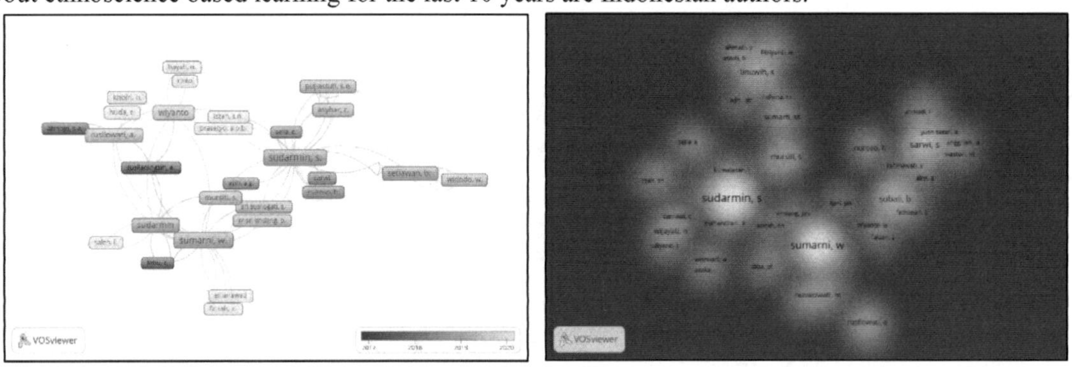

Figure 8. (a) Overlay visualization co-authorship on ethnoscience-based learning (Scopus database) and (b) Density visualization co-authorship on ethnoscience-based learning (Google scholar database)

3.3 Recommendations for future ethnoscience research in physics learning

The trend of research that has not been carried out but has the potential to be carried out is visible in the word variable (concrete or abstract) which has a little circle in the bibliometric analysis shown by VOSviewer. If the term variable does not exist, it will be a novelty for future researchers who want to research ethnoscience in learning and its contribution to physics learning. Based on the bibliometric visualization analysis of ethnoscience-based learning results, the recommended research trends in the Scopus and Google Scholar databases are physics learning about ethno-stem, development of ethnoscience-based physics teaching materials ethnoscience-based physics learning to train students' scientific literacy and problem-solving skills. The literature review, in this case, is qualitative and relies on personal judgment [31].

4. Conclusion

Based on the results of the analysis and discussion, the research trend of ethnoscience-based learning for the last 10 years (2011-2020) from the Scopus database consists of four clusters with the most dominant terms being students and learning. Meanwhile, the Google Scholar database consists of eight clusters with the most dominant terms being effects and ethnoscience approaches. Ethnoscience-based learning greatly contributes to physics learning, both at the elementary school (physics science) to high school and college. Ethnoscience can be integrated with learning innovations in schools and can also train/improve students' thinking skills. Besides being able to contribute to science learning (physics), ethnoscience also dominates social science. The recommended research trends in the Scopus and Google Scholar databases are physics learning about ethno-stem, development of ethnoscience-based physics teaching materials ethnoscience-based physics learning to train students' scientific literacy and problem-solving skills.

References

[1] Hidaayatullaah H N, Dwikoranto, Suprapto N, Mubarok H, and Wulandari D 2020 *J. Phys. Conf. Ser* **1491** 012053
[2] Suprapto N, Prahani B K, and Cheng T H 2021 *J. Pendidik. IPA Indones.* **10** 69
[3] Hikmawati, Suastra I W, and Pujani N M 2021 *J. Phys. Conf. Ser* **1816** 012105
[4] Parmin P and Fibriana F 2019 *J. Penelit. Pembelajaran IPA* **5** 2
[5] Lubis S S, Sahyar S, and Derlina 2021 *J. Phys. Conf. Ser.* **1811** 012081
[6] Silvini Y, Derlina, and Ginting E M 2020 *J. Phys. Conf. Ser* **1485** 012048
[7] Zidny R, Sjöström J and Eilks I 2020 *Sci. Educ.* **29**
[8] Dwikoranto, Jatmiko B, Hariyono E, Lestari N A, Prahani B K, and Suyidno 2021 *J. Phys. Conf. Ser.* **1805** 012006
[9] Haspen C D T and Syafriani 2020 *J. Phys. Conf. Ser* **1481** 012056
[10] Maftukhin A, Mundilarto, and Kuswanto H 2019 *Int. J. Recent Technol. Eng.* **8** 1C2
[11] Nurcahyani D, Yuberti, Irwandani, Rahmayanti H, Ichsan I Z, and Rahman M 2021 *J. Phys. Conf. Ser.* **1796** 012094
[12] Pradipta D D, Madlazim, and Hariyono E 2021 *IJORER Int. J. Recent Educ. Res.* **2** 3
[13] Atmojo S E, Lukitoaji B D, and Muhtarom T 2021 *J. Phys. Conf. Ser* **1823** 012001
[14] Kurniawan D A, Asrial, Syahrial, Salsabilah W S, Kurniawati E F, Anandari Q S, Perdana R, Lumbantoruan A, Nasih N, Samosir S C, and Dewi U P 2019 *Univers. J. Educ. Res.* **7** 2789
[15] Khoiri A, Nulngafan N, Sunarno W, and Sajidan S 2019 *J. Ilm. Pendidik. Fis. Al-Biruni* **8** 2
[16] Suprapto N, Prahani B K, and Deta U A 2021 *Libr. Philos. Pract.* 5599
[17] Sudarmin, Sumarni W, Mursiti S, and Sumarti S S 2020 *J. Phys. Conf. Ser* **1567** 022037
[18] Kurniawan R and Syafriani 2021 *J. Phys. Conf. Ser* **1876** 012067
[19] Sotero M C, Alves Â G C, Arandas J K G, and Medeiros M F T 2020 *J. Ethnobiol. Ethnomed.* **16** 23
[20] Linnenluecke M K, Marrone M, and Singh A K 2020 *Aust. J. Manag.* **45** 2
[21] Nimsai S, Yoopetch C, and Lai P 2020 *Sustain.* **12** 18
[22] Todeschini R and Baccini A 2016 *Handbook of Bibliometric Indicators: Quantitative Tools for Studying and Evaluating Research* (Weinheim, Germany: Wiley-VCH)
[23] Yanuarti E A and Suprapto N 2021 *Stud. Philos. Sci. Educ.* **2** 7
[24] Novia N, Permanasari A, and Riandi R 2021 *J. Phys. Conf. Ser.* **1806** 012209
[25] Masitoh P N A, Latifah S, Saregar A, Aziz A, Suharto, and Jamaluddin W 2021 *J. Phys. Conf. Ser.* **1796** 012009
[26] Paltrinieri A, Hassan M K, Bahoo S, and Khan A 2020 *Int. Rev. Econ. Finance* **69** 389

[27] Ayudha C F H and Setyarsih W 2021 *J. Pendidik. Fis. Undiksha* **11** 16
[28] Setyaningsih I, Ciptono W S, Indarti N, Iman N, and Kemal V 2019 *E3S Web Conf.* **120** 01001
[29] Irwandani, Rinaldi A, Pricilia A, Sari P M, and Anugrah A 2020 *J. Phys. Conf. Ser.* **1467** 012006
[30] Vallaster C, Kraus S, Merigó Lindahl J M, and Nielsen A 2019 *J. Bus. Res.* **99** 226
[31] Liu Z, Yin Y, Liu W, and Dunford M 2015 *Scientometr.* **103** 135

The Effect of ARICESA on Achievement Motivation and Understanding of Basic Science Concepts for Prospective Teacher Students

D A Putra[1], R Setiani[2], Dwikoranto[3]*, and Widiasih[4]

[1]Informatics Engineering, Faculty of Engineering, Universitas Negeri Surabaya, Indonesia
[2]Universitas Bhinneka PGRI, Indonesia
[3]Physics Education, Faculty of Science and Mathematics, Universitas Negeri Surabaya, Indonesia
[4] Physics Department, Faculty of Teacher and Training, Universitas Terbuka, Indonesia

*Email: dwikoranto@unesa.ac.id

Abstract: The aim of this research to determine effect of the application ARICESA on student teacher achievement motivation and understanding of basic concepts. The subjects in this study were 34 students in the Department of Elementary School Teacher Education, Universitas Bhinneka PGRI Indonesia. One group pretest-posttest design was used in this study. The achievement motivation of prospective teachers is measured by a motivational questionnaire and understanding of basic concepts is measured by a conceptual test of concept understanding as a pre-test and post-test. Data analysis with t-test and prerequisite test and N-gains core. The results showed a significant increase in motivation effect for 5 indicators of achievement motivation with moderate criteria with an N-gain value range of 0.62 to 0.66 and the normalized mean N-gain is 0.65 in the medium criteria. There is a significant effect of increasing understanding of basic science concepts with a range of N-gain coefficients from 0.72 to 0.77 and a normalized average N-gain of 0.75 with high criteria.

1. Introduction

The role of science subjects in elementary schools is very important, because it becomes the scientific foundation to further explore natural phenomena [1]. Therefore, students must get the correct science concepts in learning in elementary schools. The figure who is very influential on the achievement of students' needs for the scientific basis is the elementary school teacher. This means that prospective teachers must have the correct understanding of concepts and have high motivation to continue to develop their competencies. The fact states that some elementary school teacher candidates still have low motivation to learn science and low concept understanding in learning science [2,3].

Based on the analysis of Trends in International Mathematics and Science Study (TIMMS) data in 2011 from 26 countries as a sample, it is stated student's learning motivation is in low criteria, specifically in the Southeast Asian region [4]. The data was updated in the 2015 TIMMS survey which specifically involved Indonesia, and it was stated that the level student's learning motivation in Indonesia was in low criteria [5]. Middle school students have low motivation to learn [6,7] and their learning motivation tends to decrease in the span of one academic year, especially for classes with conventional teacher-centered learning [8,9].

Prospective science teachers who become students in science lecture programs are also detected to have a tendency to decrease their motivation to learn in the span of one semester running [10-12]. The survey conducted on primary school teachers in Turkey also showed low teacher motivation to continue learning and self-actualization [13], which was thought to be caused by a lack of knowledge and skills of prospective teachers in terms of performance management and self-motivation strategies. Prospective teachers who are not able to motivate themselves so that they are not motivated themselves will have difficulty motivating students to learn. This will have an impact on the low quality of learning motivation. Prospective teachers must have superior competencies with various 21st century skills. The teaching and assessment [ATC21s] into four categories, one of which is ways of thinking [14]. The categories of ways of thinking consist of critical thinking, innovation, creativity [15], also metacognition [14]. Motivation, confidence, thinking category to succeed are the capital to achieve these skills. Understanding the concept is also the basic capital to reach the category level of ways of thinking. This shows that learning motivation and understanding of concepts are supporting target of 21st century skills.

Learning independence and motivation have a positive relationship which is one of the 21st century skills and graduate competencies that refer to the Indonesian Qualification Framework (IQF). Motivation is the main factor for independent learning [16,17]. Students with high motivation, they will use a lot of time studying independently [18,19]. Extrinsic and Intrinsic motivation is adaptive which encourages and maintains independent learning. The higher the student's learning motivation, it is suspected that the higher the level of learning independence will be.

The ARICESA model that has been developed is a learning model related to learning motivation and understanding science concepts for students. The ARICESA model is the development of Keller's ARCS motivation model, with the addition of several aspects, namely Inquiry (I), Enjoyment (E), and Self-Assessment (A). The Inquiry aspect in this model consists of several stages, namely: confronting students with problems, collecting experimental data and hypotheses, observing, interpreting data, presenting results and verifying results. The aspects that make up this model are outlined in the model syntax or learning flow, namely Attention (A), Relevance (R), Inquiry (I), Confidence (C), Enjoyment (E), Satisfaction (S), and Self-Assessment (SA). Each phase has goals that must be achieved in a complete series of learning processes, namely: (a) Attention phase to focus attention and arouse students' enthusiasm when starting the learning process, (b) Relevance phase to show learning objectives and the relevance of the material to the context of life, (c) Inquiry phase to actively involve students to construct conceptual understanding, (d) Confidence phase to generate confidence and confidence to succeed in learning in students, (e) Enjoyment phase to make students feel comfortable and happy in learning. the learning process, (f) the Satisfaction phase to make students feel valued and able to appreciate the effort and results in learning, and (g) the Self-Assessment phase to give students the ability to be able to assess, reflect, and develop further learning plans.

Based on the foregoing, purpose this research to examine the effect of ARICESA in overcoming the learning motivation of prospective teachers and their conceptual understanding of basic science. Because it is very urgent to apply when teaching in elementary schools after graduating from college in Elementary School Teacher Education.

2. Method

One group pretest posttest research design was used on 34 prospective science teacher students in the School Teacher Education at the University of Bhinneka PGRI Indonesia. To see the effect of applying the achievement motivation of prospective science teacher students, it is done in 2 ways, namely by using the Wilcoxon Signed Rank test and N-gain. N-gain used to determine the level of increase in achievement motivation. Wilcoxon Signed Rank Test used to determine the significance effect of the increase [20].

The description of understanding of basic science concepts for prospective teacher students uses N-gain and paired t-test. The use of N-gain is intended to determine the criteria for increasing understanding of the basic science concept of effects from the application of ARICESA. While the use of paired t-test is intended to determine the significance of the increase. The significance of increasing concept understanding between the results of the pre-test and post-test illustrates the significant positive effect of the results of the application of ARICESA. Measurement of understanding of basic science concepts is carried out with pre-test instruments before learning and post-test after learning is carried out. The tests carried out included six learning topics that had been implemented, namely quantities and units, simple machines, temperature, electricity, magnetism, and life processes.

3. Results and Discussion

3.1 ARICESA effect on achievement motivation

The results of the analysis of achievement motivation on 34 prospective students for basic science teachers at the University of Bhinneka PGRI Indonesia are described statistically in Table 1.

Table 1. Description of achievement motivation results

No	Description	Initial Motivation	Final Motivation
1	N	34	34
2	*Mean*	1.9062	3.9171
3	*Median*	1.8900	3.8900
4	*Mode*	1.5500	3.8100
5	*Variance*	0.1150	0.0560
6	*Skewness*	0.3760	0.2020
7	*Kurtosis*	-0.7830	-0.6380
8	*Range*	1.1900	0.9300
9	*Minimum*	1.3800	3.5000
10	*Maximum*	2.5700	4.4300
11	*Sum*	64.810	133.18

Based on Table 1, the average initial motivation and final motivation are 1.91 and 3.92, respectively, which indicates an increase in the average student learning motivation. The mean value of learning motivation data shifted from 1.89 as initial motivation to 3.89 and as late motivation. The data mode of learning motivation is 1.55 for initial motivation and 3.81 for final motivation, respectively. The mean, median, and mode data show a measure of central tendency that is close to each other between the three, both initial and final motivation.

The results differences in initial and final achievement motivation for basic science teacher candidates obtained the diversity of data on initial motivation and final motivation, respectively, 0.115 and 0.056. Skewness shows a value of 0.376 as initial motivation and 0.202 as late motivation, both of which indicate the slope of the data distribution to the left or the mode is to the left of the median and mean. Kurtosis data motivation, initial and final motivation are -0.783 and -0.638, both of which fall into mesokurtic criteria or have a homogeneous tendency [11]. Early learning motivation has a value range of 1.19 from a minimum value of 1.38 to a maximum value of 2.57. The final learning motivation has a value range of 0.93 from a minimum value of 3.50 to a maximum value of 4.43.

While the results of N-gain are based on 5 indicators of achievement motivation [4], namely: (1) tasks of choice*, (2) effort*, (3) persistence*, (4) self-confidence*, and (5) achievement*, to 34 student teacher candidates in basic science courses the results are presented in Table 2.

Table 2. N-gain* indicators of achievement motivation

No	Indicator	Initial Score	Final Score	N-gain*	
				<g>	Criteria
1	*Tasks of choice* *	1.920	3.950	0.660	Medium
2	*Effort**	1.860	3.940	0.660	Medium
3	*Persistence**	1.890	3.880	0.640	Medium
4	*Self-Confidence**	1.970	3.960	0.660	Medium
5	*Achievement**	1.890	3.830	0.620	Medium
	Average	**1.910**	**3.910**	**0.650**	Medium

Based on Table 2, it shows that the results initial and final motivation measurements showed an increase in learning motivation for the 5 motivation indicators with moderate criteria with an N-gain value range of 0.62 to 0.66. Meanwhile, the normalized N-gain average is 0.65 in the medium criteria. This means that there is an effect of implementing ARICESA on the achievement motivation of the 34 student teacher candidates.

The results of the Wilcoxon signed rank statistic between the initial motivation and the final motivation were 34 which showed that all students experienced an increase in learning motivation. The test results also have a p-value [significance] of 0.0001 [<0.05]. The significance value of the test results is less than 0.05, the conclusion is that there is a significant difference between the initial and final motivation of prospective basic science teacher students.

3.2 Effects of ARICESA on understanding basic science concepts

The results of ARICESA's effect on understanding basic science concepts are measured by test questions for understanding basic science concepts consisting of topics on quantities and units, simple machines, temperature, electricity, magnetism, and life processes. The test questions in the form of essays with a total processing time of 120 minutes as Table 3.

Table 3. Analysis of test values for basic concepts

Topic	Criteria Understanding	*Pre-test*			*Post-test*		
		∑Student*	%	*X*	∑Student*	%	*x*
Quantity and Unit	High	0	0.000	17.53	31	91.18	78.88
	Medium	2	5.880		3	8.820	
	Low	32	94.12		0	0.000	
Simple plane	High	0	0.000	15.65	31	91.18	76.59
	Medium	0	0.000		3	8.820	
	Low	34	100.00		0	0.000	
Electricity	High	0	0.000	16.41	32	94.12	80.53
	Medium	1	2.940		2	5.880	
	Low	33	97.06		0	0.000	
Magnets	High	0	0.000	17.65	33	97.06	79.76
	Medium	2	5.880		1	2.940	
	Low	32	94.13		0	0.000	
Temperature	High	0	0.000	17.94	33	97.06	80.18
	Medium	2	5.880		1	2.940	
	low	32	94.12		0	0.000	
Life Process	high	0	0.000	19.59	33	97.06	81.06
	medium	4	11.77		1	2.940	
	low	30	88.25		0	0.000	

Based on Table 3, it shows the test scores understanding basic concepts for prospective teacher students based on the criteria high, medium, and low concept understanding on each topic of discussion. In addition, it also shows the average value of understanding basic concepts on each topic of discussion. shows that the application of the ARICESA model can increase the average test scores for understanding basic science concepts on 6 topics of discussion.

Students before joining the learning process with ARICESA more than 30 were still in low comprehension criteria on each topic. There are only a maximum four students who enter criteria moderate understanding, no students enter the criteria high understanding. The effect of implementing ARICESA can increase the number of students meet high criteria. Almost all students after participating in the study have an understanding basic science concepts in high criteria. There are no students have low concept understanding criteria.

Based on the results of the pre-test and post-test, the effect of implementing ARICESA in class can increase a higher level of understanding on each topic [5]. Improved understanding of basic science concepts on 6 subjects can be determined by determining value of N-gain. The value of the N-gain understanding of basic science concepts from prospective teacher can be seen in Table 4.

Table 4. N-gain* understanding basic science concepts

No	Topic*	*N-gain**	
		Coeficient*	**Information***
1	Quantity and Unit	0.740	high
2	Simple plane	0.720	high
3	Electricity	0.770	high
4	Magnets	0.750	high
5	Temperature	0.760	high
6	Life Process	0.760	high

Based on Table 4, the ARICESA effect on increasing understanding of basic science concepts for prospective teacher students has high criteria with an N-gain coefficient range from 0.720 to 0.770. If averaged, the N-gain value is 0.750 and has high criteria.

The results of the t-test prerequisite test by testing the normality of data on understanding basic science concepts which overall topic has a value greater than 5%, meaning that the data on understanding basic science concepts follows the distribution function. normal for all topics. Since the data normality requirements have been met, it is possible to test the mean difference both of them for all topics using paired t-test.

The results of the statistical value of the paired t-test between the pre-test and post-test understanding of basic science concepts in each class have a negative t-count, meaning that the post-test result is greater than the pre-test. The significance or the p-value of paired t-test statistics for all pairs pre-test and post-test on each topic of discussion is less than 5% so it can be stated that there is a significant difference between understanding the concept basic science between pre-test and post-test for all topics. The overall difference in the mean shows a negative value, which means that the mean value of the pre-test is smaller than the post-test or there is an increase in the post-test value of understanding basic science concepts.

4. Conclusion

The effect of ARICESA application on achievement motivation and understanding of basic science concepts is summarized as follows: There is a significant increase in motivation effect for 5 indicators of achievement motivation of prospective teacher students with moderate criteria with an N-gain value range of 0.62 to 0.66. and the normalized mean N-gain is 0.65 in the medium criteria. There is an effect of increasing understanding of basic science concepts for prospective teacher students, which is significant with high criteria with a range of N-gain coefficients from 0.72 to 0.77 and a normalized average N-gain of 0.75 with high criteria. This research was conducted on first year students. Furthermore, it can be tested on different subjects and different majors.

References

[1] Rhodes H and Feder M A 2014 *Literacy for Science: Exploring the Intersection of the Next Generation Science Standards and Common Core for ELA Standards: A Workshop Summary* (Washington D C: National Academies Press)
[2] Goodhew L M and Robertson A D 2017 *Phys. Rev. Phys. Educ. Res.* **13** 010106
[3] Hull M M, Lindsey B, Archambault M, Davey K, and Liu Y A 2016 *Phys. Rev. Phys. Educ. Res.* **12** 010101
[4] Liou P Y 2017 *Int. J. Educ. Res.* **81** 83
[5] Mullis I, Martin M, Foy P, and Hooper M 2016 *TIMSS 2015: International Results in Mathematics and Sience* (Boston: TIMSS & PIRLS International Study Center)
[6] Fortus D and Vedder-Weiss D 2014 *J. Res. Sci. Teach.* **51** 497
[7] Unesco Education Sector 2010 *Current Challenges in Basic Science Education* (Paris: Unesco, Education Sector)
[8] Williams D, Brule H A, Kelley S S, and Skinner E A 2018 *Int. J. STEM Educ.* **5** 8
[9] Zhu Y A N and Leung F K S 2011 *Int. J. Sci. Math. Educ.* **9** 1189
[10] Young A M, Wendel P J, Esson J M, and Plank K M 2018 *Int. J. Sci. Educ.* **40** 1016
[11] Plenty S and Heubeck B G 2013 *Educ. Psychol.* **33** 14
[12] Pintrich P R and Zusho A 2002 *Student Motivation And Self' Regulated Learning In The College Classroom in Higher Education: Handbook of Theory and Research* (Dordrecht: Kluwer Academic Publishers)
[13] Gokce F 2010 *Sch. Leadersh. Manag.* **30** 487
[14] Griffin P, McGaw B, and Care E 2012 *Assessment and Teaching of 21st Century Skills* (London: Springer)
[15] Wagner T 2010 *The Global Achievement Gap* (New York: Basic Books)
[16] Mahmoodi M. H, Kalantari B, Ghaslani R 2014 *Procedia Soc. Behav. Sci* **98** 1062
[17] Zhang W 2017 *Front. Educ. China* **12** 261
[18] Kim K J 2009 *J. Interact. Learn. Res.* **20** 317
[19] Tagawa M 2008 *Kaohsiung J. Med. Sci.* **24** 380
[20] Fortus D and Vedder-Weiss D 2014 *J. Res. Sci. Teach.* **51** 497

Profile of Students' Creative Thinking Skills on Global Warming Material: Gender Perspective in Physics Learning

D Wulandari*, **E Hariyono, N Suprapto, H N Hidaayatullaah, and B K Prahani**

Science Education Postgraduate Program, Universitas Negeri Surabaya, Indonesia

*Email: desi.20010@mhs.unesa.ac.id

Abstract. Students need creative thinking skills, critical thinking, collaborative, and communicative in the 21st century in learning physics. In predicting further research, it is necessary to analyze the initial ability of students' creative thinking skills. This study aims to analyze the profile of physics learning on the students' creative thinking skill of high school students' on global warming material and analyze the differences in students' creative thinking abilities based on gender. Research method uses 100 samples of class eleventh grade in senior high school students'. Profiles of creative thinking skills of senior high school students' on global warming were analyzed quantitatively descriptively. The results showed that the physics learning profile of students' creative thinking skills on global warming material was relatively low, and there was a significant difference between the results of male and female students' creative thinking skills. The results of this study can be used as empirical evidence that the creative thinking skills of high school students' physics learning on global warming material still need to be improved. Further research is needed to overcome the low creative thinking skills of high school students'.

1. Introduction

In the process of learning physics, both teachers and students are required to think creatively. Creative thinking is one of the skills in the 21st century, especially physics learning. Skills that can be involved in learning physics in the 21st century such as 4C are creative thinking skills, critical thinking, collaboration, and communication [1,2]. CTS is creative thinking Skill. One of the outputs of CTS is shown by creating new ideas during the learning process, and the teacher creates an interactive learning atmosphere. To make it easier for students to convey ideas, fun learning conditions are needed.

In physics learning, CTS is needed. CTS is a skill to solve problems by creating unique, different, and logical ideas. Indicators of creative thinking skills are fluent thinking, flexible thinking, elaboration, and originality [3-5]. Students who can think creatively are shown to dare to give arguments of different types from physical phenomena, produce many different views that are not similar, can create ideas, can find relationships between variables of physical quantities, and tend to have high imagination power while applying physics concepts [6,7]. CTS is measured by three levels, namely high, medium, and low. The main principle in CTS is included in the realm of higher-order thinking skills so that with these abilities students will be trained to compete in the global world.

From the before CTS explanation, students who are creative in understanding problems are a big hope for teachers. Students who are creative in offering solutions will be trained in higher-order thinking, but students' CTS in solving problems is still relatively low. This is shown by the global index data in 2010. The data states that the CTS of students in Indonesia is ranked 81 out of 82 countries in the world [8-10]. From the data, it is stated that students have difficulty in implementing ideas. Difficulty in implementing ideas because students do not understand the material being studied, including understanding global warming material. So, the data contradicts the expectations of skills in the 21st century. Where teachers and students must have good CTS. In creating change, high integrity is needed in physics education, so students need to train their thinking skills [11-13]. In order to analyze appropriate learning in training students' physics CTS level, it is necessary to analyze the physics CTS profile, one of which is on global warming material.

Through the phenomenon of low creativity of students in physics, so necessary to analyze the CTS of students specifically in the subject of global warming. Observing the importance of CTS in learning physics for students, this study aims to look at the profile of students' CTS levels, especially the material on global warming. The results of this study are expected to be empirical evidence for education in Indonesia to continue to innovate in improving students' creative ideas.

2. Method

2.1 General background

The design of this study was pre-experimental. Main objective of the study was to analyze students' thinking skills on the subject of global warming in physics learning. The second objective is to analyze students' CTS each gender in physics learning.

2.2 Sample of research

The subjects of this study were 100 students of class XI at senior high school in Indonesia. Sampling with purposive sampling technique.

2.3 Procedure of research

The research procedure consists of 1) Researchers conducting literature studies related to creative thinking skills 2) Developing instruments for CTS related to global warming material in physics learning 3) Providing instruments developed by three experts to be validated 4) Research conducting revisions related to input and suggestions from validators 5) Researchers measure creative thinking instruments in 100 students of class XI 6) researchers process data related to creative thinking instruments 7) researchers analyze and conclude research results based on the level of CTS in physics learning.

2.4 Instruments

The use of creative thinking instruments in research is to determine the profile of students' CTS level. The research instrument is 7 questions about global warming material that have been integrated with CTS. In physics learning, CTS have 4 indicators, namely fluency, flexibility, originality, and elaboration. The CTS instrument was evaluated by 3 experts based on 5 statements. The first statement is related to the suitability of the questions with the material indicators of global warming, the second statement, the questions given are by the skills being trained, third, the questions made have the right proportions with the material being taught, the four questions use good and correct language, the five questions are made not to cause multiple interpretations. This instrument has been said to be valid and reliable. The results of the validation of the CTS instrument in Table 1.

Table 1. Results of the validation of CTS instrument

Statement to-	V1	V2	V3	Modus	Validity	Cronbach's Alpha (a)	Reliability
1	4.00	4.00	3.00	4.00	Valid		Reliable
2	4.00	3.00	3.00	3.00	Valid		Reliable
3	4.00	4.00	4.00	4.00	Valid	0.63	Reliable
4	3.00	3.00	4.00	3.00	Valid		Reliable
5	4.00	3.00	4.00	4.00	Valid		Reliable
Average	3.80	3.40	3.60	3.60	Valid	0.63	Reliable
Conclusion: Instruments of CTS that can be further processed in research							

The instrument about CTS to determine the level of creative thinking of class XI high school students on global warming material was declared valid and reliable. Improvements and suggestions by the physicist validator regarding the instrument are reducing questions from 10 questions to 7 essay questions, improving grammar, and writing correctly according to Bloom's taxonomy. From the calculation results obtained the percentage of validity of 90%.

2.5 Data Analysis
The profile CTS of class XI high school students on global warming material in physics learning was analyzed quantitatively descriptively. The results of the study were to determine the initial profile of student's CTS on global warming material. The level of student's CTS is categorized as high if the score is > 8.00; medium if 6.10<s<7.90; and low if s<6.00. While the percentage of students' CTS is said to be high if > 80%. Analyze the differences in the results of students' CTS between men and women were analyzed using the ANOVA test. In the next stage, it will be used as a tool in the application of physics learning related to global warming material.

3. Results and Discussion

3.1 Level of students' CTS
After test instrument was declared valid by 3 experts physics, then instrument was tested initially on 100 high school class XI students who taught physics. The test results of students' CTS are presented in Table 2.

Table 2. The results of students' CTS on global warming material

Personal Student	Group 1		Group 2		Group 3	
	Value	Criteria	Value	Criteria	Value	Criteria
P1	4.60	Low	3.70	Low	2.30	Low
P2	4.50	Low	4.00	Low	2.60	Low
P3	4.20	Low	5.10	Low	4.00	Low
P4	6.40	Medium	5.50	Low	3.10	Low
P5	5.50	Low	2.80	Low	3.50	Low
P6	5.60	Low	3.80	Low	3.70	Low
P7	6.40	Medium	4.40	Low	4.50	Low
P8	4.50	Low	5.50	Low	4.10	Low
P9	3.80	Low	4.30	Low	4.40	Low
P10	5.80	Low	2.60	Low	4.60	Low
P11	3.50	Low	5.00	Low	4.80	Low
P12	6.30	Medium	5.40	Low	4.80	Low
P13	4.10	Low	2.40	Low	5.00	Low
P14	3.60	Low	4.80	Low	2.60	Low
P15	5.40	Low	5.40	Low	7.60	Medium
P16	5.50	Low	5.30	Low	4.40	Low
P17	3.40	Low	6.30	Medium	6.40	Medium
P18	2.80	Low	2.10	Low	6.00	Low
P19	3.00	Low	3.80	Low	5.20	Low
P20	4.30	Low	3.60	Low	6.00	Low
P21	6.10	Medium	5.30	Low	5.00	Low
P22	610	Medium	6.40	Medium	2.60	Low
P23	3.00	Low	4.00	Low	2.60	Low

Personal Student	Group 1		Group 2		Group 3	
	Value	Criteria	Value	Criteria	Value	Criteria
P24	5.70	Low	5.10	Low	4.50	Low
P25	5.40	Low	5.50	Low	7.00	Medium
P26	5.50	Low	3.30	Low	5.80	Low
P27	5.80	Low	6.30	Medium	5.50	Low
P28	4.50	Low	6.50	Medium	5.00	Low
P29	6.30	Medium	6.20	Medium	5.10	Low
P30	5.60	Low	7.70	Medium	4.20	Low
P31	6.00	Low	7.00	Medium	4.50	Low
P32	3.70	Low	3.90	Low	5.20	Low
P33	5.70	Low	6.50	Low	4.20	Low
P34					5.00	Low
Average	4.93	Low	4.83	Low	4.58	Low

Table 2 shows that students' CTS on global warming material in each group are stated low. A total of 84 students have low-level creative thinking skills as indicated by a score of <6.00, while the other 16 students have medium-level creative thinking skills with 6.10<score r<7.90. Most students are not able to achieve high-level skills for all indicators. These results indicate that the level of CTS in physics learning of class XI high school students is still relatively low.

3.2 Percentage CTS of each indicator

Overall indicators of students' CTS on global warming material are in a low category. The indicators physics learning of students' CTS consist of fluency, flexibility, originality, and elaboration [14,15]. Percentage results of the initial test of students' CTS are presented in Figure 1. While each percentage of the initial results of student's CTS on global warming material is presented in Figure 2.

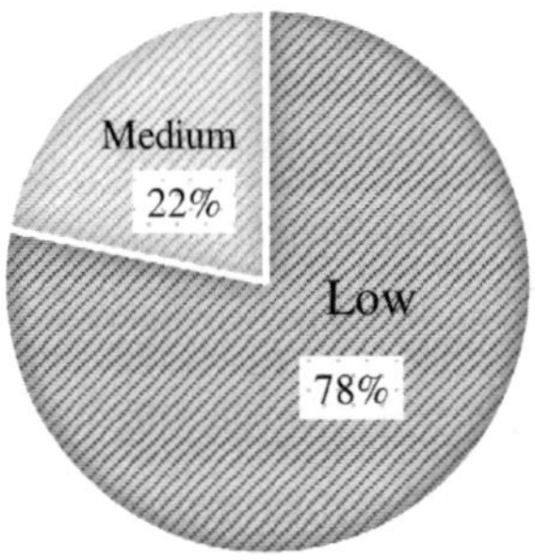

Figure 1. Percentage of trial results

Based on Figure 1 shows that the level of CTS of students at the low level is 78% and the medium is 22%. This shows that the overall CTS of students are in a low category. Furthermore, the percentage graph of each student indicator is presented in Figure 2.

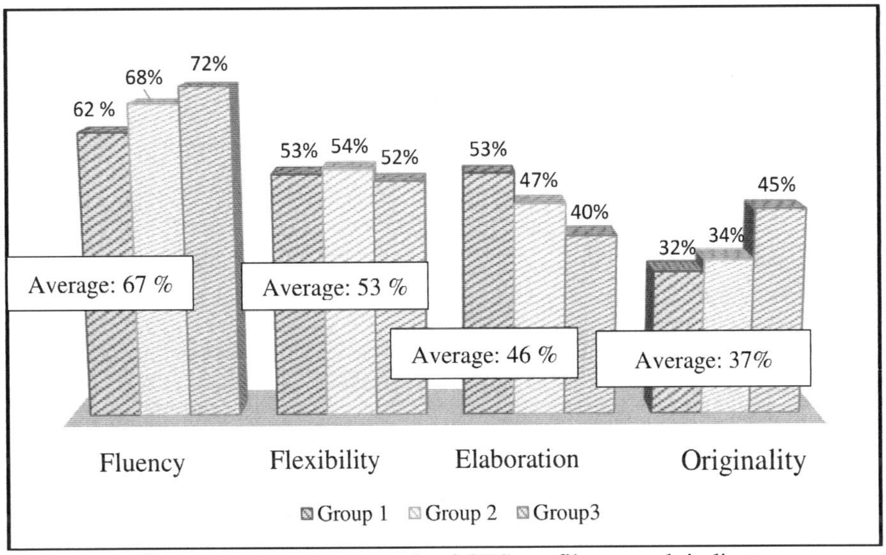

Figure 2. Percentage graph of CTS profile on each indicator

Figure 2 shows that the percentage of CTS indicator in physics learning gets the largest percentage on the fluency indicator, then the lowest CTS indicator on originality. Fluency indicators have an average of 67%, 53% flexibility, 46% elaboration, and 37% originality. In this case, it shows that of the four indicators, there are 3 indicators that are still low. So it shows that the profile of student's CTS at physics learning global warming material is still low. The lack of reading and students' material physics in everyday life is the cause of the low creativity of students [16]. In this case, innovative applications are needed to increase students' CTS [16], especially global warming material.

3.3 Results of differences in students' CTS on gender and groups

After looking at the profile of students' CTS as a whole and based on indicators, then the differences in students' CTS based on gender were analyzed. The results of the differences in gender are presented in Table 3.

Table 3. Results differences by gender

Variabel Independent and Dimentions	df	Mean Square	F	Sig. (2-tailed)
Male*female	1	53.730	50.995	.000*

*significant at p < .05

Based on Table 3 shows that the results of sig 0.000. This shows that there is a significant difference between the results of the initial CTS of male and female students in physics learning. The result of this significance is that women have higher creativity than men in physics learning. With a female average of 5.236 and a male of 3.603. Furthermore, the results of the differences in CTS in each group are presented in Table 4.

Table 4. Results of differences groups

Variabel Independent and Dimentions (I)		Mean Difference (I-J)	Sig.
Group 1	Group 2	.09394	.951
	Group 3	.34492	.506
Group 2	Group 1	.09394	.951
	Group 3	.25098	.696
Group 3	Group 1	.34492	.506

Variabel Independent and Dimentions (I)		Mean Difference (I-J)	Sig.
	Group 2	.25098	.696

*significant at p < .05

Based on Table 4, shows that there is no significant difference in the results of students' CTS between group 1, group 2, and group 3. This is indicated by a significant value > 0.05. Each group got a low and commensurate score because students had not received previous global warming lessons. In this case, it shows that this research can be continued by applying innovative learning and building students' creativity.

Students' fluency thinking skills at physics learning global warming material get a larger percentage than other indicators. In the indicator of fluency thinking students get many similar ideas. Students can respond smoothly to a problem obtained from everyday life. Problems with fluency indicators direct students to stimulate ideas in answering questions [17,18]. So that the fluency indicator gets a medium percentage. Then at physics learning material global warming with indicators of flexibility, elaboration, and originality directs students to analyze the mechanism of the greenhouse effect. The questions of the three indicators obtained a low percentage. In the research, students have difficulty in giving different arguments that are not similar [19]. So that students can give many different and dissimilar arguments, it is necessary to read a lot of literacy on material topics [20]. Furthermore, students have not completed the questions with indicators of elaboration and originality. Students cannot give original ideas in answering questions. Students cannot provide arguments for the entire contents of the table and do not give a unique view in responding to the phenomenon of the greenhouse effect. This is because students are not trained in honing creative thinking skills. They can provide unique arguments if they are accustomed to sharpening their original skills by generating new ideas [21]. All of these indicators give the result that students' CTS in physics learning are very low. In line with [22] states that students have high creativity when trained with project-based physics learning.

Solutions to improve students' CTS by giving different learning physics. Learning physics that actively involves students will create an independent learning atmosphere [23,24]. Student activity during the learning physics process can be created with the help of problem-based learning [25-27]. Learning physics by giving a variety of cases can make students have many arguments and ideas [28]. Then learning based on projects and problems will make students create original ideas [29-31]. Learning strategies by making students the center of learning will facilitate mastery of concepts physics. In line with [32,33] states that problem-based learning strategies are able to train students' creativity. Problem-based learning physics will make it easier for students to create ideas that are unusual and not thought of by others. In this study, the female gender had higher CTS than the male gender. This is in line with [34,35] mentioning that women have higher creativity. Besides gender differences affect students' creativity and other factors such as the management of affective learning attitudes also affect students' creativity [36].Through the indications of the results of this study, it can be used as further research in comparing the results of student creativity based on gender in a wider and different respondent. Analytical thinking is able to train students' creativity [37-39].

4. Conclusion

In the case of this study, the profile of student's creative thinking skills (CTS) at physics learning global warming material is low, and there is a significant difference between the results of male and female students' creative thinking skills (CTS). The results of this study can be used as empirical evidence that the CTS of grade XI students in the subject of global warming in physics learning still need to be improved. Further research is needed to be able to overcome the low CTS of students in high school.

References

[1] Trisnayanti Y, Ashadi, Sunarno W, and Masykuri M 2020 *J. Phys. Conf. Ser.* **1511** 012072
[2] Ida M, Sri Y, and Suryadi B D 2019 *Sci. Educ.* **8** 119
[3] Torrance E P 1965 *Daedalus* **94** 663

[4] Munandar U 1999 *Mengembangkan Bakat dan Kreativitas Anak Sekolah : Penuntun Bagi Guru dan Orang Tua* (East Jakarta: Grasindo)
[5] Sugiyanto F N and Masykuri M 2018 *J. Phys. Conf. Ser.* **1006** 012038
[6] Tri S, Mustaji, and Miftakhul J 2020 *Int. J. Emerg. Technol. Learn. (iJET)* **15** 245
[7] Mawarni N, Edy S, and Edi S 2017 *Int. J. Adv. Res. Innov. Ideas Educ.* **3** 2146
[8] Madyani I, Yamtinah S, Utomo S B, Saputro S, and Mahardiani L 2020 *AP Proc. Int. Conf. Learn. Innov. Qual. Educ.* **397** 957
[9] Ghufron N and Rini R S 2017 *Teori-Teori Psikologi* (Pamekasan: Ar-Ruzz media)
[10] Nurhamidah D, Masykuri M, and Dwiastuti S 2018 *J. Phys. Conf. Ser.* **1006** 012035
[11] Cayirdag N 2017 *Educ. Sci. Theory Pract.* **17** 6
[12] Ndiung S, Dantes N, Ardana I, and Marhaeni A A. I 2019 *Int. J. Instr.* **12** 731
[13] Suratno, Komaria N, Yushardi, Dafik, and Wicaksono I 2019 *Int. J. Instr.* **12** 133
[14] Treffinger D J, Young G C, and Selby E C 2002 *Nat. Res. Cent. Gift. Talent.* **4** 7
[15] M Tohir, and Abidin Z 2018 *J. Phys. Conf. Ser.* **1008** 012072.
[16] Nazula N H 2019 *J. Phys. Conf. Ser.* **1211** 012109.
[17] Saprudin S, Liliasari S, Prihatmanto A S, and Setiawan A 2019 *J. Phys. Conf. Ser.* **1157** 032030
[18] Siburian J, Aloysius D C, and Murni S 2019 *Eur. J. Educ. Res.* **19** 81
[19] Murtianto Y H, Rahmawati N D, and Apriana D 2020 *J. Phys. Conf. Ser.* **1663** 012013
[20] Jusuf R, Sopandi W, Wulan A R, and Sa'ud U S 2019 *Int. J. Learn. Teach. Educ. Res.* **18** 70
[21] Sari D K A, Permanasari, and Supriyanti F M 2019 *J. Pendidik. IPA Indones.* **6** 71
[22] Ersoy E 2014 *Proc. Soc. Behav. Sci.* **116** 3494
[23] Wahyu W, Kurnia, and Eli R N 2016 *AIP Conf. Proc.* **1708** 040008
[24] Puspitasari L, In'am A, and Syaifuddin M 2018 *Int. Electron. J. Mat. Educ.* **14** 49
[25] Komaria N and Wicaksono I 2019 *Int. J. Instr.* **12** 33
[26] Wulandari D and Ermawati F U 2019 *Pros. Sem. Nas. Fis.* **3** 60
[27] Hidaayatullaah H N, Suprapto N, Mubarok H, and Wulandari D 2020 *J. Phys. Conf. Ser.* **1491** 012053
[28] Khoiriyah A J and Husamah H 2018 *J. Pendidik. Biol. Indones.* **4** 151
[29] Sunarti T, Hariyono E, Setyarsih W, Prahani B K, and Suyidno S 2020 *Prisma Sains J. Pengkaj. Ilmu Pembelajaran Mat. IPA* **8** 91
[30] Wartono W, Diantoro M, and Bartlolona J R 2018 *J. Pendidik. Fis. Indones.* **14** 32
[31] Birgili B 2015 *J. Gift. Educ. Creat.* **2** 71
[32] Rudibyani R B 2019 *J. Phys. Conf. Ser.* **1155** 012049.
[33] Prahani B K, Deta, U A, Lestari N A, Yantidewi M, Jauhariyah M N, Kelelufna, V P, and Suyidno, S 2021 *J. Phys. Conf. Ser.* **1760** 012020
[34] Fatmawati A, Zubaidah S, and Mahanal S 2019 *J. Phys. Conf. Ser.* **1417** 012070
[35] Anjarwati P G P, Sajidan S, and Prayitno B A 2018 *J. Biol. Biol. Educ.* **10** 313
[36] Deta U A, Suparmi, Widha 2013 *J. Pendidik. Fis. Indones.* **9** 28
[37] Maskur R 2020 *Eur. J. Educ. Res.* **9** 375
[38] Marleny L and Aloysius D C 2017 *J. Phys. Conf. Ser.* **795** 012071
[39] Hidayati N, Zubaidah S, Suarsini E, and Praherdhiono H 2019 *J. Turkish Sci. Educ.* **14** 35

A Profile of Scientific Literacy of Senior High School Students on Physics Learning

F Azura*, B Jatmiko, M Ibrahim, E Hariyono, and B K Prahani

Science Education Postgraduate Program, Universitas Negeri Surabaya, Surabaya, Indonesia

*Email: filza.20027@mhs.unesa.ac.id

Abstract. In physics learning, students need to know the extent to which scientific concepts have been studied. Scientific literacy is the ability to make decisions based on scientific considerations, solve problems and apply the knowledge gained in everyday life. This study aims to scientifically analyse the scientific literacy ability of senior high school students on the kinetic theory of gases. The sample of this research is 90 high school students. Data analysis used descriptive qualitative. The results study show that high school students' ability was very low, many students did not have scientific literacy ability on physics learning. The implications of this study can be used as empirical evidence that the scientific literacy of senior high school students on physics learning still needs to be improved.

1. Introduction

In the 21st century education is more based on learning by learning to think logically and rationally oriented to knowledge, learning to behave oriented to how to solve problems [1]. In the 21st century students are required to have scientific literacy ability (SLA) [2]. The SLA achievement of high school students can be seen from the results of an international survey, namely PISA. According to PISA, SLA is divided into four aspects, namely knowledge, context, attitudes, and indicators. PISA results show that Indonesian students' SLA are low, including using scientific facts, identifying scientific problems, and understanding life. Attitudes are aspects of SL that support to analyze how students are interested in science, support science and students' motivation to be responsible for natural resources and the surrounding environment. The results of PISA 2018 showed that Indonesian high school students have low level of SLA. Indonesia's average SLA score is ranked 75th out of 80 participating countries. Indonesia's score is below the international average value [3] in science subjects from 2000 to 2018. This means that the SLA of students in Indonesia is still low [4].

SLA are very important to face development and changing times. SLA does not only train students to read, but students must learn to understand the content of to explore current scientific issues, and students must be active in the learning process. SL is an obligation for every individual. Research on students' SLA on an international scale is organized by the Organization for Economic Co-Operation and Development (OECD) through PISA. SLA are used to identify questions, use scientific knowledge, make decisions about nature through human activities by drawing conclusions based on the available evidence [5]. SLA is a person's ability to use and apply scientific concepts in everyday life, describe and explain scientific phenomena based on data and scientific evidence [6]. Literacy is related to the competence to think and process information. So, it's not just reading or spelling skills.

Someone with a high level of literacy has the ability to reason and solve problems in various fields, including science, numeracy, and finance. To find out the extent of students' literacy of the science concepts that have been studied, it is very important to measure SLA [7]. Therefore, as individuals and as social beings, they must master SLA to solve problems in the world and interact in the environment and society [8]. Students can easily solve real today and in the future and meet the needs of life through understanding positive attitudes towards science and SLA [9]. Aspects of SLA indicators are aspects that students achieve to have SLA. Aspects of SLA consist of several indicators that become a benchmark for students to have SLA. There are three indicators must be considered in physics literacy according to PISA [3]. These indicators include being able to: (1) explain scientific phenomena means being able to explain scientific phenomena, which means students can recognize, identify and predict and explain phenomena scientifically and their uses (2) evaluating and designing scientific investigations means being able to evaluate and design scientific investigations, which means students are able to explain and interpret scientific investigations and propose ways to deal with scientific questions (3) interpret scientific data and evidence means being able to interpret scientific data and evidence, which means students are able to analyze, transform and evaluate scientific data, claim in various presentations of scientific phenomena to draw conclusions.

In learning physics, students need to develop SLA. If students have high SLA, then these students can master the concept of the material. Thus, physics learning must be carried out optimally. One of the materials in physics that is considered difficult by students is the kinetic theory of gases. The basic competence 3.6 explains the kinetic theory of gases and their characteristics in a closed space and 4.6 presents works related to the kinetic theory of gases and their physical meaning. These competences require students to have SLA. SLA is useful to be applied in real situations and problems related to science. The things that are central to students' SLA are knowledge of science, scientific students' understanding, scientific processes, and the development of scientific attitudes towards science, so that students are expected to make decisions based on scientific considerations and apply science skills in solving problems in life. In addition, students are expected to be able to apply the knowledge gained in everyday life. So that students can understand the meaning of the subject matter they are studying.

In order to analyze of SLA for senior high school students, so there is a need for research that focuses on looking at the profile of scientific literacy of senior high school students', especially in the kinetic theory of gases. The next study is needed as an alternative solution of SLA that is very low category. It is hoped that the research results can be empirical evidence of SLA in the development of physics learning research.

2. Method

2.1. Background

The purpose of this research is to analyze the profile of SLA of senior high school students on physics learning. This research is pre-experimental research.

2.2. Sample

The subjects in this study were senior high school students consisting of 90 students of class XI high school, this study using purposive sampling in determining sample.

2.3. Procedure

The procedure of this research consists of: (1) researchers conduct a literature study related to SLA (2) researchers focused on finding assessments and indicators of SLA (3) the researcher developed a SLA test instrument to determine the profile of the SLA of senior high school students on gas kinetic theory material (4) the researcher conducted a limited trial (5) the researcher revised the test results (6) the researcher used a SLA test instrument to measure 90 students in determining the profile of senior high school students' SLA on the kinetic theory of gases (7) researchers conduct conclusions and analysis based on the results of the test instruments.

2.4. Instruments of Research
This study uses a SLA test instrument in determine the SLA profile of senior high school students on physics learning which consists of three indicators: explain scientific phenomena, evaluate and design scientific investigations, interpret data and scientific evidence. This instrument has been declared reliable based on the Cronbach's alpha reliability test using SPSS. Validation results of scientific literacy instrument in Table 1.

Table 1. Results of scientific literacy ability validity

Components	**The validity of Scientific Literacy Ability Instruments**							
	Construct Validity				**Content Validity**			
	ICC	**V**	**α**	**R**	**ICC**	**V**	**α**	**R**
Scientific Literacy Ability	0.17	Valid	0.67	Reliable	0.17	Valid	0.67	Reliable
Conclusion: The scientific literacy instrument that has been developed is valid. It can take as an instrument to measure determine scientific literacy ability of students on physics learning								

Note = ICC: Interclass Correlation Coefficient, α: Cronbach's Alpha, V: Validity, R: Reliability

The test instruments of scientific literacy ability to determine a profile of scientific literacy of senior high school students on physics learning by statistic analyse with Cronbach's alpha. This instrument valid (ICC = 0.17) and Reliable (α = 0.67).

2.5. Data Analysis
A profile of SLA of senior high school students on physics learning was analysed using qualitative descriptive. To determine the scientific literacy ability measured by the average percentage score on each indicator according to the criteria of 86-100 very good, 76-85 good, 60–75 enough, 55-59 low and $\leq$ 54 very low [10]. The results of the analysis of this study are used to obtain an initial preliminary data on the SLA of high school on the kinetic theory of gases so that it can be used as evidence in the development of research in the field of education. Then next, we will use the profile as a recommendation in learning science, especially physics.

3. Result and Discussion
A pre-test on 90 high school students shows that the kinetic theory of gases test instruments is valid and reliable. The results obtained from the SLA of senior high school students on the kinetic theory of gases are presented in Table 2.

Based on Table 2, it can be seen that students' SLA on the kinetic theory of gases are included in the very low category in all classes. A total of 90 students did not have a score above 54. It can be seen that all students have not achieved completeness on the indicators of SLA in the kinetic theory of gases. The results of the average pre-test score show that students' SLA in the aspects of knowledge and competence in the three classes are in the very low category. The SLA test instrument consists of three indicators: explain scientific phenomena, evaluating and designing scientific investigations, interpret scientific data and evidence. All indicators are still in the low category which can be seen in the percentage shown in Figure 1.

Based on Figure 1 it can be seen that the percentage of students' SLA is still very low on each indicator. In class 1 indicator 1 of 56.51% and indicator 2 of 54.23% in low category, indicator 3 of 26.44% in very low category. In class 2 indicator 1 of 71.95% and indicator 2 of 62.74% in low category, indicator 3 of 53.64% in very low category. In class 3 indicators 1 of 73.69%, indicator 2 of 59.54% in low category, indicator 3 of 51.14% in very low category. The average percentage of SLA on indicators explaining scientific phenomena is 45.73%, on indicators evaluating and designing scientific investigations is 62.77%, on indicators of interpret scientific data and evidence is 61.46%. These results indicate that the profile SLA of senior high school students on gas kinetic theory is low. SLA in this case students only have scientific knowledge to explain the context or make conclusions based on simple investigations. However, students are less able to explain scientific components in various complex life situations and evaluate scientific evidence that is correct to respond to situations.

Table 2. Scientific literacy ability of students on the kinetic theory of gases

Student Initial	Class 1		Class 2		Class 3	
	Score	Category	Score	Category	Score	Category
U1	15	VL	22	VL	40	VL
U2	33	VL	30	VL	27	VL
U3	22	VL	28	VL	34	VL
U4	23	VL	29	VL	30	VL
U5	13	VL	28	VL	40	VL
U6	20	VL	37	VL	48	VL
U7	15	VL	37	VL	30	VL
U8	15	VL	44	VL	50	VL
U9	27	VL	32	VL	20	VL
U10	10	VL	25	VL	25	VL
U11	25	VL	37	VL	25	VL
U12	21	VL	35	VL	44	VL
U13	22	VL	24	VL	33	VL
U14	23	VL	30	VL	23	VL
U15	22	VL	35	VL	30	VL
U16	12	VL	20	VL	30	VL
U17	35	VL	44	VL	45	VL
U18	20	VL	22	VL	50	VL
U19	10	VL	38	VL	35	VL
U20	22	VL	20	VL	46	VL
U21	20	VL	27	VL	50	VL
U22	22	VL	39	VL	29	VL
U23	10	VL	25	VL	23	VL
U24	24	VL	20	VL	50	VL
U25	28	VL	23	VL	20	VL
U26	13	VL	28	VL	37	VL
U27	20	VL	49	VL	40	VL
U28	20	VL	20	VL	50	VL
U29	24	VL	20	VL	35	VL
U30	23	VL	32	VL	40	VL
Average	20.3	VL	30	VL	36.3	VL

Note = VL: Very Low

In maintaining, improving the quality of life, and developing public policies, PISA assessments are carried out so that students know that science has important value for society and individuals optimally achieved [11]. Literacy skills are needed by students. If students have high SLA, then these students can master the concept of the material. In addition, SLA is useful to be applied in real situations and problems related to science. Only some students who have a high of ability master indicators in explain phenomena scientific, evaluate and design scientific investigations, interpret data and evidence scientific. Not all students can achieve these indicators. Related previous studies have supported these findings. Lack of scientific attitude makes students' SLA very low so that research with a scientific approach is needed on students [12]. This is also supported by Budiarti who found that the SLA of students about the concept of motion of living things is still low in Papua [13]. In line with the reason when students do not have an android smartphone with an internet connection [14]. SLA with a scientific approach as learning requires strategies that must be achieved by students to have the ability conceptual understanding and can help students solve scientific problems, make it easier to understand the material being taught, make it easier to connect scientific concepts with everyday phenomena. Make it easier for students to evaluate and design scientific investigations, encourage collaboration in groups, increase responsibility in group learning, facilitate interpretation of scientific data and evidence. Better learning can be created if students have high SLA [15]. The solution for students who have not achieved mastery in SLA indicators is through effective remedial learning in active learning, which means learning ability. In line with the research results of Gunel et al [16].

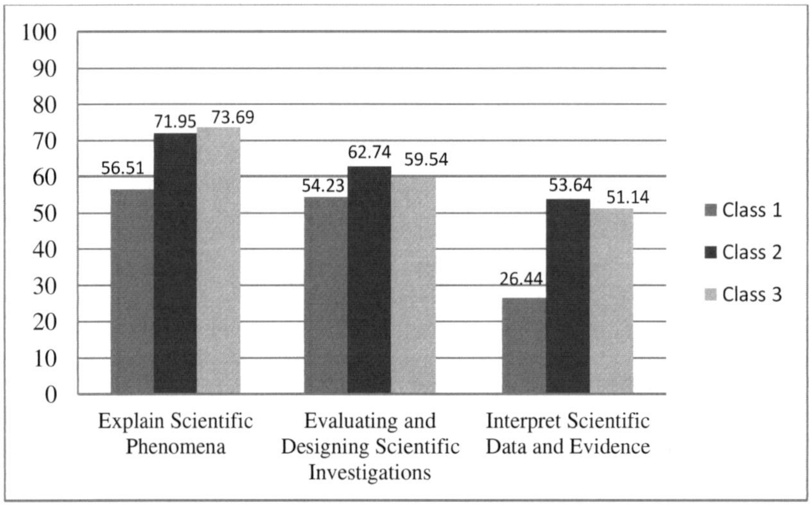

Figure 1. Data Percentage Indicator of Scientific Literacy Ability

Students' low SLA is influenced by several factors, namely low reading interest, evaluation tools that have not led to the development of scientific literacy, and the teacher's lack of knowledge about SLA [17]. The cause of the low SLA of Indonesian students is due to several other things, namely: teacher centered learning, the low positive attitude of students in understanding science, there are several basic competencies that are not liked by students related to process, context and content. This is caused by the material being tested has never been studied, Questions that use discourse are not used to being done by students, and students' SLA are not supported by the learning process. A learning model is needed to improve SLA. Scientific processes that are not involved in the learning process can lead to low SLA [18]. The same thing was also revealed that the SLA of high school students was still low [19]. Diana [20] concluded that the SLA of Class X SMA students is still relatively low due to differences in learning targets applied in schools with the demands of PISA. Science learning aims to assist students in developing more understanding of the nature of science and making students scientifically educated citizens, it is considered an important outcome that must be obtained from science learning [21]. Teachers need to apply effective science learning to improve students' SLA. Based on the opinions expressed by some of these experts, it can be concluded that science learning must be able to develop SLA [22].

Science learning must be balanced with the provision of evaluation questions that encourage students to develop thinking skills and develop students' reasoning in the given situation [23]. The achievement of competence in science learning leads to the formation of science attitudes. The process of forming this scientific attitude must be based on the mastery of knowledge and skills [24]. The results of Huryah's research [25] concluded that students who are not accustomed to working on questions that require analysis are one of the factors that cause students' low SLA. The idea of SLA still requires a more in-depth study, either through discussions, seminars, or applications to make it more applicable, namely the assessment tools can be developed [26]. However, a person's SLA can develop along and that ability in a person can be very high in certain fields but can be very low in other fields [27]. Even according to Surpless, et al [28] Geology Physics students in Texas also still do not have SLA adequate.

Teachers' knowledge about integrated science learning and conceptual literacy are important assets to improve the quality of teaching and learning in the classroom [29]. This is the reason why most students find physics difficult and confusing. Teachers must be able to overcome these problems, namely by linking the kinetic theory of gases with phenomena that exist in life. So that students are expected to be able to understand physics because it will be very useful to apply and develop. The education system, curriculum, learning facilities and infrastructure, the selection of teaching models and methods by teachers, and teaching materials cause the low SLA of Indonesian students [30]. Apart from these factors, most students only memorize concepts and are less able to think deeply about the relationship of what they have learned with its application in new situations [31]. The results of the study show that the students' SLA profile is in a sufficient category [32].

4. Conclusion

In the case of this study in Indonesia, a profile of SLA of senior high school students on physics learning is generally still at a very low category. This can be seen from the results of the SLA test in the form of 10 items that have been carried out by high school students. The average percentage of indicator aspects with indicators in explain phenomena scientific, evaluate and design scientific investigations, interpret data and evidence scientific are 45.73%, 62.77%, and 61.46% respectively. SLA in this case students only have scientific knowledge to explain the context or make conclusions based on simple investigations However, students are less able to explain scientific components in various complex life situations. The implications of this study can be used as empirical evidence that the SLA of high school students on physics learning still need to be improved. The limitation of this research is that it still uses 90 senior high school students. Further research is needed to overcome the problem of high school students' SLA level on physics learning.

References

[1] Yuliati 2017 *Cakrawala Pendidik.* **3** 21
[2] Dewi 2018 *J. Inov. Pendidik. Fis.* **7** 3
[3] OECD 2016 *PISA Result 2015: Assessment and Analytical Framework PISA 2015 Assessment and Analytical Framework* (Paris : OECD Publishing)
[4] OECD 2018 *PISA Result 2018: The Most Comprehensive and Rigorous International Assessment of Student Learning Outcomes To Date* (Paris : OECD Publishing)
[5] Hadi W P, Munawaroh F, Rosidi I, and Wardani W K 2020 *J. IPA Pembelajaran IPA* **4** 178
[6] Bybee R, McCrae B, and Laurie R 2009 *J. Res. Sci. Teach.* **46** 865
[7] Holbrook and Rannikmae 2007 *Int. J. Sci. Educ.* **29** 1347
[8] Rosana D, Widodo E, Setianingsih W, and Setyawarno D 2020 *J. Pendidik. Sains Indones.* **8** 292
[9] Vieira and Tenreiro V 2016 *Int. J. Sci. Math.* **14** 659
[10] Purwanto 2008 *Metodologi Penelitian Kuantitatif* (Yogyakarta: Pustaka Belajar)
[11] Bieber T and Martens K 2011 *Eur. J. Educ.* **46** 101
[12] Kusumah R G T 2019 *Indones. J. Integr. Sci. Educ.* **1** 71
[13] Budiarti, Indah S, and Tanta 2021 *J. Pendidik. Sains Indones.* **9** 36
[14] Indarti S 2019 *Indones. J. Integr. Sci. Educ.* **1** 100
[15] Rini E S 2020 *Indones. J. Integr. Sci. Educ.* **2** 108
[16] Gunel M, Hand B, and Gunduz S 2006 *Sci. Educ.* **90** 1092
[17] Sutrisna N 2021 *J. Inov. Penelit.* **1** 2683
[18] Anggraini G 2014 *Proc. Math. Sci. Forum* **2** 161
[19] Rizkita L 2016 *J. Pendidik Biol.* **1** 771
[20] Diana S, Rachmatullah A, and Rahmawati E S 2015 *Sem. Nas. XII Pendidik. Biol. FKIP UNS* **1** 285
[21] Roberts D A 2007 *J. Sci. Educ.* **1** 729
[22] Pantiwati Y and Husamah 2014 *Konferensi Ilm. Tah. HEPI 2014* 158
[23] Gormally C, Brickman P, and Lutz M 2012 *J. Life Sci. Educ.* **11** 364

[24] Sari K and Nurwahyunani A 2016 *J. Pendidik. IPA.* **6** 249
[25] Huryah F, Sumarmin R, and Effendi J 2017 *J. Eksakta Pendidik.* **1** 72
[26] Asyhari A 2015 *J. Ilmiah Pendidik. Fis. Al-Biruni* **4** 179
[27] Shwartz Y, Ben-Zvi R, and Hofstein A 2006 *Chem. Educ. Res. Pract.* **7** 203
[28] Surpless B, Bushey M, and Halx M 2014 *J. Geosci. Educ.* **62** 244
[29] Rubini B, Ardianto D, Pursitasari I D, and Permana I 2016 *J. Pendidik. IPA Indones.* **5** 299
[30] Kurnia F 2014 *J. Inov. Pembelajaran Fis.* **1** 43
[31] Rustaman N, Firman H, and Kardiawarman 2004 *Analisis Hasil PISA Bidang Literasi Sains* (Bandung: Puspendik)
[32] Lestari D and Setyarsih W 2021 *J. Phys. Conf. Ser.* **1805** 012040

A Critical Thinking Skill Profile of Science Education Undergraduate Student in Basic Physics

R Wulandari*, B Jatmiko, M Budiyanto, E Hariyono, N A Lestari, and B K Prahani

Science Education Postgraduate Program, Universitas Negeri Surabaya, Indonesia

*Email: reni.20013@mhs.unesa.ac.id

Abstract. An undergraduate student should have ability to make the right decision based on data analysis and information able to provide instructions in choosing various alternative solutions independently and in groups. This is an important point in the Indonesian National Qualification Framework at the undergraduate level. One of the skills that must be possessed by students who are part of 21st century skills is critical thinking skills. The aim of this research is to analyze the critical thinking skills profile of science education undergraduate students in basic physics. Subject of this research were undergraduate students with sample of 109 undergraduate students determined by purposive sampling. The instruments have fulfilled the validity requirements ($r_\alpha = 0.26$) and reliability ($\alpha = 0.99$). Result of this research that the profile of science education undergraduate students is very low level. The students' critical thinking skills need to be improved based on that level.

1. Introduction

Basic physics is the basic science that must be mastered by science education undergraduate student. In basic physics there are various sciences related to science, technology, pedagogy and knowledge. Lecturers often find it difficult to convey basic physics materials to be easily accepted by student. The basic physics lecture has more effort in the class everyday: they must achieve that their undergraduate students extract the major possible learning results in the class. Therefore, the lecture must motivate the students in order that they learn and take part in class [1]. Many undergraduate students are not interested in studying and understanding basic physics. Based on the opinion of this student is only a requirement to study basic physics and the younger generation who like it [2]. But lecturers strive to train and improve a variety of 21st century skills for undergraduate student, one of which is critical thinking skills. Critical thinking skills involve the ability to explain important questions, give relevant information, reach well-reasoned conclusions, make accurate decisions assessing the credibility of sources, identify causes that affect relationships and effectively communicate with others in search of solutions [3].

The facts show that indicators of critical thinking skills are almost very difficult to find in classroom learning. Indicators of critical thinking skills have not been developed as competencies that must be honed among students. It is rare to find students proactively addressing problems in learning. Similarly, very few students dare to answer the problems posed. As a result, students almost never offer solutions to the problems found. This can be used as an indicator that the critical analytical power of students is still low. Some students are still trapped in memorization and mere memory. This causes the information received by students to be very easily forgotten and disappeared. As prospective teachers, students need to do activities that hone critical thinking skills. The thinking of prospective teachers about critical thinking does not change and less noticeable differences over time. Teacher or lecturer can make decisions with expression references and answer the questions [4]. This is certainly not in accordance with the objectives in the National Qualification Framework

Critical thinking is also known as a metacognitive process, which includes a number of sub-skills i.e. analysing, evaluating, and concluding that are used to improve the logic of inferring the opportunity of creation toward an argument or solution. The low critical thinking skills of learners can be seen from the lack of participation of learners in asking, arguing, commenting, and explaining. Critical thinking skills are concerned with deeper and higher thinking as the process of considering a topic or problem from a variety of perspectives [5-7].

Undergraduate students need to be trained and developed their critical thinking skills, which are analysis, explanation, evaluation, interpretation, inference and self-regulation ability, now we called them critical thinking skill's indicators [8]. But, the critical thinking skills' indicators to be observed in this research are analysis, evaluation, interpretation, and inference.

In order to analyse critical thinking skills for science education undergraduate students, so there is a need for research that focuses on looking at the profile of science education undergraduate students' especially in basic physics and use this result as empirical data to developing critical thinking skill. Furthermore, the next study is needed as an alternative solution of critical thinking that is very low level.

2. Method

2.1 Background

This pre-experiment research is to analyse the profile of science education undergraduate students' critical thinking skills on basic physics.

2.2 Sample

The sample is science education undergraduate students consist of 109 students of first, second, third and fourth year. Purposive sampling is used to determine of the sample in this research.

2.3 Procedure

This research procedure that researcher to be done: (1) studying the literature that relate with critical thinking skills; (2) finding indicators; (3) finding critical thinking skills instrument to determine the profile of science education undergraduate students on basic physics (4) analysing validity and reliability instrument; (5) Measuring 109 students using the critical thinking instrument to determine the profile of science education undergraduate students critical thinking skills on basic physics; (6) analysing and making conclusions based on the results of the critical thinking skills instrument.

2.4 Instruments of this research

Using the critical thinking instrument to determine the profile of science education undergraduate students critical thinking skills on basic physics that developed by Jatmiko et al. [9]. This instrument consists of four indicators: analysis, evaluation, interpretation, and inference. This instrument has been declared valid and reliable. Validation results of critical thinking instrument in Table 1.

Table 1. Result of critical thinking skills validity

Components	The Validity of Critical Thinking Skills Instruments											
	Construct Validity						Content Validity					
	Cohen's Kappa	R	r_α	V	α	R	Cohen's Kappa	R	rα	V	α	R
Critical Thinking Skills Test	1.00	Reliable	0.26	Valid	0.99	Reliable	1.00	Reliable	0.26	Valid	0.99	Reliable
Conclusion: The critical thinking skills instrument that has been developed is valid. It can take as an instrument to measure determine critical thinking skill of science education undergraduate students on basic physics												

Notes: r_α = Single measure interrater coefficient correlation; α = Cronbach's alpha; *R* = Reliability; *V* =Validity

The instrument of critical thinking skill test to determine the profile of science education undergraduate students critical thinking skills on basic physics by statistic analyse with Cronbach's alpha. This instrument valid ($r_\alpha = 0.26$) and reliable ($\alpha = 0.99$).

2.5 *Data Analysis*

The profile of science education undergraduate student's critical thinking skill in basic physics. Qualitative descriptive is used to analysed the data. The results will be used to obtain the profile of science education undergraduate's critical thinking skill, especially on basic physics. After that, this result will be used to recommendations in basic physics learning based on critical thinking skill.

For the results of students answers in completing written tests in order to get a profile of undergraduate students critical thinking skills, they are categorized according to the highest scoring criteria for each criterion. The methods used in completing the study are written in this section.

3. Results and Discussion

The critical thinking skill instrument was valid and reliable based on test of validity, then it done by 109 undergraduate students whose took basic physics subject. The criteria of critical thinking skill based on Table 2 [10].

Table 2. Criteria of critical thinking skill

Interval	Criteria
$81.25 < x \leq 100$	Very High
$71.5 < x \leq 81.25$	High
$62.5 < x \leq 71.5$	Medium
$43.75 < x \leq 62.5$	Low
$0 < x \leq 43.75$	Very Low

The result of analysis undergraduate student critical thinking skill in basic physics that presented in Table 3.

Table 3. Completeness of indicators and undergraduate students' critical thinking skill on basic physics

Student Initial	Group I		Group II		Group III		Group IV	
	Value	Criteria	Value	Criteria	Value	Criteria	Value	Criteria
U1	55	Low	43	Very Low	10	Very Low	23	Very Low
U2	35	Very Low	48	Low	35	Very Low	33	Very Low
U3	40	Very Low	30	Very Low	28	Very Low	36	Very Low
U4	43	Very Low	31	Very Low	31	Very Low	46	Low
U5	40	Very Low	47	Low	37	Very Low	35	Very Low
U6	35	Very Low	49	Low	26	Very Low	36	Very Low
U7	32	Very Low	36	Very Low	28	Very Low	37	Very Low
U8	28	Very Low	31	Very Low	27	Very Low	28	Very Low
U9	38	Very Low	29	Very Low	24	Very Low	35	Very Low
U10	40	Very Low	45	Low	26	Very Low	8	Very Low
U11	37	Very Low	38	Very Low	26	Very Low	40	Very Low
U12	33	Very Low	29	Very Low	57	Low	33	Very Low
U13	38	Very Low	34	Very Low	40	Very Low	11	Very Low
U14	51	Low	39	Very Low	43	Very Low	34	Very Low
U15	45	Low	32	Very Low	28	Very Low	33	Very Low

Student Initial	Group I		Group II		Group III		Group IV	
	Value	Criteria	Value	Criteria	Value	Criteria	Value	Criteria
U16	33	Very Low	34	Very Low	36	Very Low	35	Very Low
U17	33	Very Low	57	Low	41	Very Low	32	Very Low
U18	32	Very Low	36	Very Low	34	Very Low		
U19	39	Very Low	41	Very Low	8	Very Low		
U20	44	Low	48	Low	37	Very Low		
U21	39	Very Low	27	Very Low	47	Low		
U22	44	Low	44	Low	47	Low		
U23	36	Very Low	34	Very Low	24	Very Low		
U24	43	Very Low	22	Very Low	39	Very Low		
U25	36	Very Low	28	Very Low	41	Very Low		
U26	43	Very Low	35	Very Low	37	Very Low		
U27	34	Very Low	46	Low	46	Low		
U28	36	Very Low	28	Very Low	38	Very Low		
U29	34	Very Low	28	Very Low	31	Very Low		
U30	35	Very Low	42	Very Low	42	Very Low		
U31	23	Very Low	33	Very Low				
Average	38	Very Low	37	Very Low	34	Very Low	31	Very Low

Table 2 shows test outcome of critical thinking skill. In all classes, critical thinking skill of science education undergraduate student in basic physics subject is in the very low level generally. As many as 109 undergraduate student's score is under 62.5. It also shows that undergraduate student not complete in all indicators. In group I and group II which are first year science education undergraduate students show only 13 with low level and 49 in very low level. In group III, which are second year science education undergraduate students have just 4 in low level, and 26 in very low level. Last in group IV, which are third and fourth year science education undergraduate students just 1 in low level, almost them in very low level. This result was surprised. Because the students are still unknown the test of critical thinking skills in basic physics.

This research used first, second, third and fourth grade undergraduate students. The critical thinking skills in each grade show with Figure 1.

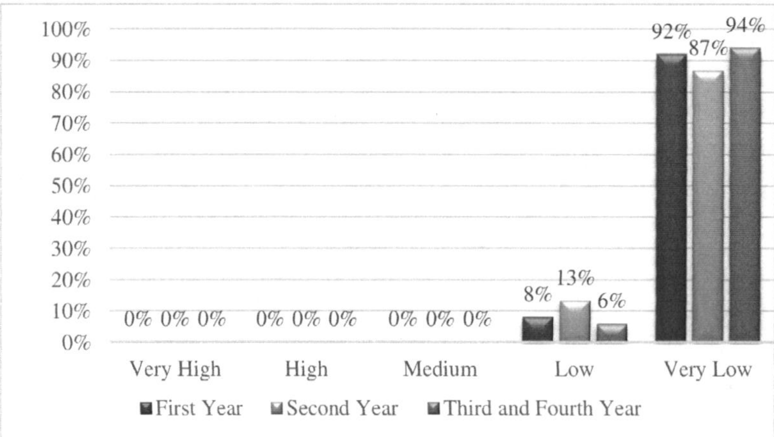

Figure 1. Graph of critical thinking skill undergraduate students in each year

Figure 1 shows criteria of critical thinking skill of undergraduate students in each year. Based on graph, generally undergraduate students on very low level all of year. Specifically, 92% first year, 87% second year and 94% third and fourth year in very low level. Just 8% first year, 13% second year and 6% third and fourth year in low level. There are no one in medium, high and very high level.

Instrument of critical thinking skill consist of four indicators: analysis, evaluation, interpretation and inference. Analysis is ability to identify relationships intended for questions, statements, concepts, description or other representation intended to express judgment, information or opinion. Evaluation is ability to review the credibility of answer the questions or other representations in the form of reports or description of perceptions, experiences, situations, judgment, or opinion, and interpreting.

Logical of the relationship or representation other means. Interpretation is ability to understand expressing or interpreting the meaning of some experiences, events, data, or situations. Inference is the ability to identify and obtain elements necessary to logically in conclusion, making assumptions and hypotheses, consider relevant information and infer the consequences of data and questions other [11]. Critical thinking skill is one of 21th century skills of undergraduate students should have for succeed in today's world. Undergraduate students need some skills such as creativity, critical thinking, innovation, ability of problem solving, good communication and also collaboration [12-16]

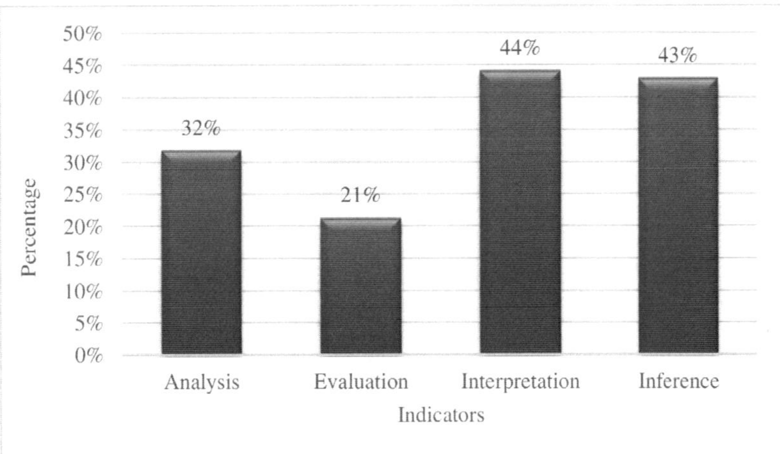

Figure 2. Graph of achieving critical thinking skill indicators

Figure 2 shows achieving critical thinking skill indicators. Four indicators of critical thinking skill, analysis with 32% its mean low, evaluation with 21% means low, interpretation with 44% means low and inference with 43% means low. All indicators in low level.

Figure 3 shows achieving critical thinking skill indicators of undergraduate students in each year. Based on that graph, the first-year undergraduate students have lowest level in each indicator. But, the second, third and fourth-year undergraduate students almost at same point. Generally, all of the undergraduate students in low level.

Increasing of undergraduate students' critical thinking skills is closely related to the understanding of the student's physics concepts. Highlights the importance of critical skills in the framework of educational concepts [17]. In his research mentioned that critical thinking skills and high-level thinking skills must develop among students in the educational process. Furthermore, to practice the critical thinking skills one of them is by angling students to argue. Reveals the influence of scientific arguments on the development of student critical thinking. In his research it is known that the critical thinking skills of students improve significantly by using scientific argument strategies [18]. This is in line with four indicators of critical thinking skills providing basic clarification as well as advanced clarification.

Critical thinking skills along with scientific argumentation contribute to understanding concepts in different spell models [19].

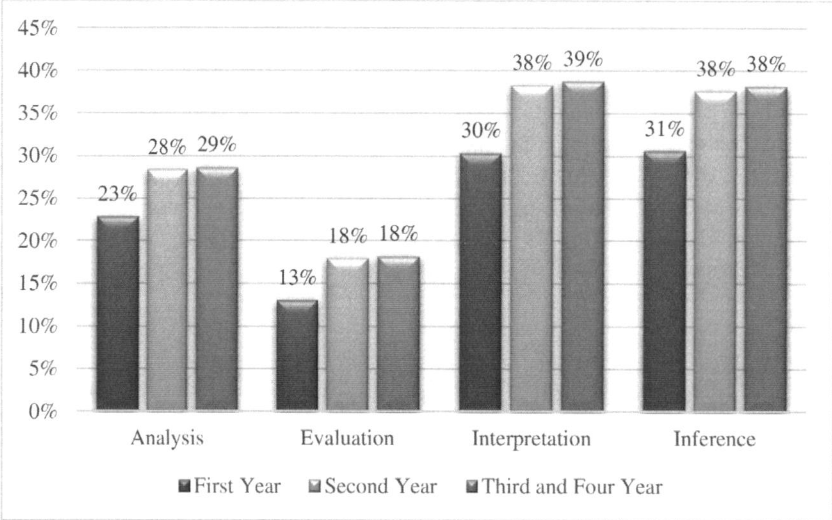

Figure 3. Graph of achieving critical thinking skill indicators of undergraduate students in each year

An essential skill that must have of undergraduate student is critical thinking skill, it is from some studies result about science teacher [20-23] which showed that assessment and learning about critical thinking skill are indispensable and driven by the needs of students at the college level and work environment.

Based on table and figures above the result of this research showed that undergraduate students are very difficult in applying critical thinking skill in basic physics subject, especially in solving the test which are need deeply of analysis, evaluation, interpretation and inference. So, it is needed to find an alternative solution to solve this problem. Learning with experience could be try to this problem.

Nevertheless, the student's critical thinking skills cannot be acquired in a short period of time. An ongoing process is needed to practice critical thinking skills in students. In line with this, research by Grussendorf and Rogol [24] reflect on critical thinking skills at a time. In the results of his reflection obtained that the initial jump in critical thinking skills after training for one semester, but at the end of the study it takes a longer time to score above average.

Critical thinking skills can be achieved one of them by practicum-based learning with workbook laboratory [25]. One of way to improve student's critical thinking skills with clear and simple instruction was by practicum-based learning [26]. In research of Balta N [27] stated that the results of the survey of instructional design science laboratory needs obtained a very high average value in the sense that laboratory instruction can improve the learning process of science in laboratory settings. In some scopes of education, science practicum learning receives less attention, as expressed in research laboratory instruction as part of the reform of science does not fully have the support of the relevant parties [28].

In line with this, it is necessary to develop learning tools that can accommodate the needs of students' knowledge and skills. Learning tools based on practicum activities with simple and easy to understand patterns. Learning tools developed with Inquiry Based Laboratory Instruction are expected to be a means for students to improve critical thinking skills [29]. Inquiry Based Laboratory Instruction have six syntax: (1) Present contextual issues, (2) Planning, (3) Implementation, (4) Analyse and interpretation, (5) Communicating/Presenting, and (6) Make follow up [30]. This syntax of Inquiry Based Laboratory Instruction estimated improved critical thinking skill of science education undergraduate students. The Inquiry Based Laboratory Instruction learning model is believed to be the most relevant teaching model for promoting scientific concepts and scientific processes as well as developing research skills, including asking research questions, formulating hypotheses, and compiling tests of hypotheses [31]

4. Conclusion

From this result, critical thinking skill profile of science education undergraduate students in basic physics is still at a very low level generally. The achieving level of indicators competence also at low. Impact of this research can be used as empirical evidence that critical thinking skill of science education undergraduate students in basic physics still needs to be improved. The next research needs as an alternative solution to solve the very low level of critical thinking skills of science education undergraduate students in basic physics.

References

[1] Coca M and Slisko 2013 *Eur. J. Phys. Educ.* **4** 17
[2] Prahani B K, Deta U A, Lestari N A, Yantidewi M, Jauhariyah M N R, Kelelufna V P, Siswanto J, Misbah M, Mahtari S, and Suyidno S 2021 *J. Phys. Conf. Ser.* **1760** 012020
[3] Maknum J 2020 *Int. Educ. Stud.* **13** 117
[4] Özelçi S Y, Çalışkan G 2019 *Int. J. Eval. Res. Educ.* **8** 495
[5] Dwyer, Michael H, and Ian S 2015 *Elsevier J.* **16** 11
[6] Afrizon R, Ratnawulan, and Fauzi A 2012 *J. Penelit. Pembelajaran Fis.* **1** 1
[7] Lloyd M and Bahr N 2010 *Int. J. Scholarsh. Teach. Learn.* **4** 1
[8] Facione P A 2013 *Critical Thinking: What It Is and Why It Counts* Retrieved from: https://www.insightassessment.com/article/critical-thinking-what-it-is-and-why-it-counts
[9] Jatmiko B, Prahani B K, Munasir, Supardi Z A I, Wicaksono I, Erlina N, Pandiangan P, Althaf R, and Zainudin 2018 *J. Baltic Sci. Educ.* **17** 300
[10] Setyowati A 2011 *J. Pendidik. Fis. Indones.* **7** 89
[11] Facione P A 2015 *Critical Thinking New Jersey* (USA: Prentice Hall)
[12] Jatmiko B, Prahani B K, Munasir, Supardi Z A I, Wicaksono I, Erlina N, Pandiangan P, Althaf R, and Zainuddin 2018 *J. Baltic Sci. Educ.* **17** 1
[13] Griffin P and Care E 2015 *Assessment and teaching of 21st century skills* (New York: Springer)
[14] Jatmiko B, Widodo W, Martini, Budiyanto M, Wicaksono I, and Pandiangan P 2016 *J. Baltic Sci. Educ.* **15** 441
[15] Pandiangan P, Sanjaya M, Gusti I, and Jatmiko B 2017 *J. Baltic Sci. Educ.* **16** 651
[16] Suyidno, Nur M, Yuanita L, Prahani B K, and Jatmiko B 2018 *J. Baltic Sci. Educ.* **17** 136
[17] Padmanabha C H 2018 *J. Educ. Psychol.* **11** 45
[18] Giri V and Paily MU 2020 *J. Sci. Educ.* **29** 673
[19] Hasnunidah N, Susilo H, Irawati M, and Suwono H 2020 *J. Univ. Teach. Learn. Pract.* **17** 6
[20] Atabaki A M S, Keshtiaray N and Yarmohammadian M 2015 *Int. Educ. Stud.* **8** 93
[21] Birgili B 2015 *J. Gift. Educ. Creativity* **2** 71
[22] Kivunja C 2015 *Creative Educ.* **6** 380
[23] Kalelioğlu F and Gülbahar, 2014 *J. Educ. Technol. Soc.* **17** 248
[24] Grussendorf J and Rogol 2018 *J. Political Sci. Educ.* **14** 151
[25] Nainggolan B, Hutabarat W, Situmorang M, and Sitorus M 2020 *Int. J. Instr.* **13** 895
[26] Zeiner M, Viehauser P, Steiner-Friedman C 2019 *J. Chem. Educ.* **96** 2518
[27] Balta N 2015 *J. Math. Sci. Technol. Educ.* **11** 957
[28] Cherbow K, McKinley M, McNeill K, and Lowenhaupt R 2020 *Sci. Educ.* **104** 446
[29] Gudyanga R and Jita L C 2019 *Issues Educ. Res.* **29** 715
[30] Irwanto, Saputro A D, Rohaeti E, and Prodjosantoso A K 2019 *Eurasian J. Educ. Res.* **80** 151
[31] Tatar N 2012 *J. Baltic Sci. Educ.* **11** 248

Profile of Students' Scientific Literacy in Physics Learning during COVID-19 Pandemic

V Melinda*, E Hariyono, E Erman, and B K Prahani

Science Education Postgraduate Program, Universitas Negeri Surabaya, Indonesia

*Email: vina.20012@mhs.unesa.ac.id

Abstract. Online learning during the COVID-19 pandemic that is not optimal has an impact on students' Scientific Literacy (SL). Facts in Indonesia, online learning has not been fully run optimally. In this case, teachers and students are still less adaptable to changes in learning methods. This certainly has an impact on aspects of students' SL. The main objective is to analyze the initial level of students' SL during COVID-19 pandemic. Research use 105 senior high school students in Surabaya. Based on the result we can conclude that the profile of students' SL in Physics learning during COVID-19 pandemic is still at low category, that's percentage are 28.38% (level 2). It's means students can perform basic epistemic and procedural knowledge, contain knowledge in daily, basic procedural knowledge to give identification of scientific explanations, interpret scientific data, and draw conclusions from simple experimental data. The implication of this research can be used as empirical evidence stating that the SL of Senior High School students still need to be improved.

1. Introduction

Indonesia is faced with a fairly serious problem, namely the outbreak of the corona virus, which is now known as the pandemic of COVID-19. It has a negative impact on various sectors of many countries, especially in the education sector of Indonesia. The Indonesian government made a policy by changing the face-to-face learning series from being online learning. There are still many teachers and students who complain because of difficulties in the learning process, especially technical problems in teaching, this can hinder the learning process [1]. While flexibility in time and location is an advantage of online learning, it can also present problems. The manner of students who are not serious about the online learning process can cause many problems. Moreover, many students often do not participate in online learning due to internet accessibility constraints [2].

The results according of interviews with several students, some students said that they felt uncomfortable when studying online; this led to increased frustration and confusion [3]. Many students complain because the teacher does not explain the topic in a coherent manner, and the teacher tends to only give independent assignments continuously. This is a challenge to develop learning content that does not only include the curriculum but must involve students [4]. The components of learning content, technology, students' psychology need to be considered, so that the online learning process can be balanced. If the components in the learning process are not balanced, it will have an impact on students' Scientific Literacy (SL). In this case, understanding of science and the implication is very important to learn, so science must be a feature of the education of every young generation [5,6].

The results of the last scientific literacy assessment conducted by PISA in 2018 seen by the Organization for Economic Cooperation and Development (OECD) stated that the SL aspects of Indonesian students before pandemic COVID-19 were still low. Indonesia is ranked 70th out of 78 OECD member countries with an average SL score of 396 [7]. This is an illustration that Indonesia's SL are still low, which is at level 1a. It means a student is able to use basic concepts and knowledge that associated lives with everyday life. They can conduct a scientific investigation of no more than two variables, the students also to identify simple correlations and they draw graphical and visual data that require low cognitive comprehension [7].

In addition, many studies in Indonesia conducted prior to COVID-19 stated that students' SL were still relatively low [8-12]. SL are influenced by several factors they are teaching materials, learning models, learning medias, students' worksheets, evaluation instrument based on SL [12]. Three main categories of inventions were identified as consisting of issues related to online learning, teachers, and content development. Learner-related issues are including students' expectations, identities, readiness, and participation in online learning. Meanwhile, content-related issues are including the role of teachers in content development, integration of multimedia in content, strategic roles of teaches in content development, and considerations for content development. The problems that faced by teachers are the transition from face to face to online learning, the role of agencies, the time management, and the teaching methods [4]. The low SL of students is possible because of these factors.

The SL measured by PISA includes the Physics subject matter. Therefore, the measurement of students' SL in the physics topic is very necessary. Meaningful learning can be obtained through students' SL which is useful for problem solving in everyday life. One of the physics topics is the kinetic theory of gases. Based on the questionnaire, students think that the kinetic theory of gases topic is difficult to understand, because of its abstract nature related to microscopic things. Student in explaining an abstract concept is a main to receive knowledge [13]. Therefore, a great reasoning is necessary to understand abstract concept. In this case, teachers must have many ideas to make learning more meaningful, and need to adjust so that students' scientific performance can still be carried out even though learning is carried out online.

Based on the description, this research aims to describe the profile of students' SL on the kinetic theory of gases topic during the COVID-19 pandemic. The results of this research will be used as a reference for efforts to improve students' SL.

2. Method

2.1 General Background

The design of this research was pre-experimental. The main objective of this research was to analyze the initial level of SL of senior high school students during the COVID-19 pandemic on kinetic theory of gases topic.

2.2. Sample of Research

The subjects of this research were Senior High School students in Surabaya (East Java, Indonesia) that consist of 105 eleventh grade students in Surabaya. Purposive sampling is used in this research.

2.3. Procedure

The procedure of this research consists of: (1) researchers conduct a literature about SL; (2) researchers focused on finding indicators and assessments of several SL in the physics learning; (3) researchers develop SL test instrument to determine the initial level of SL of senior high school students during the COVID-19 pandemic; (4) researcher gave instruments to three experts, to validate the SL test Instruments; (5) Researchers revise related on suggestions from three validator experts; (6) Researchers tested 105 students using a question instrument to determine the profile of high school students' SL on the topic of Kinetic Theory of Gases; and (7) researchers conducted analysis and conclusions the results of the instrument on SL.

2.4. Instrument

In this research, the SL question instrument was developed to determine the initial level of SL of high school students during the COVID-19 pandemic on kinetic theory of gases topic. Indicators of SL questions consist of (1) Explaining phenomena scientifically; (2) Evaluating and designing scientific investigations; and (3) Interpreting scientific data and evidence. The results of instrument validation on SL test are shown in Table 1.

Table 1. The results of SL test instruments' validation and reliability

Aspect	**Indicators of Validation**	**Assessment (validator)**			**Score**	**Criteria**	**R (Cronbach's Alpha)**
		V1	V2	V3			
Content	The suitability of the item with the topic	4.00	3.00	3.00	3.33	Valid	Reliable
	The suitability of the items with the indicators	4.00	3.00	3.00	3.33	Valid	Reliable
	The items presented are able to reveal the profile of students' SL	4.00	3.00	3.00	3.33	Valid	Reliable
Construct	The clarity of instructions on how to do the questions	4.00	3.00	3.00	3.33	Valid	Reliable
	The clarity of questions and answer boundaries	4.00	3.00	4.00	3.67	Valid	Reliable
Language	The structure of communicatives' sentence	4.00	3.00	4.00	3.67	Valid	Reliable
	The use of question sentences does not cause double interpretation	4.00	3.00	4.00	3.67	Valid	Reliable
	Use of good and correct language	4.00	3.00	3.00	3.33	Valid	Reliable
Average		4.00	3.00	3.38	3.46	Valid	Reliable
Conclusion: The Instrument of SL test is valid and reliable							

The development of SL test instrument of senior high school on the kinetic theory of gases by experts is 3.46 valid. The instrument validation sheet for SL test based on the statistical reliability of Cronbach's alpha, which is 0.956, is categorized as very reliable.

2.5 Data Analysis

The initial level of high school students' SL during the COVID-19 pandemic on kinetic theory of gases topic analyzed using quantitative descriptive. The results of this research show the initial level of SL for Senior High School students on the kinetic theory of gases topic.

Data analysis of students' SL was carried out by:

a. Giving weight or score to each question that has different literacy competencies. SL cognitive competence adapted from PISA
   1. The point 1, maximum scores: 5, questions with low cognitive level
   2. The point 2, maximum score: 10, questions with moderate cognitive level
   3. The point 3, maximum score: 15, questions with high cognitive level
b. Determine the score for each question with the description assessment rubric that has been made.
c. Total score of students:

$$\text{Total Score} = \frac{score\ obtained}{Maximum\ score}\ x\ 100 \quad (1)$$

d. Determine the achievement score for each competency:

$$\text{Total Score} = \frac{score\ obtained}{Maximum\ score\ x\ sample}\ x\ 100 \quad (2)$$

e. The score obtained is converted into a level score based on PISA [7] based on Table 2.

Table 2. Score and level of scientific literacy

Level (Score of PISA)	Conversion Indonesian's Score
6 (708)	86-100
5 (633)	73-85
4 (559)	58-72
3 (484)	43-57
2 (410)	28-42
1a (335)	13-27
1b (261)	0-12

3. Results and Discussion

The SL test instrument that has been valid and reliable was tested on 105 eleventh grade Senior High School students in physics learning. The questions of test were developed by researchers and the categorization of SL levels adapted from PISA 2018. The result of students' SL test based on levels of SL shows in Table 3.

Table 3. The results students' SL in level of PISA 2018

Levels of scientific literacy	Average score of scientific literacy	Percentage (%)	Students who can answered questions on level
1b	-	-	-
1a	31.59	39.05	41
2	23.31	60.95	64
3	-	-	-
4	-	-	-
5	-	-	-
6	-	-	-
Percentage of all students'SL	28. 36 %	Students	105

Based on Table 2, we can see that students' SL in online learning during the COVID-19 pandemic are still low, namely dominant at level 2. This means that students' SL has not changed before and after the pandemic COVID-19, that's in low category [10,11]. Table 3 shows that percentage of all students' SL in aspects of competence and knowledge in 105 students are 28.36%, it can be interpreted into the level 2 [7]. In this case, the students of eleventh grade in physics learning can demonstrate the basic of epistemic knowledge. They can use their contain knowledge in daily, basic procedural knowledge to give identification of scientific explanations, interpret scientific data, and draw conclusions from simple experimental data [7]. This result shows one level higher than the result which states that the average SL of Indonesian students is low, it's show at level 1 a [7]. There are several causes of low SL during the COVID-19 pandemic, including learning media, textbooks, learning models, worksheets, and evaluation instruments based on SL [13]. Improving the SL of Senior High School students can be done by using science textbooks that are developed appropriately. Textbooks, worksheets, learning models, and evaluation instruments who develop based on SL can improve the students' SL [14].

The students are not used to learning more about scientific issues. Students also do not understand the problem, because the reading level is low [12]. There are still many students in Indonesia who experience misconceptions, especially on kinetic theory of gases topic, so in this case, teachers need to prepare learning well so that the concepts conveyed can be accepted by students clearly. They also increase self-confidence and improves the relationship between teaching and learning process with the environmental context where students live [15]. Teachers should try to choose the right model and method to be applied to the online learning process during pandemic COVID-19. In addition, learning topic on the kinetic theory of gases tends to be abstract and related to microscopic. In this case, the teacher can integrate technology, especially in online learning. The development of high-quality computer-based learning resources on kinetic theory of gases topic effectively improves students' understanding, use and quality, and has received good responses from students [16]. During the pandemic COVID-19 media that can help in the virtual lab, namely interactive PhET. The function of PhET interactive simulation is also effective in increasing students' motivation to experiment, facilitating students to collect experimental data very well [17]. Developing inquiry-based teaching materials effectively improves SL in learning. [18].

Based on several research, it is concluded that several models that can improve SL are problem-based learning [19-21]. STEM approach in learning can also improve students' scientific literacy [22-24]. If viewed from the achievement of students' SL competence, the result of SL test shown in Table 4.

Table 4. The percentage of indicator SL competencies

Indicator of Scientific Literacy	Question number	Percentage results (%)
1. Explanation of scientific phenomena	1 and 6	59.81
2. Evaluation and designation of scientific investigations	2, 4, and 5	20.87
3. Interpretation of scientific evidence and data	3 and 7	29.57

Table 4 shows the percentage of SL competencies on each question. The highest percentage is shown in the first scientific literacy indicator, which is explaining phenomena scientifically, while the lowest is shown in the indicator of evaluating and designing scientific experiments. Based on Table 4 it can be obtained as many as 40.19 % of students are still not able to understand the concept of kinetic theory of gases on indicator 1 clearly. This shows that students have not been able to use and relate the concept of the kinetic theory of gases to issues in the surrounding environment. Several factors can cause this to happen, students assume that the phenomena around them have nothing to do with physics concepts [25], students often think that physics is a difficult lesson, it's just identical to formulas that can make students confused and experiments that can only be done in the laboratory. This means that the level of reading literacy in Indonesia is still low category [26]. Student still use traditional literacy in their learning activities [27]. Based on the analysis, the information was obtained that most Indonesian students still have limited scientific knowledge that can only be applied to a few situations. They are only able to provide clear scientific explanations and follow explicit evidence. Other than that, many students are less sensitive to the phenomena around them, even more, sensitive to the virtual world, especially in online learning. In the modern era that is increasingly advanced, many students prefer to play online games than observing scientific phenomena around them. So, the impact of misused internet use can affect students' academic achievement and social life [28].

It can be obtained on Table 4, as many as 79.13 % of students still do not understand the concept of evaluating and designing scientific experiments, the fact that in the classroom there are still many teachers who only teach using conventional methods so that learning becomes boring and weakens students' thinking patterns to propose ideas, investigate scientific questions, and solve problems around them. When online learning students tend only writes the topics explained by the teacher [29]. The results of students doing less of laboratory activities when, the solution for that condition is the teacher can integrate STEM in their learning. It's can improve students' understanding of concepts than learning with conventional methods [14,30]. Problem-solving that occurs in everyday life can be overcome if students already have a good understanding of science [31]. Other factor show that the weakness of online learning is the obstacle to internet accessibility, students who do not have an internet network will find it difficult to access the material provided by the teacher [2,29,32,33], this results in students cannot learn optimally.

Based on Table 4, it can be known as many as 70.43% of students still weak in terms of interpreting scientific data. The reasons are the teaching materials provided by the teacher only use full text, it's less interactive. This is because the teaching materials used by the teacher are only in the form of textbooks, containing formulas, concepts, materials that are packaged classically, so that they are easily understood by students. Good teaching materials should be prepared with various attractive representations and can motivate students in learning physics. The development of unique, creative, and innovative teaching materials by teachers needs to be done with the hope of increasing students' understanding of concepts, interactive e-modules can help students learn to understand material independently anywhere [34]. The development of physics instruments can improve students' scientific attitudes [35]. To improve SL teacher can develop SL-based teaching materials [36,37].

To determine whether gender affects the results of students' SL literacy, the data were analyzed using two-way analysis of variance. The assumption of homogeneity variance was tested using Levene test. The results of the Levene test are shown in table 5, the value of the Levene test p 0.764 > 0.050 indicates the homogeneity of the variance between the gender.

Table 5. The results of Levene test for equality of variances

Dependent variable	F	df1	df2	Sig.
Scientific literacy	3.288	1	103	.764

The results of Table 6 show that the mean SL values of male and female students are 27.30 and 29.06, respectively. According to the result, we can conclude that the average SL of female students is higher than male students [7,9]. That supported by the PISA 2018 results, the average of female students (399) than boys (393) [7].

Table 6. Students' mean score and standard deviation of scientific literacy based on gender

Gender	Number of participants	Mean	Standard deviation
Male	42	27.30	4.89
Female	63	29.06	4.84

Table 7 shows the independent t-test, the sig value of 0.073 > 0.050 its mean there is no statistically significant difference of the average scores of females and males' SL. The mean difference value shows the difference between the average scientific literacy score of female and male students, namely -1.753. It can be concluded that gender does not significantly affect the SL scores of senior high school students in learning physics in the COVID-19 pandemic.

Table 7. As a result of independent sample t test with factors of gender

Source	df	Sig.	Mean difference
Gender	103	0.073	-1.753

However, the researchers' findings prove that there is no statistically significant difference between the two. The average score of women and men on SL was statistically proven that there was no significant difference between them. It can be concluded that the gender of students does not affect the score of their SL [38].

4. Conclusions

In this case of the research, the profile of students' scientific literacy in Physics learning during COVID-19 pandemic is still at low category, that's percentage are 28.38%. The students' SL based on PISA is still at level 2. It's means students can perform basic epistemic and procedural knowledge, contain knowledge in daily, basic procedural knowledge to give identification of scientific explanations, interpret scientific data, and draw conclusions from simple experimental data. The students' SL for each indicator are: (1) 59.81% can give the explanation of scientific phenomena; (2) 20.87% can give evaluation and designation of scientific investigations; and (3) 29.57% can give interpretation of scientific evidence and data. This research will be used as empirical evidence on how to improve the scientific literacy level in physics learning.

References

[1] Favale T, Soro F, Trevisan M, Dargo I, and Mellia M 2020 *Comput. Netw.* **176** 107290
[2] Dwijuliani R, Rijanto T, Munoto, Nurlaela L, Basuki I, and Maspiyah 2021 *J. Phys. Conf. Ser.* **1810** 012072
[3] Dhawan S 2020 *J. Educ. Technol. Syst.* **49** 5
[4] Kebritchi M, Angie L, and Lilia S 2017 *J. Educ. Technol. Syst.* **46** 4
[5] COSCE 2011 *Informe ENCIENDE, Enseñanza de las Ciencias en la Didáctica Escolar paraedades tempranas en España, Confederación de Sociedades Científicas de España, Madrid* Retrieved from: https://www.cosce.org/pdf/Informe_ENCIENDE.pdf
[6] National Research Council 2012 *A Framework for K-12 Science Education: Practices, Crosscutting Concepts, and Core Ideas* (Washington, D.C: The National Academies Press)
[7] OECD 2019 PISA 2018 *Assessment and Analytical Framework* (Paris: OECD Publishing)
[8] Widowati A, Widyastuti L, and Widodo E 2019 *J. Phys. Conf. Ser.* **1321** 032070
[9] Nur'aini D, Rahardjo S B, and Susanti V H E 2018 *J. Phys. Conf. Ser.* **1022** 012016
[10] Suroso J, Indrawati, Sutarto, and Mudakir I 2021 *J. Phys. Conf. Ser.* **1832** 012040
[11] Mukti W R, Yuliskurniawati I D, Noviyanti N I, Mahanal S, and Zubaidah S 2019 *J. Phys. Conf. Ser.* **1241** 012043
[12] Purwani L D, Sudargo F and Surakusumah W 2018 *J. Phys. Conf. Ser.* **1013** 012019
[13] Rusilowati A, Nugroho S E, Susilowati E S M, Mustika T, Harfiyani N, and Prabowo H T 2019 *J. Phys. Conf. Ser.* **1170** 012028
[14] Parlindungan S, Ida K, and Andhy S 2017 *J. Turk. Sci. Educ.* **14** 92
[15] Udompong L and Suwimon Wongwanich 2014 *Procedia Soc. Behav. Sci.* **116** 5091
[16] Pratidhina E 2019 *J. Phys. Conf. Ser.* **1157** 032004
[17] Rustana C E, Andriana W, Serevina V, and Junia D 2020 *J. Phys. Conf. Ser.* **1567** 022011
[18] Aulia E V, Poedjiastoeti S, and Agustini R 2018 *J. Phys. Conf. Ser.* **947** 012049
[19] Wicaksono R S, Susilo H, and Sueb 2019 *J. Phys. Conf. Ser.* **1227** 012005
[20] Husna F R, Sukaesih S, and Setiati N 2017 *J. Biol. Educ.* **6** 357
[21] Nurtanto M, Nurhaji S, Widjanarko D, Burhan M R, Wijaya, and Sofyan H 2018 *J. Phys. Conf. Ser.* **1114** 012038
[22] Afriana J, Permanasari A, and Fitriani A 2016 *J. Pendidik. IPA Indones.* **5** 261
[23] Khaeroningtyas N, Permanasari A, and Hamidah I 2016 *J. Pendidik. IPA Indones.* **5** 94
[24] Amiruddin B, Budi A S, and Sumantri M S 2021 *J. Phys. Conf. Ser.* **1869** 012176
[25] Al Faizah W A, Suparmi, and Aminah N S 2019 *J. Phys. Conf. Ser.* **1155** 012025
[26] Pratama W A, Hartini S, and Misbah 2019 *J. Inov. Pembelajaran Fis.* **6** 9
[27] Brown C 2013 *Literacy Boost Indonesia Endline Report* (Indonesia : Save the Children)
[28] Soegoto E S and Tjokroadiponto S 2018 *IOP Conf. Ser. Mater. Sci. Eng.* **407** 012176
[29] DebarghaBanerjee, Das D, Pal S, Paul S R, Debnath A, and Reza M 2021 *J. Phys. Conf. Ser.* **1797**

012066
[30] Thahir A, Anwar C, Saregar A, Choiriah L, Susanti F, and Pricilia A 2020 *J. Phys. Conf. Ser.* **1467** 012008
[31] Salamah P N and Rusilowati A 2017 *Unnes Phys. Educ. J.* **6** 3
[32] Sulisworo D, Wulandari Y, Effendi M S, and Alias M 2021 *J. Phys. Conf. Ser.* **1783** 012117
[33] Akmam, Hidayat R, Mufit F, Jalinus N, and Amran A 2021 *J. Phys. Conf. Ser.* **1876** 01208
[34] Maison and Wahyuni I 2021 *J. Phys. Conf. Ser.* **1876** 012089
[35] Gunada I W, Sanida, and Sutrisno H 2015 *J. Pendidik. Fis. Teknol.* **1** 38
[36] Hatta, Miterianifa, and Octarya Z 2021 *J. Phys. Conf. Ser.* **1842** 012045
[37] Thaybah, Mastuang, and Mahtari S 2018 *J. Phys. Conf. Ser.* **1097** 012032
[38] Sema Altun-Yalçın, AçıŞlı S, and Turgu U 2011 *Procedia Soc. Behav. Sci.* **15** 783

Suplementary File:

SCIENTIFIC LITERACY TEST INSTRUMENT (SLTI)

Answer the following questions honestly and thoroughly based on your own ability, please!

Indicator 1:

Pay attention to the following phenomena!

Figure 1A. Hot air balloon fly early hours of the morning

Hot air balloon is a technology of the first flight by humans, invented by the Montgolfier brothers in Annonay, France in 1783. You need to know, the first flight was carried out in Paris November 21, 1983 by Pilâtre de Rozier and the Marquis d'Arlandes. Hot air balloons can be controlled and not only carried by the wind known as thermal airships or airships. The original idea behind the hot air balloon we know today has been around for a long time. Archimedes, an ancient Greek mathematician, described the principle of floating more than 2000 years ago.

Hot air balloons can fly in the air because of their buoyancy. Initially, the air pumped was hydrogen. Due to the risk of explosion, now the noble gas helium is used as the propulsion medium. "Hot air balloons are often flown in the early hours of the morning. In theory, hot air balloons were easier to fly at that time".

1. Explain your opinion, why hot air balloons are theoretically easier to fly in the early hours of the morning? ***(max. score: 5)***

Indicator 2:

2. **Pay attention to the following phenomena!**

Figure 2A. Dented tennis ball

One day your friends, Dini and Bella were playing table tennis, during the game you accidentally step on a table tennis ball and the ball becomes dented. At that time there was no replacement and the game wanted to continue, then what would you do? Design scientifically, how do you restore a dented ball to its original shape and why do you use that method? *Hint: relate your answer to the concept of an ideal gas!* ***(max. score: 15)***

Indicator 3:

A student conducts an experiment using PhET media to investigate the relationship between temperature and pressure. From these experiments the data obtained shows on the table A1.

Table A1. The experimental data.

V (nm^3)	T (K)	P (Atm)
9	400	0.70
	500	0.91
	600	1.08
	700	1.26

3. Based on the data obtained, interpret the magnitude of the pressure (P) at a temperature (T) of 800 K. Please, analyze the data and make a graph that fits the data! ***(max. score: 10)***

Indicator 2:

The heating of an ideal gas looks like the figure A4.

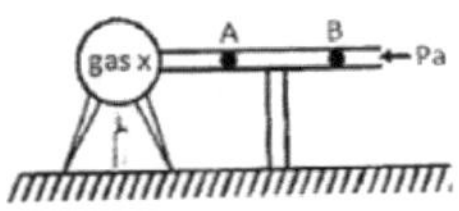

Figure 4A. Heated gas

Journal of Physics: Conference Series **2110** (2021) 012031 doi:10.1088/1742-6596/2110/1/012031

Due to gas heating:

a. Particle speed increases
b. The momentum of the particle increases
c. The kinetic energy of the particles increases
d. Gas pressure increases

4. If due to rising temperature drops A moves to B, make a design of figure 4A and then evaluate what will happen to gas X! *Hint: choose the option appropriate!* ***(max. score: 15)***

5. A hairspray can fill with isobutane gas at constant volume reads a warning that "Store Only at Temperatures below 49^0 C".
   a. The gas in the can initially has a temperature of 23 °C and 360 kPa with a volume of 340 ml, a can is placed in a motorbike seat storage, the temperature reaches 45 °C on a hot day. Evaluate the new pressure in the can!
   b. If the hairspray can accidentally catch fire, evaluate what will happen to the hairspray can and design what steps should be taken to prevent this from happening! ***(max. score: 15)***

Indicator 1:

Figure A5. Divers are diving in the sea of different depths

Divers wear diving equipment containing compressed air (nitrogen-oxygen) when they dive in the deep ocean. When the diver swims in the deep sea, the bubbles come out of the diver's mouth and these bubbles rise. It is observed that the volume of the bubbles increases gradually. These bubbles rise and become several times larger than they were before. There is no change in the chemical construction when the bubble rises.

6. Explain why did the volume of the bubble at that time change? (Assume that the temperature in seawater remains constant the same at all points in the ocean) ***(max. score: 5).***

Indicator 3:

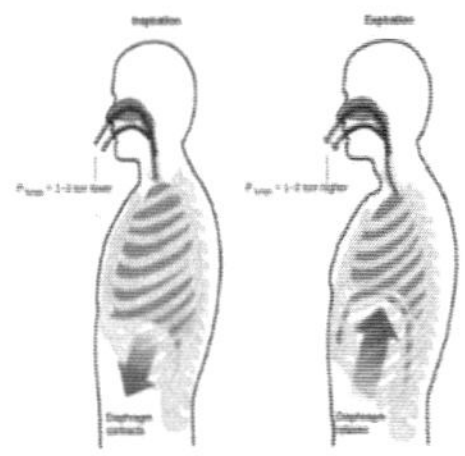

Figure A6. The human respiration

The lungs are made of spongy and elastic tissue, so that when you breathe, they contract and expand. When humans inhale, the intercostal muscles and diaphragm contract, expanding the chest cavity so that the volume of the human lungs is larger. This volume increase causes a decrease in pressure. In this case the air flows into the lungs (from high pressure to low pressure). When humans exhale, the diaphragm and rib muscles relax, as opposed to inhaling. The chest cavity will contract, and the volume of the lungs will decrease, so the pressure increases.

7. From the process of human respiration, make a table of assumptions about the relationship between volume, pressure, at constant temperature and interpret the statement into a graph that allows the situation! Which ideal gas theory fits this situation? ***(max. score: 10)***

Electromagnetic Wave Equation Approximation using FDTD method on Conductivity Material

R A Firdaus[1]*, M Khoiro[1], A Asnawi[1], M A Bustomi[2], and J Annovasho[3]

[1] Physics Department, Faculty of Mathematics and Natural Sciences, Universitas Negeri Surabaya, Indonesia
[2] Institut Teknologi Sepuluh Nopember, Indonesia
[3] Tadris Fisika, IAIN Palangka Raya, Indonesia

*Email: rohimfirdaus@unesa.ac.id

Abstract. FDTD is a method that is applied in the simulation of electromagnetic waves. This study aims to simulate the propagation of electromagnetic waves on a material with conductivity and permittivity on the plate. The approximate form of Maxwell's equations can be used to describe discrete electromagnetic waves. Signal analysis in the form of electromagnetic waves using position domains for magnetic field H and electric field E. By taking into consideration boundary conditions, stability, and boundary conditions, the proposed research employs the basic concept of differential equation method. The simulation results show that materials with high conductivity will cause the waves to decay. Under certain conditions, the relationship between the shape of the field to changes in conductivity and permittivity of the material is needed in the analysis process.

1. Introduction

Electromagnetic theory is a theory that explains the relationship between electric and magnetic fields, which cause electromagnetic waves to propagate. This theory was put forward by James Clark Maxwell in 1865. According to Maxwell's, shifting electric field will cause a magnetic field. Meanwhile, Faraday argued that a changing magnetic field creates an electric field. Thus, electromagnetic waves form as result of the relation between the magnetic and the electric field. In other terms, electromagnetic waves are mixed waves between electric and magnetic fields caused by accelerated moving electric charges.

Modeling of electromagnetic waveforms is difficult to visualize so an appropriate method is required to overcome this problem. One of the suitable methods to solve the problem of visualizing electromagnetic waves is the Finite Difference Time Domain (FDTD) method. The FDTD had utilized to solve problems including propagation, radiation, and electromagnetic wave propagation. Kane Yee initially proposed FDTD method in 1966 to evaluate electromagnetic fields [1]. It can analyze problems based on integral equations much easier which are very difficult to solve with the Moment Method and others. Applying this method does not require the basics of deep knowledge [2]. The FDTD method has been used as a numerical solution since 1920, and is widely used in various fields. There are many fields that had utilized the methodology, for example are fields that study electromagnetic waves, seismology, radiation and scattering and the propagation of sound waves in an acoustic space [3].

Previous research related to FDTD had mostly been done on regional hyperthermia [4], absorption of dispersive media [5], and the electromagnetic waves effect on materials of biology [6]. FDTD also analyzed how the dielectric effect in the case of electromagnetic wave propagation [7,8] and frequency dependent dispersive media [9]. Further, in the term of divergence, it was also utilized for ADI-R-FDTD 3-D modelling [10]. FDTD could also simulate the frequency shifted using radar cross-section (RCS) and CPML [11,12]. FDTD was also utilized in the material industry to estimate and calculate bands in photonic crystal [13], as well as in the telecommunications industry for acoustic and antennas analysis [14,15].

According to previous study, majority of researchers utilized FDTD method to identify material properties. However, FDTD had not been developed and calculated to explain how permittivity and conductivity variations influenced field patterns on metal plates. The simulations can be used for materials selection and physical phenomena detection associated using electromagnetic waves. Furthermore, It can be used to compute gains and return losses of antennas, especially microstrip antenna for telecommunications development. In this study, FDTD method was used to identify the influence of conductivity and permittivity for the electromagnetic wave patterns on the metal plate.

2. Method

The mathematical relation between the time-dependent electromagnetic radiation field and current as well as charge density was generally introduced by Maxwell's. On the time-dependent equations of Maxwell's equation [16], differential equations of propagating wave in one dimension may be explained and developed in Equations (1) and (2).

$$\frac{\partial \vec{B}}{\partial t} = -\nabla \times \vec{E} - \vec{M} \tag{1}$$

$$\frac{\partial \vec{D}}{\partial t} = \nabla \times \vec{H} - \vec{J} \tag{2}$$

Equations (1) and (2) describe the propagation of the electric field, magnetic field, and wave in the directions of the x-axis, y-axis, and z-axis, respectively. The following equation is used to solve time and space dependent differential equations.

$$\overline{E}_x^{n+1/2}(k) = \overline{E}_x^{n-1/2}(k) - \frac{1}{\sqrt{\varepsilon_0 \mu_0}} \frac{\Delta t}{\Delta x}\left[H_y^n(k+1/2) - H_y^n(k-1/2)\right] \tag{3}$$

$$H_y^{n+1}(k+1/2) = H_y^n(k+1/2) - \frac{1}{\sqrt{\varepsilon_0 \mu_0}} \frac{\Delta t}{\Delta x}\left[E_x^{n+1/2}(k+1) - E_x^{n+1/2}(k)\right] \tag{4}$$

The relation between electric and magnetic fields as a function of time and position is shown in Equations (3) and (4).

The differential equations can be modified in different approximations for derivation of spatial and temporal dependent electric and magnetic fields to simulate electromagnetic wave propagation in the medium having both permittivity and conductivity [16]. The approximation equation can be written as in Equation (5).

$$\overline{E}_x^{n+1/2}(k) = \frac{\left(1 - \dfrac{\Delta t.\sigma}{2\varepsilon_r \varepsilon_0}\right)}{\left(1 + \dfrac{\Delta t.\sigma}{2\varepsilon_r \varepsilon_0}\right)} \overline{E}_x^{n-1/2}(k) - \frac{1/2}{\varepsilon_r \left(1 + \dfrac{\Delta t.\sigma}{2\varepsilon_r \varepsilon_0}\right)}\left[H_y^n(k+1/2) - H_y^n(k-1/2)\right] \tag{5}$$

In programming language, Equation (5) was modified in Equation (6).

Ex(k) = A(k) * Ex(k) + B(k) * (Hy(k - 1) - Hy(k)) (6)

where

E = dt * Sigma / (2 * epsilon0 * epsilonr) (7a)
A(i) = (1 - E) / (1 + E) (7b)
B(i) = 0.5 / (epsilonr * (1 + E)) (7c)

The equation for the constant propagation and attenuation of electric and magnetic fields propagating on the plate is defined as in Equations (8) and (9).

$$\alpha = \frac{\omega}{c_0}\sqrt{\frac{\varepsilon_r}{2}}\left[\sqrt{1+\left(\frac{\sigma}{\omega\varepsilon_0\varepsilon_r}\right)^2}-1\right]^{1/2} \tag{8}$$

$$\beta = \frac{\omega}{c_0}\sqrt{\frac{\varepsilon_r}{2}}\left[\sqrt{1+\left(\frac{\sigma}{\omega\varepsilon_0\varepsilon_r}\right)^2}+1\right]^{1/2} \tag{9}$$

Because the velocity of electromagnetic waves in a vacuum could not be faster than the light's, it is necessary to determine time interval using FDTD stabilities. The "Courant Condition" in n-dimensions in Equation (10) can be used to solve the problem [17].

$$\Delta t \leq \frac{\Delta x}{\sqrt{n}c_0} \tag{10}$$

The theory suggested boundary condition absorption, which was necessary to avoid reflection on the electric field (E) and magnetic field (H). For the fundamental of the FDTD method, the E field calculation was obtained by considering the encircling H field value [18]. The theory was known as Theory of Absorbing Boundary Condition (ABC Theory). In the boundary conditions, it was important to consider the values in both sides [19]. However, according to the theory, it will more advantageous if so, the boundary field must spread outside due to there is no source on the outside of boundary. Both of the facts were utilized to calculate the value in the end, which was based on the value in the area. When $k = 0$, the wave will propagate to the boundary at the speed of light, just as it would in free space. While travelling, it generated one time step of the FDTD algorithm.

$$\text{displacement} = \frac{\Delta x}{2} \tag{11}$$

Calculation of FDTD simulation is constrained by computer resources. If wave generated from a points source was simulated as in free space, the unexpected reflections inside will be generated. It was impossible to distinguish between incident waves and reflected waves. During FDTD process, ABC theory became the appropriate solution of the complicated problem. Previous researchers developed the Perfect Match Layer (PML) to improve the flexibility and efficiency of ABC theory [19].

3. Results and Discussion

Figure 1 describes the visual appearance of the FDTD program that had been made to illustrate electromagnetic waves. The procedure for making the proposed electromagnetic wave FDTD program includes determining the input parameters, the number of N-Steps, calculating the attenuation and propagation constants either manually or based on simulations. In addition, it is also necessary to determine the variation of the thickness of the conductor plate and the type of pulse used.

The input parameters present the position domain length, time domain, wave pulse position, permittivity, conductivity, and plate thickness. Pulse variations can use gaussian or sinusoidal wave pulse types. Program-bound variables include the domain length of position 620, frequency of 700 MHz, the initial position of the wave pulse at position 1. While the independent variables that we use include plate thickness, permittivity (epsilon), and conductivity (sigma).

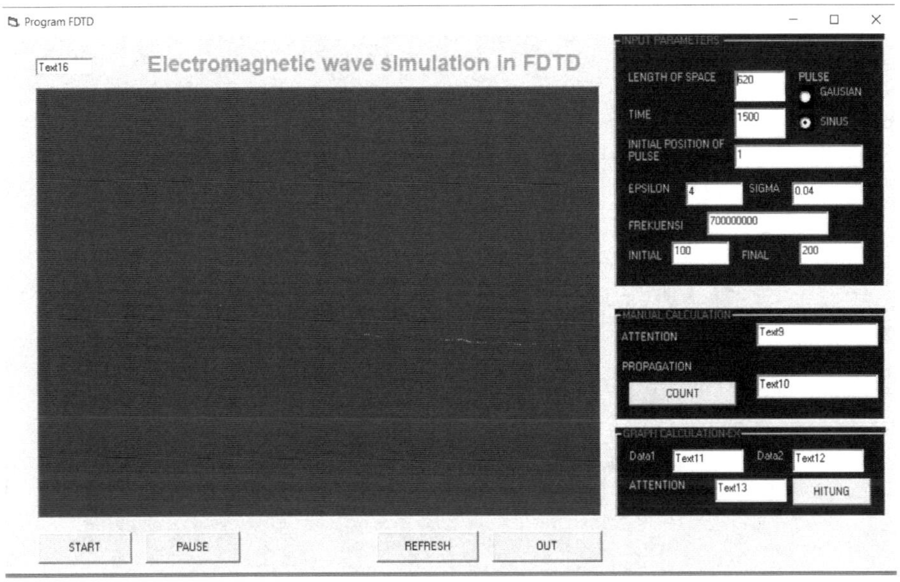

Figure 1. FDTD program display

Figure 2 shows the symptoms of electromagnetic waves when passing through a conductor plate. The relative permittivity of the plate is 4 and the conductivity value is 0.04 and the plate thickness is 100. The sinusoidal electromagnetic wave pulses E_x (red) and H_y (yellow) decay with B values ranging from 0.12383450 Equation (7c) and A from 0.98135203 in Equation (7b). Based on the simulation results, the E_x-value decays first and then H_y is followed according to Equation (5).

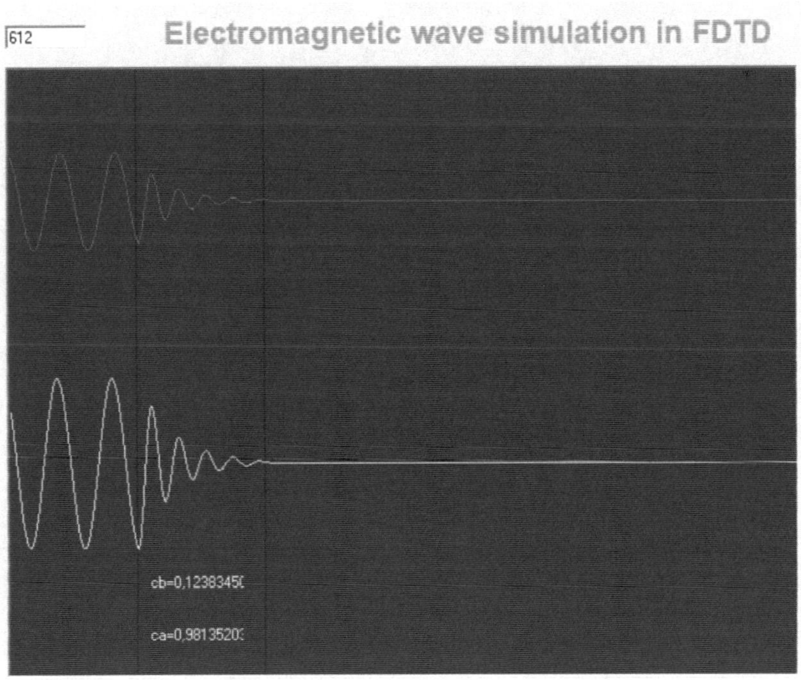

Figure 2. Electromagnetic wave simulation results on the plate

The calculation results are taken at N-step 620 with the attenuation value in Equation (8) showing a value ranging from 3.7344977 Np/m while the simulation results show 4.1632605 Np/m, this difference is due to the position factor of the location of the test point taking the E_x value in the initial position and end. While the value of the propagation constant according to Equation (9) shows 29.54364880 rad/m.

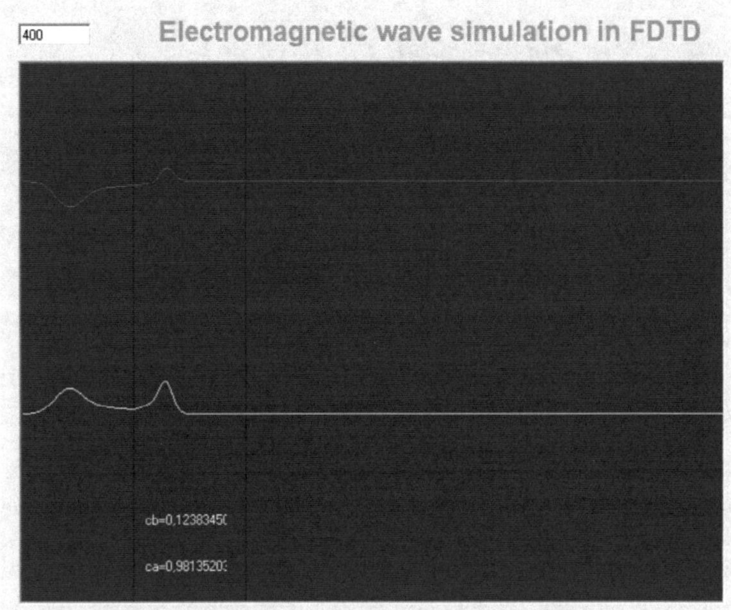

Figure 3. Simulation using Gaussian wave pulse

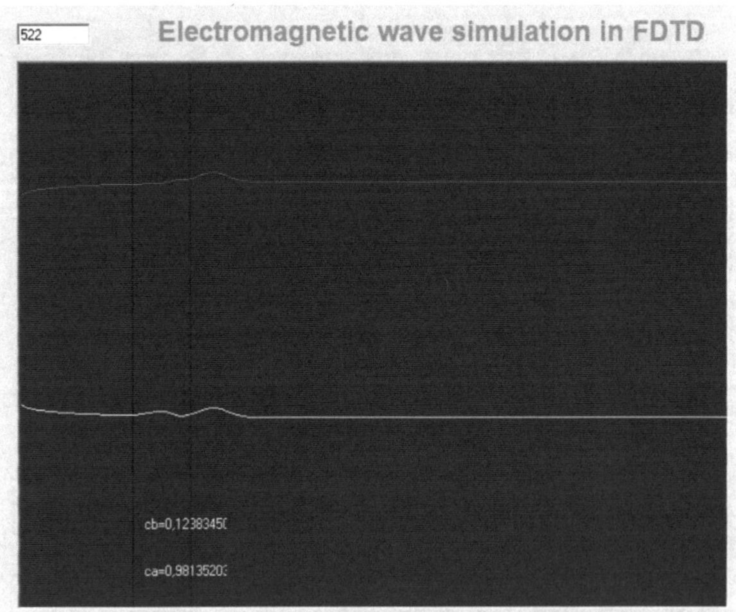

Figure 4. Simulation using Gaussian wave pulse

Figure 3 illustrates how gaussian pulses form in electromagnetic waves. The simulation results show that the reflection results of the E_x and H_y field components experience different wave phases, while the transmission results after passing through the plate show the same phase and experience attenuation so that the waveform decays. In addition, the values of B and A on N-step 400 showed constant values. The decay of electric and magnetic fields is caused by a decrease in the intensity or energy of the wave due to the absorption of energy by the permittivity of the plate material. Previous studies have not shown this physical phenomenon [19].

Figure 4, which shows the plate change to 50, shows that for the permittivity value of 4 and the conductivity of 0.04 the wave pulse still penetrates the conductor plate. While attenuation and wave propagation are not affected by plate thickness. These results indicate that what affects the attenuation and wave propagation values is the permittivity and conductivity of the material [8].

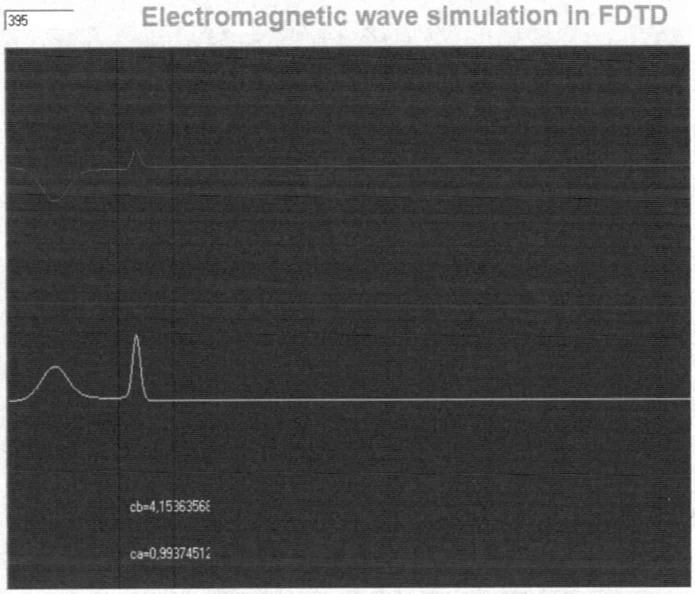

Figure 5. Simulation using Gaussian wave pulse with different permittivity

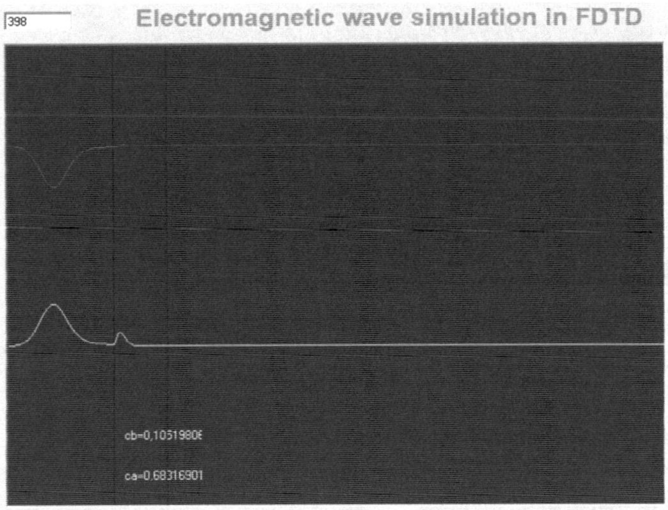

Figure 6. The Simulation uses Gaussian wave pulses with different conductivity

Figure 5, shows the change in permittivity of the material to 12 at a thickness of 50. The simulation results show that changes in the permittivity value cause the wave superposition in the E_x field to decrease while the H_y field component increases. This is different from Figure 6, which shows an increase in the conductivity value to 0.08 causing the E_x and H_y components to rapidly decay so that no wave pulses are transmitted.

4. Conclusion

The conclusion that can be drawn based on the simulation results is that the permittivity and conductivity properties affect the wave propagation on the plate, both reflected and transmitted. Changes in the thickness of the plate have no effect on attenuation and wave propagation. Meanwhile, changes in the permittivity value cause the field components E_x and H_y to have different wave superposition results in the destructive E_x and constructive E_y components. Meanwhile, the change in the conductivity value causes the E_x and H_y field components to decay faster before passing through the plate.

References

[1] Yee K 1996 *IEEE Trans. Antennas Propag.* **14** 302
[2] Kunz K S and Luebbers RJ 1993 *The Finite-Difference Time Domain for Electromagnetic, Boca Raton* (Florida: CRC Press)
[3] Ishimaru A 1991 *Elektromagnetic Wave Propagation, Radiations, and Scaterring: From Fundamentals to Applications* (New Jersey: Prentice Hall)
[4] Sulivan D M 1991 *IEEE Trans. Microw. Theory Tech.* **39** 862
[5] Joseph R M, Hagnes S C, and Taflove A 1991 *Optic. Lett.* **16** 1412
[6] Johnson C C and Guy A W 1972 *Proc. IEEE* **60** 692
[7] Stuchly M A and Stuchly S S 1980 *J. Microw. Power* **15** 16
[8] Winarno N, Firdaus R A, and Afifah R M A 2019 *Mater. Phys. Mech.* **42** 617
[9] Gandhi O P, Gao B Q, and Chen Y Y 1993 *IEEE Trans. Microw. Theory Tech.* **41** 658
[10] Jin X H, Chen Y W, and Zhang P 2011 *Adv. Mater. Res.* **317** 1172
[11] Mao Y F, Huang P H, and Ma L G 2014 *Adv. Mater. Res.* **989** 1869
[12] Wang L F, Wu Z Y, and Meng Z J 2014 *Appl. Mech. Mater.* **635** 1502
[13] Yan Z Z 2014 *Appl. Mech. Mater.* **575** 108
[14] Yang Z R and Yang J 2014 *Appl. Mech. Mater.* **602** 3702
[15] Thiele E and Taflove A 1994 *IEEE Trans. Antennas Propag.* **42** 633
[16] Sulivan D M 2013 *Electromagnetic Simulation Using the FDTD Method. 2nd Edition* (Hoboken: John Wiley and Sons)
[17] Taflove A and Brodwin M E 1975 *IEEE Trans. Microwav. Theory Tech.* **23** 623
[18] Mur G 1981 *IEEE Trans. Electromagn. Compat.* **23** 377
[19] Berenger J P 1994 *J. Comput. Phys.* **114** 185

AUTHOR INDEX

Institute of Physics
Dirac House, Temple Back
Bristol BS1 6BE UK

ISSN: 1742-6588
ISBN 978-1-7138-5081-6

CURRAN ASSOCIATES INC.
proceedings.com